MICROWAVE LINE OF SIGHT LINK ENGINEERING

MICROWAVE LINE OF SIGHT LINK ENGINEERING

PABLO ANGUEIRA and JUAN ANTONIO ROMO
University of the Basque Country
E.T.S.I. Bilbao
Bilbao, Spain

A JOHN WILEY & SONS, INC., PUBLICATION

Published by John Wiley & Sons, Inc., Hoboken, New Jersey.
Published simultaneously in Canada.

Library of Congress Cataloging-in-Publication Data
Angueira, Pablo, 1972–
Microwave line of sight link engineering / Pablo Angueira, Juan Antonio Romo.
p. cm.
Includes bibliographical references.
ISBN 978-1-118-07273-8
1. Microwave communication systems. 2. Line-of-sight radio links. I. Romo, Juan Antonio, 1958– II. Title.
TK5103.4833.A54 2012
621.382–dc23

2012007155

Printed in the United States of America

10 9 8 7 6 5 4 3 2 1

To my parents, Gloria and Juan Antonio, and to my family: Amaia, Leire, and Itziar.
—Juan Antonio Romo

To my parents, Dolores and Juan Carlos, and to my family: Jone and Sira.
—Pablo Angueira

CONTENTS

PREFACE

For decades, microwave line of sight (LOS) links have been one of the basic technologies used to build telephone networks. Until 1980, the fast rollout of high capacity transport networks and deployment of links in areas with challenging geographic characteristics could not be understood without this technology.

During the 1980s, different voices from academy and industry forecasted a decline of point-to-point radio technologies and an increase of the deployment of optic links and satellite fixed service systems as the basic technologies to build transit, backbone networks. The evolution during the following decades has confirmed the success of optical and satellite technologies, but has not confirmed the decline of microwave radio. Owing to their intensive use in mobile cellular access networks, the number of microwave links installed worldwide has increased significantly, where fast and flexible link deployment at reduced costs is required. Quite the contrary.

Today, it is expected that this tendency continues and high capacity microwave LOS links in bands above 40 GHz will be one of the techniques used for deploying 4G mobile networks. Also, it should not be forgotten that microwave LOS links are still a cheap and reliable option for point-to-point links in areas with difficult geographical features. In summary, microwave LOS links will be in use for a few decades more and the required frequency regulatory work has been accomplished by the ITU-R during the last few years with frequency allocations for fixed services above 100 GHz.

This book aims at being a complete guide for design, bringing into service, and operation of microwave LOS systems. It contains material to understand and accomplish all the tasks involved in the design and planning of a LOS link. The book is organized according to design work areas, showing the relations between different design procedures. Throughout the different chapters, it discusses and provides recommendations for the critical decisions to be taken during the planning process such as: frequency band selection, radio channel arrangements, site selection, antenna installation, and equipment choice. In addition, the book provides a comprehensive and concise summary of the vast amount of ITU recommendations related to planning, bringing into service, and maintenance purposes of radio-relay systems in the real world.

The book is organized into nine chapters and one appendix. The first chapter contains the basic description of a microwave LOS link. This chapter also discusses the use cases and alternative technologies for point-to-point applications and ends with a summary overview of basic link budget, noise and interference concepts that, maybe familiar to some of the readers, are fundamental to understand calculations in further chapters.

The second chapter provides a description of the physical phenomena associated with the propagation of the signals involved in fixed wireless service (FWS) systems. The chapter describes the parameters that are commonly used to characterize the propagation phenomena and their influence on the perturbations suffered by the microwave LOS link signals along the transmitter–receiver path. This chapter covers the description of the phenomena, while the calculation procedures and algorithms to quantify fading, signal depolarization, and other effects of the consequence of propagation, have been compiled in Chapter 7 with the aim of including them in the context of the system design.

Chapter 3 deals with frequency planning. It describes the different aspects related to the capacity of the link and radio channel arrangement options. Here, different examples of channel arrangements in different frequency bands are provided and the standard radio channel arrangement plans for use in most cases will be described in detail. The chapter also describes the different frequency bands available for FWS in different parts of the world and discusses the criteria for selecting a specific band. The designs will depend on the kind of application the radio-relay link is used for, its capacity and the interferences that are present at the microwave LOS link operation scenario.

Chapter 4 describes the different subsystems and equipment that are combined to create a radio-relay link. Technical features and configurations of microwave LOS link equipment are explained throughout this chapter that is focused on the development of concepts and knowledge that directly affect the design of a radio-relay link. Stress is laid on the factors that a link planner should bear in mind when specifying and choosing equipment for a specific case.

Chapter 5 discusses system performance metrics and objectives for microwave LOS link design. The objective of this chapter is to explain and provide synthetic use guidelines of the concepts and objectives related to error performance and system availability according to ITU-T and ITU-R reference materials.

Chapter 6 deals with link path engineering. The path engineering analysis studies the alternatives for providing a LOS path between the transmitting and receiving antennas. A discussion of the factors for selecting the sites where stations will be installed is included in this chapter. The location of each of the nodes of the link (repeater stations), the dimensioning of each transmission site, including building and equipment room, power supply, access, towers, and antenna installation special requirements are discussed in several sections of the chapter. The chapter also describes the clearance criteria and the methods for calculating the optimum antenna heights. The impact of antenna height on reflection and multipath is also dealt with in this chapter.

Chapter 7 provides a clear and detailed description of the procedures for calculating the parameters that characterize fading that are associated with different propagation effects. The sections contain information related only to the propagation phenomena relevant to the design and operation of radio-relay links.

Chapter 8 focuses on the design according to error performance and unavailability objectives. The chapter discusses the relationship between availability and error performance objectives and system thresholds usable on the link dimensioning process. The chapter describes the steps of the link dimensioning process and discusses about the additional countermeasures to be applied on difficult propagation conditions. The final part of the chapter is dedicated to interference calculations and their impact on system design. This chapter contains a step-by-step summary of the procedures involved in the design of a microwave LOS link.

Chapter 9 provides an overview of the bringing into service (BIS) and maintenance procedures. The first part of the chapter describes the reference criteria used for system performance evaluation during BIS and contains a detailed step-by-step guide for accomplishing the BIS process successfully. The second part of the chapter deals with maintenance and describes the concepts of maintenance performance limits and associated maintenance and fault location procedures.

Finally, the Appendix contains two detailed examples of microwave LOS link design that summarize all the procedures and criteria described in previous chapters. The examples developed correspond to two of the most usual application scenarios of a microwave LOS link: a medium distance high capacity link in bands below 10 GHz and a short range link in higher frequencies for connecting different parts of a cellular mobile access network.

We would like to close this preface by acknowledging the valuable discussions and advice provided by several persons from the microwave LOS link industry. We would like to Javier Gurrutxaga and the link planning crew at *3dB Consult*, and Daniel Baldi, (*blueNEXUS*) for the discussions on real world link design practices. Also, we would like to thank the students David Esteban and Rocío Prieto for their help with the manuscript.

JUAN A. ROMO
PABLO ANGUEIRA

ACRONYMS

ACR	Adaptive clock recovery
ADM	Add/drop multiplexer
ADSL	Asymmetric digital subscriber line
AGC	Automatic gain control
AIS	Alarm indication signal
ANSI	American National Standards Institute
APO	Allocated performance objectives
AR	Availability ratio
ASTER	Advanced spaceborne thermal emission and reflection radiometer
ATM	Asynchronous transfer mode
ATPC	Automatic transmit power control
AZD	Ambiguity zone detection (modulation)
BB	Base band
BBE	Background block error
BBER	Background block error ratio
BC	Basic channel digital interface
BCM	Block coded modulation
BER	Bit error rate
BIS	Bringing into service
B-ISDN	Broadband-integrated services digital network
BISPO	Bringing into service performance objectives
BITS	Building integrated timing source
BPSK	Binary phase shift keying
BS	Base station
BSC	Base station controller

CATV	Cable television
CCDP	Co-channel dual polarization
CEPT	Conférence Européenne des administrations des Postes et des Télécommunications / European Conference of Postal and Telecommunications
CES	Circuit emulation service
C/I	Interference ratio
CLW	Cloud liquid water
CPA	Co-polar attenuation
CSMA/CD	Carrier sense, multiple access/collision detect
C-QPSK	Constant Envelope Offset-Quadrature Phase Shift Keying
DCR	Differential clock recovery
DQPSK	Differential QPSK
DRRS	Digital radio relay systems
DSL	Digital subscriber line
DTE	Digital terrain elevation
DVB-RCS	Digital video broadcasting–return channel satellite
DXC	Digital cross connect
EB	Errored block
ECC	Electronic communications committee
EDC	Error detection code
EHF	Extremely high frequency
EIRP	Equivalent isotropic radiated power
EPO	Error performance objectives
ERP	Equivalent radiated power
ES	Errored second
ESR	Errored second ratio
ETSI	European Technical Standards Institute
EV-DO	Evolution-data optimized access
FCC	Federal communications commission
FEC	Forward error coding
FET	Field-effect transistors
FS	Fixed service
Fs	Frontier Station
FSO	Fixed optical links
FSS	Fixed satellite service
FTTB	Fiber to the building
FTTC	Fiber to the curb
FTTH	Fiber to the home
FTTN	Fiber to the neighborhood
FWS	Fixed wireless system
GD	Group delay
GDEM	Global digital elevation model
GIS	Geographical information systems
HAPS	High altitude platform system

HDFS	High density fixed services
HDSL	High-bit-rate digital subscriber line
HEMT	High electron mobility transistor
HF	High frequency
HFC	Hybrid fiber cable
HP	High performance (Antennas)
HRDP	Hypothetical reference digital path
HRDS	Hypothetical reference digital sections
HRP	Hypothetical reference path
HRX	Hypothetical reference connection
HSPA	High-speed packet access
ICPCE	Inter-country path core element
IDU	Indoor unit
IEEE	Institute of electrical and electronic engineers
IETF	Internet engineering task force
IF	Intermediate frequency
IG	International gateway
IP	Internet protocol
IPCE	International path core element
ISDN	Integrated services digital network
ISI	Inter-symbol interference
ISM	In-service monitoring
ITU-R	International Telecommunication Union. Radiocommunication Sector
ITU-T	International Telecommunication Union. Telecommunication Standardization Sector
IWVC	Integrated water vapor content
LAD	Linear amplitude dispersion
LAN	Local area network
LMDS	Local multipoint distribution system
LF	Low frequency
LMS	Least mean square
LNA	Low noise amplifier
LOF	Loss of frame alignment
LoS	Loss of signal
LOS	Line of sight
LTE	Long-term evolution
MAN	Metropolitan area network
MAP	Maximum power combining (algorithms)
MARS	Multiaccess radio systems (rural telephony)
DTM	Digital terrain model
ME	Maintenance entity
MF	Medium frequency
MID	Minimum dispersion combining devices
MIFR	Master international frequency register

MIMO	Multiple input multiple output
MLCM	Multilevel coded modulation
MMIC	Monolithic microwave integrated circuits
MO	Mean outage
MODEM	Modulator demodulator
MP	Minimum phase
MPL	Maintenance performance limits
MSC	Mobile switching center
MTBF	Mean time between failures
MTTR	Mean time to repair
MULDEM	Multiplexer demultiplexer
MW link	Microwave link
NASA	National Aeronautics and Space Administration
NFD	Net filter discrimination
NGA	National geospatial-intelligence agency
NLOS	Non Line of sight
NMP	Non-minimum phase
NNI	Network-to-network interface
NPE	National path element
ODU	Outdoor unit
OI	Outage intensity
O-QPSK	Offset QPSK
PC	Primary centre
PCM	Pulse coded modulation
PDH	Plesiochronous digital hierarchy
PDV	Packet delay variation
PEP	Path end point
PHEMT	Pseudomorphic high electron mobility transistor
PLF	Performance level factor
PLL	Phase locked loop
POC	Points of concentration
PRBS	Pseudo-random bit sequence
PS	Packet switching
PSK	Phase shift keying
QAM	Quadrature amplitude modulation
QPR	Quadrature partial response (modulation)
QPSK	Quadrature phase shift keying
RAN	Radio access network
REM	Radio environment map
RF	Radio frequency
RH	Relative humidity
RLS	Recursive least square
RNC	Radio network controller
RPE	Radiation pattern envelope
RPO	Reference performance objectives

RS	Reed–Solomon
SC	Secondary centre
SDH	Synchronous digital hierarchy
SDR	Software defined radio
SES	Severely errored second
SESR	Severely errored second rate
SHDSL	Single-pair high-speed digital subscriber line
SHF	Super high frequency
SONET	Synchronous optical network
SRTM	Shuttle radar topography mission
STM	Synchronous transport module (SDH networks)
TC	Tertiary center
TCM	Trellis coded modulation
TD	Threshold degradation
TDM	Time division multiplex
TM	Terminal multiplexer
TX	Transmitter
TP	Test period
TR	Threshold report
UHF	Ultra high frequency
UHP	Ultra high performance (antennas)
UMTS	Universal mobile telecommunications system
UNI	User to network interface – ATM
UR	Unavailability ratio
VC	Virtual channel
VCn	Virtual container level n
VCO	Voltage controlled oscillator
VDSL	Very high data rate digital subscriber line
VHF	Very high frequency
VLF	Very low frequency
VP	Virtual path
VSAT	Very small antenna terminal
VSWR	Voltage standing wave ratio
WAN	Wide area network
WiMAX	Worldwide interoperability for microwave access
XIF	XPD improvement factor
XPD	Cross-polar discrimination
XPIC	Cross-polarization interference cancelation

CHAPTER 1

INTRODUCTION TO MICROWAVE LOS LINK SYSTEMS

1.1 INTRODUCTION

From a generic standpoint, a telecommunication network enables the exchange of information among users or devices that can be either fixed or mobile. This general view contains the first simple classification of telecommunication networks into fixed and mobile. Independently of the mobile or fixed nature of the target devices, the signals involved in the communication process transport digitized information that is associated with final services such as voice, pictures, video, or general data.

Every network is composed of two basic components: network nodes and transmission systems. The network nodes provide the control, access, aggregation/multiplexation, switching, signaling, and routing functions. The transmission systems enable the transport of signals either from the user devices to the network nodes or between different nodes of the network. The transmission systems can be based on different delivery media. Usually, the transmission media have been divided into wireless systems, where the information is delivered by means of electromagnetic waves that propagate through the atmosphere, and systems based on transmission lines, where the electric or optical signals propagate through a closed medium. The metallic transmission of electric signals uses lines that usually are copper pairs or coaxial cables, whereas the optical signals are sent over glass fiber cables.

Transmission systems can be found in any of the two subnetworks that compose a generic telecommunications network: access network and transit network. The

Microwave Line of Sight Link Engineering, First Edition. Pablo Angueira and Juan Antonio Romo.

access network enables the communication between the network and the user devices, whereas the transit network provides all the required functions that interconnect different access sections, including network control, signaling management, switching, interfacing with other networks, etc.

In this network context, a radio link of the fixed service (FS) [as per Radiocommunication Sector of the International Telecommunication Union (ITU-R) terminology)] is any radiocommunications link between two fixed stations based on the propagation of signals through the atmosphere at frequencies higher than 30 MHz. Currently, there is a tendency to use more the generic term of fixed wireless system (FWS), which is used to identify the telecommunication systems operated for FSs and that are used in access and transport application scenarios. Those systems are conveyed by electromagnetic wave propagation, in any form, with a limit that has been set in 3000 GHz. Terrestrial point-to-multipoint systems, terrestrial point-to-point systems, high-frequency (HF) systems, high-altitude platform systems (HAPS), and even free space optic links fall into the FWS category.

Microwave line-of-sight (LOS) links covered by this book are a subgroup of the FS or FWS general classifications. Microwave LOS links are composed of point-to-point systems between two terrestrial stations that transmit and receive signals taking advantage of the propagation of waves through the lower part of the atmosphere (troposphere). Microwave links operate in LOS condition in frequencies from 400 MHz to 95 GHz under specified availability and quality conditions. These systems are in practice referred as microwave links (MW links), LOS microwave, fixed service radio links, or simply radio links.

The frequency limits mentioned earlier are associated with the frequency band assignments that international regulatory bodies have reserved for fixed service links. Currently, a majority of the systems operate in frequency bands between 4 and 40 GHz. Higher frequency bands are used in links where the path between stations is rather short (usually less than 1 km and, in any case, no longer than a few kilometers due to availability constraints associated with rain attenuation).

A basic point-to-point microwave LOS link is composed of two nodal stations, each one at the edge of the link path, without obstacles in the propagation path that could cause blocking or diffraction, and that use antennas with high directivity, also named narrow-beam antennas.

Microwave LOS links are designed to preserve the LOS propagation path as the main propagation mechanism. This condition implies that the direct component of the space wave is well above the terrain irregularities and any diffraction effects are considered negligible under standard conditions. In practice, the LOS component coexists with additional propagation modes such as the reflection on the surface of the earth, diffraction in obstacles due to anomalous refractive conditions and multipath propagation originated both on the surface of the earth and on higher layers of the troposphere. In the design process of a MW link, the availability of accurate terrain maps, which also contain any man-made construction candidate to create diffraction, is a key requirement. Figure 1.1 shows a simplified model of the possible propagation modes in a point-to-point link.

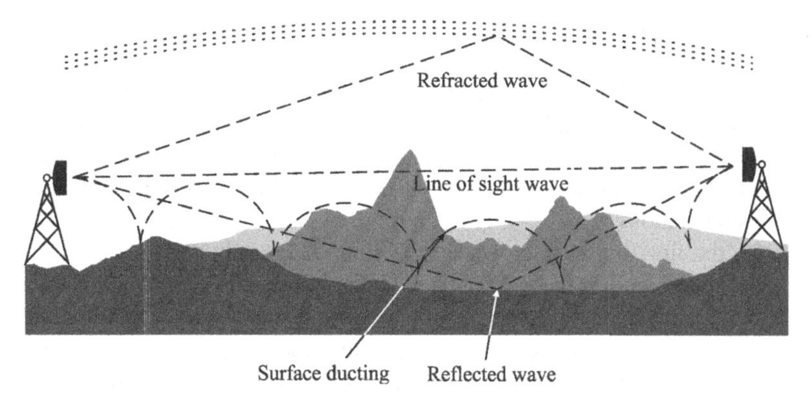

FIGURE 1.1 Propagation modes in a microwave LOS link.

In the likely event that path distance between the locations that will be communicated exceeds the LOS distance, due to terrain irregularities or simply due to the curvature of the earth, the link will be divided into concatenated shorter sections (called "hops") that are created by means of including repeater stations.

In most cases, microwave LOS links are bidirectional systems, with full duplex capacity provided by frequency division multiplex schemes. The simplest example would require two carriers, each one aimed at transporting the information in one direction. An assignment of two frequencies, each one for each direction of the communications, is called radio channel. Sometimes, LOS links can be simplex systems, transporting information in only one direction. An example of this application can be found in the transport section of terrestrial broadcast systems, where the video, audio, and data are conveyed from a production or aggregation center to the broadcast stations that will later broadcast the contents to the end users.

This chapter provides an overall view of the microwave LOS links, describing the specific terminology that will be used in this book, the most relevant characteristics of the technologies involved and identifying the most widespread application fields of microwave LOS links. The chapter contains the basic principles of the planning and design process of a microwave LOS link, starting from the definition of a link budget and identifying the main signal degradation sources that influence the fulfillment of the quality requirements of the link. The perturbation sources considered will be related to propagation through the troposphere, noise sources, and interferences. After the introductory picture given in this chapter, each one of the design procedures and modules will be covered in detail in later chapters.

1.2 HISTORIC EVOLUTION OF RADIO LINKS

The first experimental microwave LOS link was designed and installed by the Bell Labs in 1947. The system was intended to provide a two-way communication between two stations in New York and Boston. The link was an analog system in the 4 GHz band that used frequency modulation and frequency division multiplex techniques. The equipment was based on vacuum tubes. The evolution of this system led to further developments in the United States, Australia, Canada, France, Italy, and Japan during the 1950s. The preferred bands during this period were 4 and 6 GHz. In this context, in 1960, the National Long Haul Network was designed to connect the East and West Coasts of the United States of America, with a total length of 6500 km and about 125 active repeater stations.

In 1968, the first digital microwave LOS link was installed in Japan. This first digital system operated in the 2 GHz band, using phase shift keying (PSK) modulation with an equivalent capacity of 240 telephone channels. After this first digital landmark, the rollout of digital microwave relay systems starts in the 1970s and continues over the 1980s. During this period, analog systems did not disappear and coexisted with the new digital links. During the 1980s, the analog systems started being progressively replaced by equivalent digital systems, process that was generally completed by the first years of the twenty-first century.

The use of multilevel modulation schemes in high-capacity links was spread out during the first years of the 1980s. These systems were based on plesiochronous digital hierarchy (PDH) transport techniques. The inclusion of adaptive equalizers and diversity reception schemes to fight fast fading associated with multipath propagation components are also relevant milestones of the mentioned decade.

During the 1990s, the most relevant advances over the state of the art are a consequence of the new fields of application for microwave LOS links. Although the typical use for long-haul transport links in telephone networks started to decline in favor of fiber-optic links, the use of microwave LOS in access networks grows significantly, both as transport infrastructure for cellular mobile access networks and also as supporting infrastructure in fixed access networks. The exponentially growing access networks of the 1990s require new frequency bands and enhanced efficiency in bits per hertz. During the 1990s, synchronous digital hierarchy (SDH) and asynchronous transfer mode (ATM) technologies were widely adopted by transport networks, including those based on microwave LOS links.

During the first decade of the twenty-first century, there has been a convergence between mobile and fixed services and a progressive implementation of Internet protocol (IP) packet switching (PS) traffic in all networks, both in the access and transit network sections. Microwave LOS links have not been immune to this tendency and a progressive adaptation to this scenario has been put in place. The first versions consisted of interface adaptations for the coexistence of Ethernet and PDH/SDH traffics in the same links that later evolved to all IP systems. Currently, Ethernet radio equipment provides a significant flexibility to adapt the bandwidth assignments to different services carried by the MW link system. Maximum throughput values today range from several hundreds of megabits per second to a few gigabits per second if latest optimization techniques are used (dual polarized channels, high-order modulations, multiple in multiple out, etc.).

The target of new technology developments during the last years has evolved to a progressive enhancement of the spectral efficiency, following the same tendency of increasing bandwidth demands of broadband multimedia services. Nowadays, the effort focuses on increasing the capacity while maintaining performance (availability and quality), as well as a better exploitation of spectrum resources in dense frequency reuse scenarios. Following chapters will cover some of these techniques, such as adaptive modulation techniques with high-order schemes (i.e., 512 - 1024-QAM), frequency reuse channel arrangements with dual polarization or high-performance antennas.

1.3 POINT-TO-POINT FIXED COMMUNICATION TECHNOLOGIES

In order to set up a communication connection between two locations, there are several technical choices, the microwave LOS link being just one of the possible options. Among the alternatives, there is a set of choices that involve physical carriers: systems over copper pair cables, links using coaxial cables, and fiber-optic cable links. Additional alternatives are based on radiocommunication systems such as satellite links, other terrestrial point-to-point systems (i.e., transhorizon links), HF fixed systems, communication links using HAPS, free space optic (FSO) links, and point-to-multipoint wireless communication systems.

The choice of the transmission media is one of the first actions that a communications engineer must take, always at the first stages of the design of a communication system. This section will describe briefly the different choices for establishing a link between two locations, and the different advantages and drawbacks of each alternative will be discussed, always with the microwave LOS link as the comparison reference.

From a general standpoint, terrestrial microwave LOS links have inherent advantages that are a consequence of wireless propagation without the need of having a physical carrier that connects transmitter and receiver. This advantage is notorious in areas with irregular orography, zones where deploying a cable system is difficult, areas where physical access is a challenge, and cases where common infrastructures are not developed.

The microwave LOS links are usually the solutions with lowest cost in the case of access and transit network if the network rollout requires fast and flexible connection deployments in dense network scenarios, such as wireless mobile systems. The possibility of transporting physically the equipment of a microwave LOS link provides further benefits for its use in the case of emergency situations, natural disasters, or temporary backup system in severe damages suffered by fiber-optic link cables.

The major disadvantage of microwave LOS links is associated with the restriction imposed by the LOS requirements of these systems. In dense urban environments, blockage from buildings is a problem to set up links with the minimum number of hops possible. In cellular access networks, the intense reuse of frequencies provokes interference problems that require complex and careful design procedures. Moreover, the need of periodic maintenance actions in towers and stations with difficult access is one of the remarkable disadvantages of these systems. Finally, the complete dependency of the system performance upon the unstable mechanism of propagation through the troposphere is a challenge for the radiocommunications system design engineer.

1.3.1 Cabled Transport Systems

1.3.1.1 xDSL Technologies Historically, copper cable pairs have been massively used as the physical carrier in the local loop, from the telephone office to the customer premises, as well as the physical means to transport analog and digital multichannel links between offices. As a consequence, the telephone companies, most of which have become today's global telecom operators, have a wide outdoor plant infrastructure based on copper pairs.

In the case of point-to-point applications, today, there are commercial solutions that multiplex different flows and sources (PDH, Ethernet, etc.) into a single flow in a link over copper pairs (two and four wires depending on the system). These links can use some of the variations of a family of technologies called digital subscriber line (DSL), that in addition to be last mile applications, can be used as the lower layer technologies for transport systems over copper pairs. Figure 1.2 shows a block diagram with an example of DSL links.

The DSL family is a group of standards that offer different alternatives mainly targeting access networks over the copper outdoor plant. There is a variety of alternatives depending on the specific requirements as symmetry/asymmetry of the upload and download channels, maximum throughput per maximum local loop length, etc. Among the variations, asymmetric digital subscriber line (ADSL), high data rate digital subscriber line (HDSL), symmetric digital subscriber line (SDSL), single-pair high-speed digital subscriber line (SHDSL)/HDSL 2, very high speed digital subscriber line (VDSL), and VDSL2 are worth mentioning. Figure 1.3 shows a comparison among these technologies differentiating the symmetrical/asymmetrical nature of the standards as well as the maximum link length versus achievable bit rates.

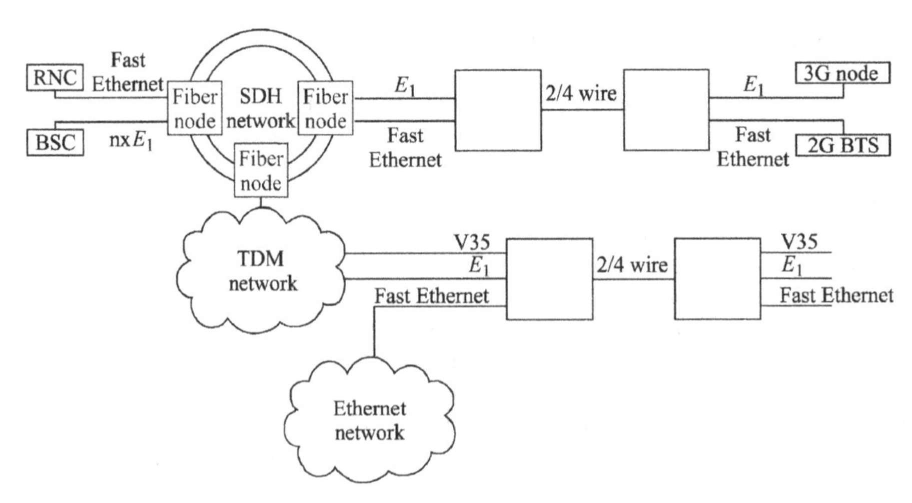

FIGURE 1.2 Point-to-point connections based on DSL technologies.

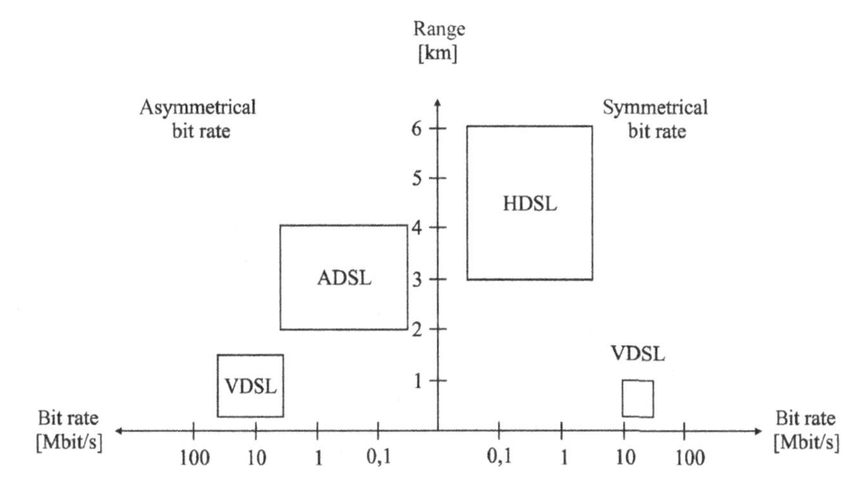

FIGURE 1.3 xDSL technologies. ADSL, asymmetric digital subscriber line; HDSL, high data rate digital subscriber line; VDSL, very high-speed digital subscriber line.

In comparison with microwave LOS links, DSL technologies are used today extensively in access networks for connecting consumer premises to the transit network (Figure 1.4). The success of DSL standards in access networks has been based on their flexibility and capacity for fast and inexpensive deployments over the existing copper outdoor plant, and, at the same time, providing bitrates high enough for wideband internet access, video distribution and in general access to networked multimedia contents. The most remarkable limitation of these systems is usually a maximum bitrate/distance limit caused by link density in urban areas and by the difficulties of propagation through the copper pair carrier. The propagation channel over copper pairs involves a considerable list of relevant impairments (interferences, crosstalk, attenuation, impulse noise, etc.). The problems in these networks are in many cases amplified by the fact that the outdoor plant is rather old.

1.3.1.2 Fiber-Optic Links Fiber optics has a long list of advantages for point-to-point link applications. Fiber-optic links have extremely high bandwidths, very low attenuation values that enable long links without repeaters and transmission quality specifications that are almost unaffected by environmental changes. Additionally, these features remain stable over time. Fibers are grouped in variable number in fiber-optic cables that have special isolation, reinforcement, and protection elements in order to preserve the integrity of the fibers. Each fiber can convey a few gigabits per second per wavelength. If wavelength division multiplex or dense wavelength division multiplex techniques are used with a cable that contains multiple fibers, for practical purposes, the transport capacity of these links is unlimited, and the network bandwidth is limited by other functions such as switching or interfacing with other networks.

Fiber-optic cables present disadvantages that are a consequence of the need for special arrangements in laying the cable. In most cases, especially in urban areas,

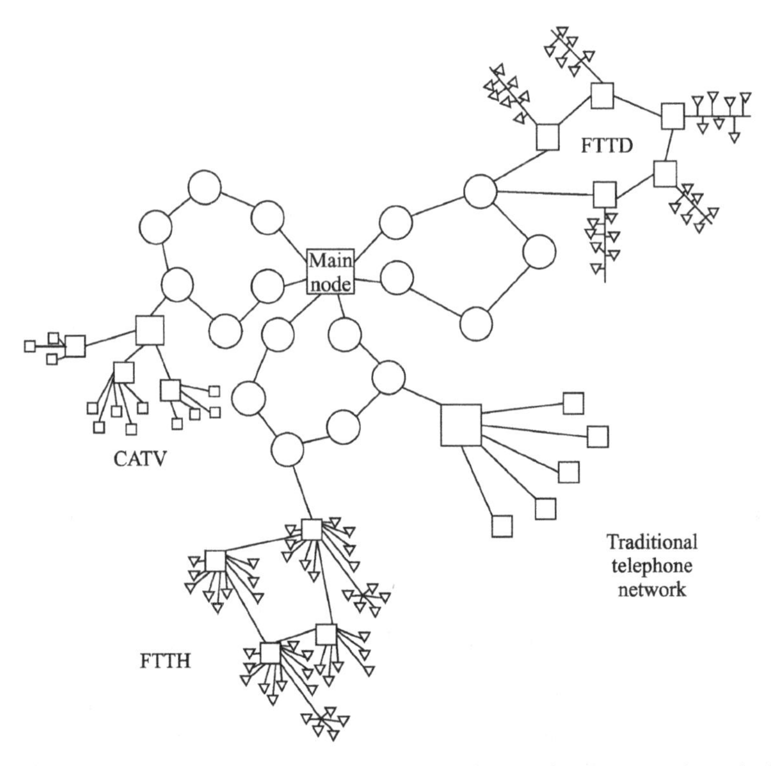

FIGURE 1.4 Network diagram showing different subnetworks (access and transit) based on fiber-optic cables.

fibers are laid in underground ducts and subducts, where special polyvinyl chloride pipes are installed previously. The construction of this underground concrete infrastructure in urban areas is expensive and time consuming, and it increases significantly the number of administrative permits required in the network deployment process. Moreover, in areas with irregular terrain, laying a fiber-optic cable is difficult and very expensive.

Fiber-optic systems can be found today in all the sections of a telecommunications network, including terrestrial long distance links, international submarine high-capacity links, high-capacity intercity and metro ring transport systems, and not forgetting the increasing number of high-speed access networks based on fiber-optic cables. Figure 1.4 shows a network diagram example with different sections in a fibre-optic network.

The application in access networks has different approaches depending on the distance between the fiber cable termination and the user premises. Figure 1.5 shows the most common schemes known as fiber to the x (FTTx), where x refers to the terminating point of the fiber: fiber to the home (FTTH), fiber to the building (FTTB), fiber to the curb, fiber to the neighborhood (FTTN).

FTTH is based on fiber-optic cables and optical distribution systems for enabling wideband services (voice, television, and Internet access) to residential and business

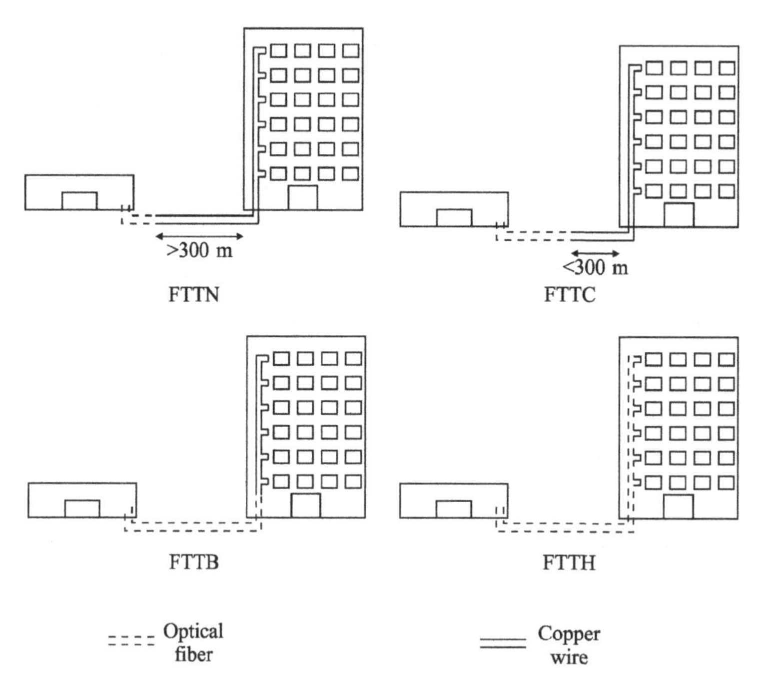

FIGURE 1.5 Fiber to the x (FTTx) technologies. Fiber to the neighborhood (FTTN); fiber to the curb (FTTC); fiber to the building (FTTB); fiber to the home (FTTH).

users. The optical distribution networks are passive optical networks (xPON). Once again there is a list of variations of the passive optical family: asynchronous transfer mode passive optical network, broadband passive optical network, gigabit-capable passive optical network, and gigabit Ethernet passive optical network.

A well-known case of FTTN is hybrid fiber coaxial (HFC) networks. These networks have been widely deployed in residential environments to provide traditional cable television and telephone services that have been complemented today with wideband Internet access. The most widespread open systems interconnection Layer 2 standard in today's HFC networks is DOCSIS (data over cable service interface specification).

1.3.2 Satellite Communication Links

Satellite communication links are a type of radiocommunication system established between two earth stations enabled by an artificial satellite that acts as a repeater station. The fixed satellite service (FSS) is defined by the ITU-R as the point-to-point link between two earth stations. FSS systems are based on satellites located on geostationary orbits. These orbits are geosynchronous. The orbit is on the equatorial plane and the satellite travels along the orbit with an equivalent orbital period, which is approximately the same as the rotation period of the earth. Geostationary satellites are thus fixed for an observer on earth's surface, independently of its latitude–longitude position (Figure 1.6).

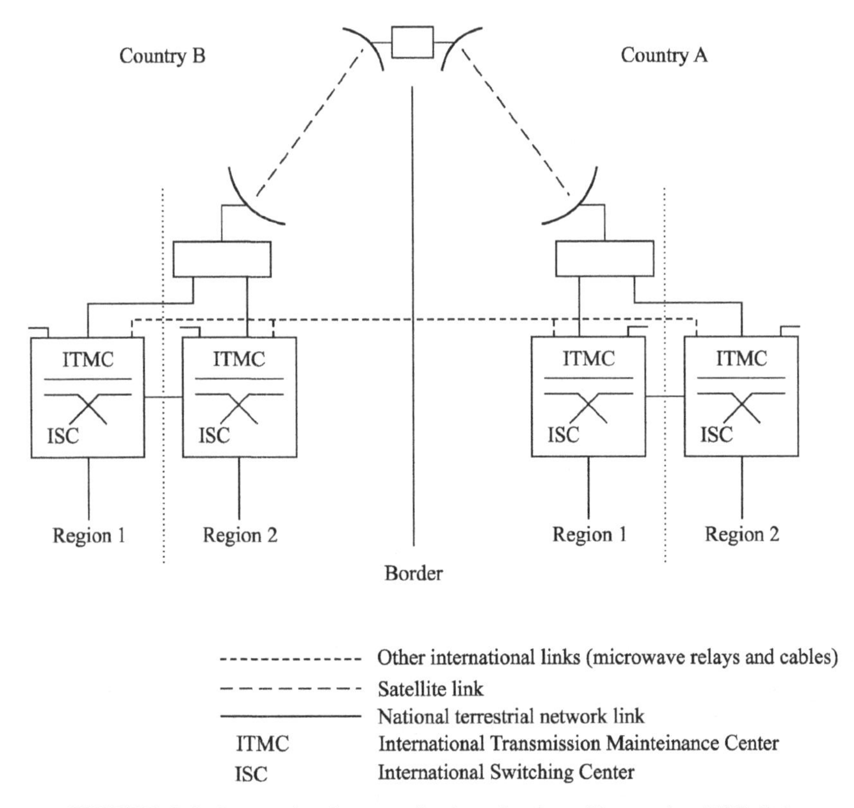

FIGURE 1.6 International connection by a fixed satellite service (FSS) link.

Satellite links have been intensively used in the past two decades as the solution for long distance point-to-point digital telephone channel transport as well as generic data, especially in the context of international communications or in countries where the distances make other solutions less adequate. This application involves earth stations with remarkable bandwidths, high equivalent isotropic radiated power (EIRP) values and Figure of Merit G/T (a parameter that quantifies the station gain vs. overall noise at the receiver, thus specifying sensibility) values also very high. These high-capacity satellite links have been progressively substituted by fiber-optic connections (land and submarine) while the satellite option is in many cases the redundant backup system for ensuring link availability in the event of a fiber link failure.

Satellite links are also adequate for point-to-multipoint networks in remote zones and over wide service areas. A common application is very small aperture terminal (VSAT) networks, where a nodal station dynamically controls the access of a high number of terminals disseminated over the coverage area. The services provided by today's VSAT networks are similar to a generic telecommunication network, providing voice, multimedia, and data network access.

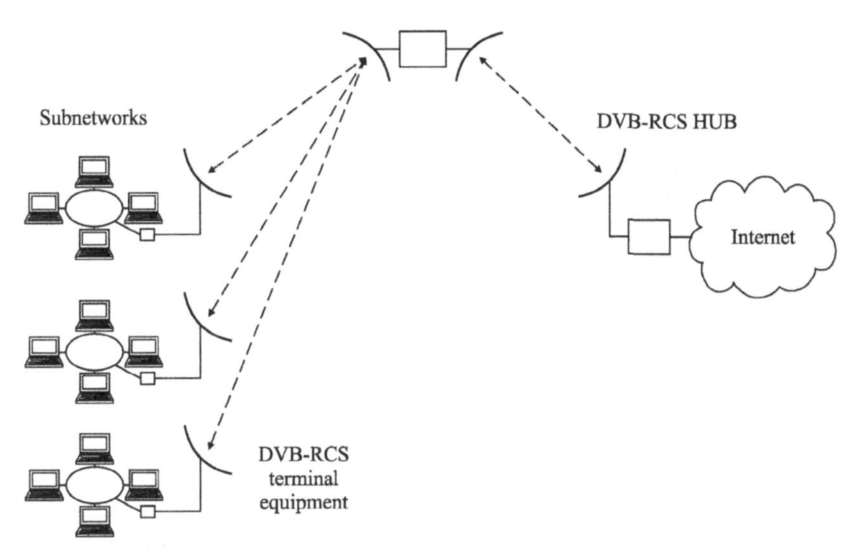

FIGURE 1.7 A very small aperture terminal (VSAT) network architecture based on DVB-RCS (digital video broadcasting-return channel satellite).

Figure 1.7 illustrates the typical architecture of a VSAT network where all the terminals are connected to the central hub through the satellite link. The central hub manages the operation of the VSAT system and acts as the gateway between the VSAT and other networks (Internet).

1.3.3 Other Fixed Wireless Systems

1.3.3.1 Free Space Optic Links FSO links are based on optical transmitters that send a narrow beam optical signal that propagates through the troposphere towards an optical receiver station. The optical nature of the energy involved is associated with the wavelength of the signals propagated (1550 nm, 780–850 nm, and 10 000 nm), in some cases very similar to the ones used by fiber-optic links. The main disadvantage of FSO links is the limited range that can be achieved today, due to propagation impairments. The limiting factor for FSO links is fog. In addition, other perturbation sources might be caused by sunlight, visibility obstructions, rain, snow, etc. Those impairments create scintillation and different degrees of fading. These perturbation sources limit the practical link ranges to 1 km. Figure 1.8 illustrates a simplified scheme of an FSO link intended to interconnect two local area networks of near buildings.

FSO links are a very interesting solution for short distances, which is complementary to microwave LOS links in dense urban environments. The major advantages if compared with LOS links are the wide bandwidths, simple equipment, and absence of interferences, which eliminates the complex frequency coordination studies from the design process.

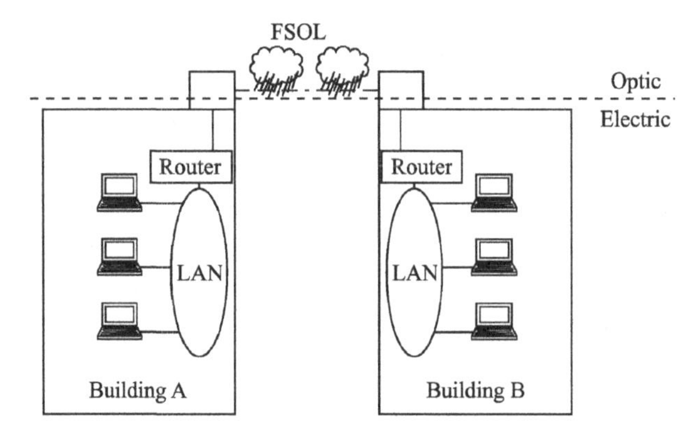

FIGURE 1.8 Free space optic (FSO) link.

1.3.3.2 Wireless Point-to-Multipoint Systems Point-to-multipoint wireless systems enable a communication between a nodal station and different stations spread out within the coverage area. Each nodal station has an associated coverage area, usually up to several tens of kilometers where substations can be installed. The network is usually composed of several nodal stations, and thus, there should be a frequency reuse coordination and interference analysis process when defining the coverage area associated with each one. This process is similar to the planning methods used by mobile cellular networks. These systems are normally used to provide wideband access to data services, both in urban residential and rural environments. The services offered today include both generic data access (Internet) and real-time services such as voice communications or television.

One of the most popular examples of point-to-multipoint systems is wireless metropolitan area networks (MANs), and different standards have been developed worldwide. The Institute of Electric and Electronic Engineers (IEEE) has developed the IEEE 802.16 standard, whereas the European Telecommunications Standards Institute has standardized the HiperACCESS and HiperMAN technologies within the broadband radio access network (BRAN) workgroup.

The radio interface of these standards can be configured for different data transmission throughputs in a variety of frequency bands in the lower part of the spectrum (2–11 GHz) and also in higher frequencies (26–31, 32, 38, and 42 GHz). The systems operating in the higher part of the spectrum are referred as high-density fixed services (HDFS).

Among all the technologies in use today, the local multipoint distribution system (LMDS) is worth mentioning. LMDS uses the frequency bands that range from 26 to 31.3 GHz and from 40.5 to 42.5 GHz. These high frequencies limit the coverage range to 5 km, due to the LOS requirements and deep fading effects associated with hydrometeors. LMDS is specified by the standard IEEE 802.16-2001 for wireless MAN environments. The system has been successfully deployed in Europe, whereas the commercial rollout in other parts of the world (Asia and North America) has not been as immediate as expected due to difficulties in finding available frequencies.

1.3.3.3 High-Frequency Links The FS systems operating in the frequencies that range from 3 to 30 MHz are an interesting alternative to point-to-point communications in applications where the link distance is very long (hundreds of kilometers) and the traffic capacity requirements are low.

These systems are especially adequate for setting up fast emergency communication systems, in cases of natural disasters that destroy or interrupt the normal operation of standard wireless or wired communication networks. In these disaster relief situations, HF links can be used as the first means to communicate (sometimes broadcast) alarm messages to communication stations or become the basic communication system to coordinate disaster relief operations.

The propagation in these frequencies is very unstable and difficult to predict. The propagation mechanism associated with these bands is the ionosphere wave. This propagation mode is based on transmitting from a station with a certain elevation angle towards the ionosphere, where the signal suffers refraction on the different layers of the ionosphere, and returns to the earth some hundreds of kilometers away from the transmitter. In certain cases, and depending on the antenna system, the frequency and the terrain soil electrical features (propagation over the sea is the most favorable case), the propagation through surface waves is also noticeable.

1.3.3.4 High-Altitude Platform Stations (HAPS) There is a special type of FS links that are based on platforms elevated at high altitudes that in theory enable a communication link between arbitrary locations of the coverage area of the HAPS platform, with high bandwidth capacities, in the same order of magnitude as satellite links. The HAPS "vehicle" is located at an altitude of 21 and 25 km and must be kept in place by complex control systems and some type of propulsion engine. HAPS systems are intended to have a cellular architecture in order to reuse the spectral resources intensively. User terminals are usually divided into three categories: urban area receivers, suburban receivers, and rural area devices.

There are frequency assignments to HAPS systems in the 800 MHz and 5 GHz band, whereas the relevant spectral resources remain at 18–32 and 47–48 GHz.

1.4 FIELD OF APPLICATION AND USE CASES

Microwave LOS link systems have played and still play a fundamental role in long-haul and high-capacity communications systems, both in transmission systems between nodes in telecommunication networks and in transport sections of broadcast networks.

Another classical application of microwave LOS links is the transport (backhaul) in mobile cellular networks, where it has become the dominant transport technology in global markets worldwide. This dominant position will be likely kept in next generation of wideband wireless communication networks.

An increasing interest field is related to access networks for licensed and unlicensed short distance links above 17 GHz, where the equipment is compact and reliable. The MW links are especially adequate for access sections in telecommunication networks, due to their economic advantages and easy deployment in practically any rollout scenario.

Next sections describe the most relevant features of the application scenarios where microwave LOS links are usually exploited.

1.4.1 Backhaul Networks

Traditional transport or backhaul Networks have used microwave LOS links that operate in frequency bands below 15 GHz. The typical hop length of these systems is in the range from 30 to 50 km and the associated bitrate capacity is equivalent to medium-to-high capacities in PDH or SDH systems (usually above 34 Mbps).

As the traffic demand increased, many service providers have deployed fiber networks that have substituted MW links as the leading technology in the mentioned network sections.

Most administrations and carriers, specially in places where the infrastructures are not well developed or in areas where topography is a challenge for deploying telecommunication networks, assume that even an increase of its use is not probable, MW links will still be used for some time in these low-frequency bands, with medium and low throughput capacities.

Some other administrations forecast a decrease in the use of microwave LOS links for high-capacity applications, shifting their field of application towards a role of backup systems that will be complementary to fiber-optic networks. Nevertheless, the same administrations envisage an intense use of microwave LOS technology in point-to-point short-distance applications, where capacity is not a relevant factor, as a means to support the increasing demand of traffic in access networks, especially in rural zones, remote areas, or areas where access is a challenge.

1.4.2 Backhaul in Mobile Networks

Microwave LOS links are the usual communication system for transport functions between base stations (BSs) (or equivalent in 3G and 4G networks), upper level control nodes (i.e., base station controllers (BSCs) in global system for mobile communications) and even with higher order nodes such as mobile switching centers (MSCs) and Packet Switching nodes. When installed in BSs, microwave LOS links share infrastructure and towers with the cellular access network equipment.

During the last two decades, second- and third-generation International Mobile Telephony 2000 networks have been deployed worldwide to serve traffic demands of voice, instantaneous messaging, and e-mail services. These standards are based on BSs furnished with E1 or T1 interface modules. These interfaces can be used directly for 2G TDM (time division multiplex) native traffic transport. In the case of 3G networks, the traffic is encapsulated into ATM and later conveyed by physical PDH/SDH interfaces. In any case, the traditional approach is a TDM link operating in different frequency bands, which are chosen by the operator depending on the link length requirements. Typical frequency allocations for this application can be found in the 10, 11.5, 18, 23, and 38 GHz bands.

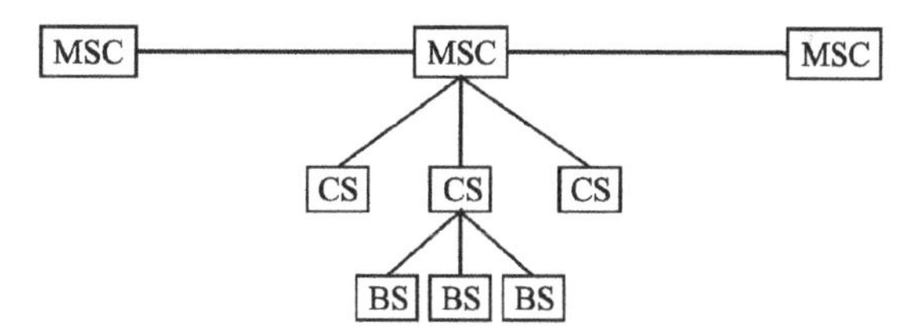

FIGURE 1.9 Interconnection topology for transport links in a mobile network.

Usually, the transport network topology is built upon different hierarchical layers that begin at the access aggregation nodes, following to the control centers (CS) and even to the switching centers (MSC). The latter link type (in MSCs) is only used in those cases where optical fibers are not available for this purpose. The links are in most cases cascaded, using multiple redundant paths, and providing high availability values even in the case of rough propagation conditions or equipment failures. Figure 1.9 illustrates the hierarchical interconnection topology from BSs to MSCs.

The situation described so far is not static. Mobile networks are experimenting a continuous change in technologies in use and also in services and traffic demands and patterns. It is remarkable that the exponential increase of data traffic demands that already deployed 3G networks are experimenting in the first years of the second decade of the twenty-first century.

Wireless access technologies are evolving to mobile wideband data access systems that transport packet mode traffic sources using direct Ethernet interfaces. This evolution has taken place in parallel to the increasing demand of basic data services, streaming, multimedia applications, and real-time mobile television, among others. All these services require higher bandwidths and, at the same time, are more sensitive to network delays. The emerging network technologies that are being used to serve the earlier-mentioned demands are known as wideband mobile systems and some technology examples are high-speed packet access, evolution-data optimized access, long-term evolution (LTE), and worldwide interoperability for microwave access. As a reference value, LTE cells are being designed with a target bitrate of 100 Mbps, with native IP BS Ethernet transport interfaces, that substitute E1/T1 connections.

The transport infrastructure of these new wideband mobile standards will have more restrictive requirements in terms of bandwidth, deployment flexibility, spectrum efficiency, and price. As the access networks evolve to optimized systems with capacities and efficiencies close to the Shannon limit, and all IP operation, the associated transport infrastructure should evolve accordingly. In any case, the capacity increase is not homogeneous over the entire transport infrastructure. Links close to access nodes will have lower capacity increase demands, while systems in higher aggregation levels will require a significant increase on traffic capacity.

The introduction of the IP on the transport section will increase the total capacity of the network at a lower operation cost per capacity unit. The evolution from the TDM traffic environment to the all IP scenario is carried out following an evolutionary approach. The intermediate links will transport a mixture of TDM and Ethernet

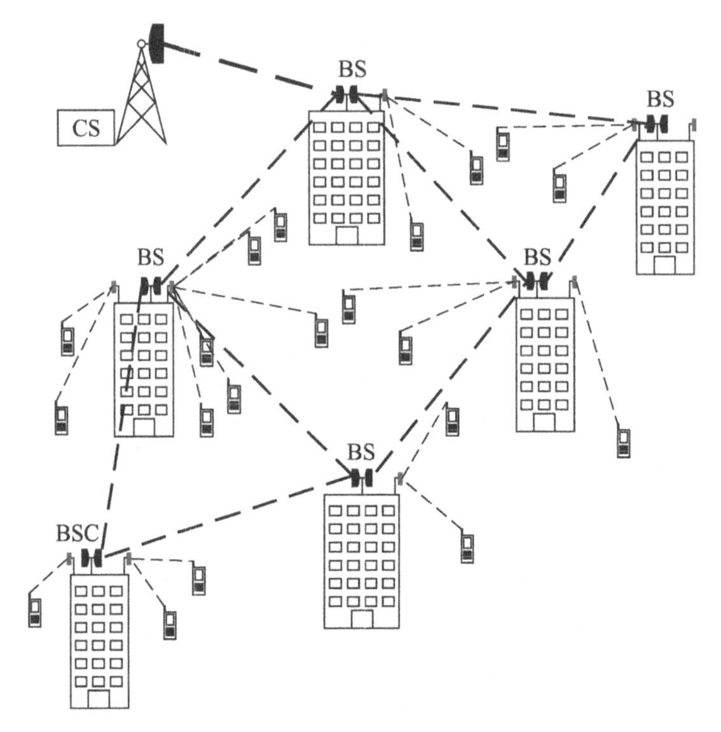

FIGURE 1.10 Microwave LOS links in a cellular mobile access network.

traffics up to the moment where the network is ready for all the IP operation. Another relevant aspect that is being studied is the adaptation of new quality measures and availability criteria, adapted to packet traffic profiles.

A paradigmatic example of the evolution described is Gigabit Ethernet Radio. This technology combines the basic features of Ethernet with high spectral-efficiency techniques that enable link throughputs around several gigabits per second, for microwave LOS links operating in usual frequency bands from 6 to 40 GHz. These high capacities are even increased if wider channeling structures are used in frequency bands around 42, 70, and 80 GHz. Figure 1.10 shows a simplified example of a cellular mobile access network with MW links associated with each access node.

1.4.3 Metro and Edge Networks

Metro networks are transport networks in urban areas based on fiber-optic rings with high capacities and usually based on synchronous optical networking SONET/SDH and Metro Ethernet standards, which transport voice, video, TV, and data traffic flows. The application of microwave LOS link in this scenario is a complementary role under certain specific conditions:

- Short-term alternative solution to fiber-optic links in cases where administrative permits for civil works delay the deployment of the fiber cable.

- Links that enable redundant paths that could reconfigure and carry traffic originally conveyed by the optic fiber ring in the cases of ring disruption. This kind of protection measure is especially relevant for high-capacity hubs, or simply nodes that are more vulnerable to accidents and dual cuts.
- Backbone extension to reach locations outside the limits of urban areas.

The microwave LOS links used in metro networks are usually designed with frequency plans in high bands, due both to the usually short link distance and the commonly high bitrate capacity requirements.

1.4.4 Fixed Access Networks

The connection of customer premises to the wideband fixed access networks is usually carried out either by means of copper pair or fiber-optic systems. This application also includes connections for LAN bridging or remote LAN connections. Microwave LOS links and high-density point-to-multipoint systems are used as alternative or complementary choices depending upon the system deployment costs if copper is not available and in cases where fast system deployment is a requirement.

Microwave LOS links used in this environment are usually high-capacity IP links, in line with the evolution tendency in access networks from ATM to IP. The frequency bands for this application are usually in the upper part of the spectrum, as the frequencies assigned to HDFS: 32, 38, 42, 52, 56, and 65 GHz. In some cases, where these bands might be saturated, links below 15 GHz are also possible. Figure 1.11 shows an example of connections by means of metro and microwave LOS links.

1.4.5 Additional Use Cases

A traditional application of microwave LOS links is the physical support to corporate networks of private companies such as utilities (electricity, gas, etc.), public security, and other industries, which might require to connect buildings and other installation facilities within an area.

Another use is LAN or personal area networks in indoor environments, where radio links are used for high-speed multimedia service connections between local devices in indoor areas, offices, etc. This application usually takes advantage of the highest bands of assigned to FSs, starting at 57 GHz. For example, in the United States, this application uses the 60 GHz band (57–64 GHz), 70 GHz (71–76 GHz), 80 GHz (81–86 GHz), 95 GHz (92–95 GHz).

An additional traditional field where microwave LOS links have been intensively used is temporary portable link installations for special events or for distress operation communications in the case of natural disasters.

Finally, it should be mentioned that microwave LOS links, in addition to their use in licensed bands, can operate in ISM bands (5 GHz), within the subband that ranges from 5.25 to 5.35 GHz and 5.725 to 5.825 GHz. If interferences to and from other services are carefully handled, these frequencies can be used to set up links in rural areas with ranges over 20 km as a means to extend urban access networks to those hypothetically isolated areas.

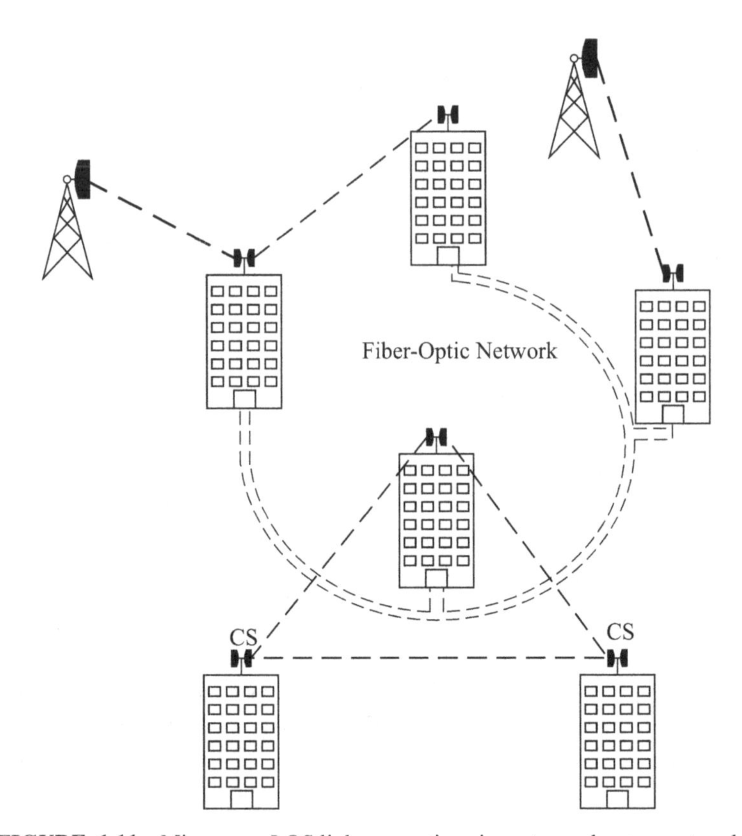

FIGURE 1.11 Microwave LOS link connections in metro and access networks.

1.5 BASIC STRUCTURE OF A FIXED SERVICE MICROWAVE LINK

Basically, a microwave LOS link is composed of a transmitting station, a receiving station, transmitting and receiving antennas, as well as the required support infrastructure, that is, a tower, to install the radiating systems.

Nodal or terminal stations of a microwave LOS link are the radiocommunication stations where the baseband (BB) payload is originated and sent to. The LOC requirement of radio links limits the maximum range between two stations. In cases where terrain irregularities or earth's curvature obstruct the LOC path between nodal stations, intermediate repeater stations will be required. There are two general types of intermediate repeater stations:

- *Passive Repeaters*: These are either simple reflecting surfaces (radioelectric mirrors) or directive antennas installed back to back through a passive transmission line. In both cases, these stations change the direction of the transmission path and are used in certain cases to avoid obstruction caused by isolated obstacles.

- *Active Repeaters*: These are radiocommunication stations that use active elements, transmitters, receivers, and radiating systems that receive, process, amplify, and transmit the signal arriving from one link hop to the next one. Depending on the processing involved, these stations can be radio frequency (RF), intermediate frequency (IF), and BB repeaters.

A hop (or link hop) in a microwave LOS link is the link section between two radio stations, either between a nodal and a repeater station or between repeater stations. A link without repeaters is a single-hop link. For obvious economical reasons, the number of hops in a link should be kept as low as possible and the hop distances should also be as short as possible.

In most application scenarios, microwave LOS links are bidirectional systems, and all the stations of the link are composed of transmitting and receiving sections. Figure 1.12 contains a block diagram of a bidirectional link with nodal and intermediate repeater stations. Figure 1.12a contains the path terrain profile illustrating how intermediate repeaters are installed to ensure LOS in all hops. Figure 1.12b shows the functional equipment blocks of the link.

The transmitter section of a link performs radio multiplexing, error correction, modulation, IF to RF frequency up-conversion, amplification and filtering functions. The receiver block, in turn, performs RF pass-band filtering, RF to IF frequency conversion, demodulation and radio de-multiplexation. Many of these functions can be integrated into common transmission and reception modules. Thus, multiplexer and demultiplexers (MULDEMs) are composed of combined radio multiplex and de-multiplexation stages; modulator demodulators combine modulation and demodulation functions and transceivers perform the frequency conversion from IF to RF and vice versa.

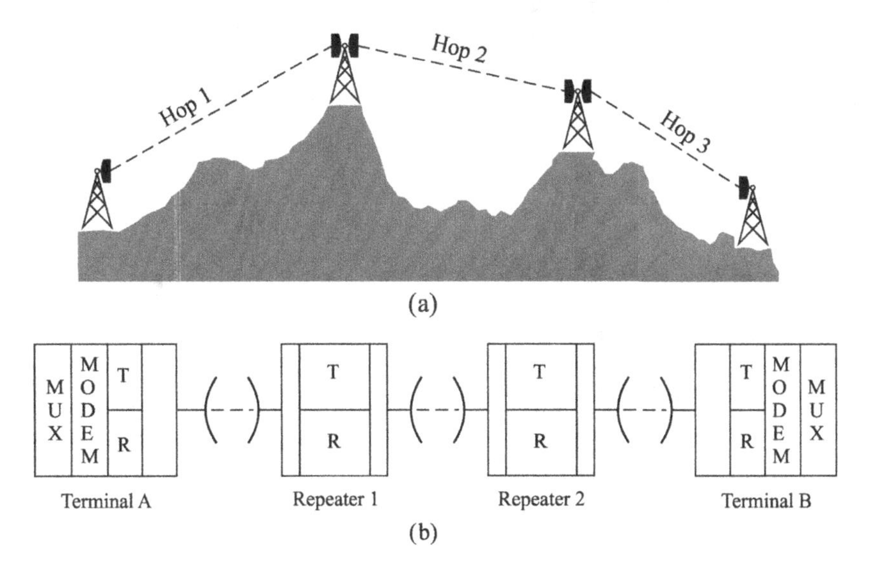

FIGURE 1.12 Building blocks of a microwave LOS link.

As illustrated in Figure 1.12, radio links are cascaded communication systems. In consequence, any interruption caused by equipment failure or severe propagation fading affects the whole link. In order to guarantee a certain availability target of the entire system, it is common practice to install redundant modules to cope with eventual equipment failures. The propagation fading occurrences cannot be avoided in most cases, but their effect can be mitigated with diversity schemes.

A microwave LOS link with M active and N reserve (backup) radio channels is designated as an M + N system. If the link has a single radio channel without any backup, it is identified as a 1 + 0 system. Reserve radio channels and associated equipment are put into service when the link suffers an interruption. The system monitoring and management is carried out by a monitoring and control system, which performs configuration, remote control, remote command (telecommand), and supervision of all the different stations, elements, and equipment that compose the microwave LOS link.

Microwave LOS links can be part of a variety of network architectures. The usual configurations are as follows (see Figure 1.13):

- *Point-to-Point*: A link which connects two terminal stations that conveys either unidirectional or bidirectional traffic.

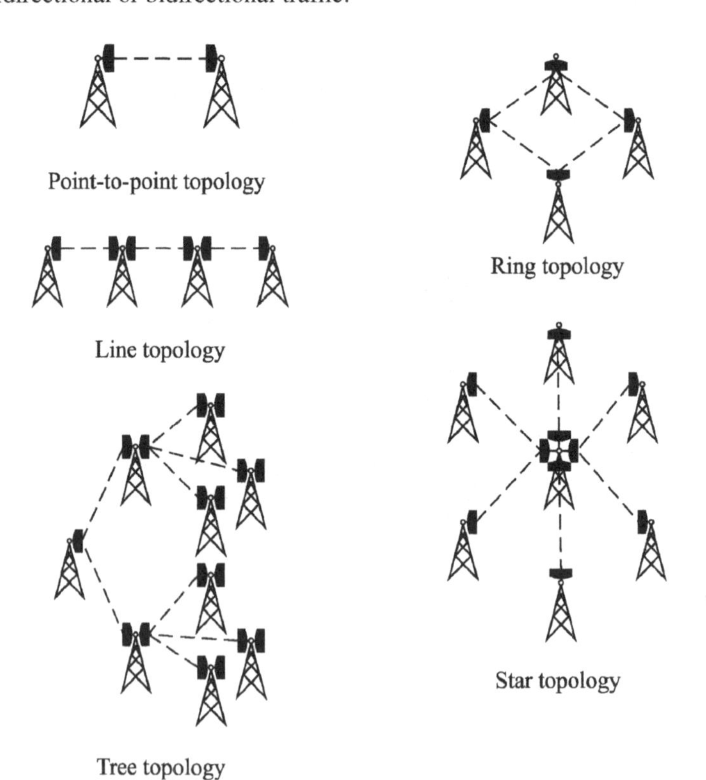

FIGURE 1.13 Microwave LOS link network topologies.

- *Line*: Each station is connected to another two (East and West), except for the nodes that are terminal stations of the entire line structure. This architecture is a cascaded composition of several point-to-point links. The nodes in this topology are usually referred as *edge nodes*.
- *Tree*: This layout is composed of a hierarchy of stations following a structure that resembles a tree. The nodes in this topology are called *aggregation nodes*, as they concentrate traffic arising from all the links connected to each node to be transmitted to the next hierarchy branch.
- *Ring*: The nodes are connected with links that form a ring. Any destination node of a ring can be reached following two paths, clockwise and counterclockwise direction.
- *Star*: All nodes of the network are connected by means of a central node.
- *Hybrid or Mesh*: This topology is a mixture of the previous layouts. Hybrid networks use a combination of other structures, so the final network architecture does not follow a dominating pattern.

1.6 SPECTRUM MANAGEMENT ASPECTS

One of the first steps when planning a microwave LOS link is the selection of the frequency band where the system will operate. The choice is not arbitrary and the radio channel plan must use resources from the available portions of the spectrum that have been allocated to the FWS by the ITU-R first and adopted with minor variations by the incumbent national or regional administration.

Usually, when requesting administrative permits for deploying a microwave LOS system, there is usually a requirement for specifying the emission bandwidth of the signals involved that, in turn, will depend on the specific channel arrangements of the selected band, the system capacity requirements, and the maximum acceptable interference levels.

It should always be kept in mind that the radio spectrum is a natural resource and as such, though reusable, is scarce. Consequently, it should be optimally used so that it can be utilized by as many as possible stations while maintaining a minimum mutual interference. For this purpose, spectrum engineering is required to manage and plan all aspects associated with the use of frequency bands for specific services. Worldwide, a relevant portion of the spectrum engineering effort is carried out by the ITU-R, which is the major international spectrum regulator. The framework for spectrum management procedures is compiled on the Radio Regulations. This document is a live compilation of the decisions taken at the World Radiocommunication Conferences. The Radio Regulations also contain Annexes, Resolutions, and Recommendations produced at the ITU-R in order to guarantee a rational, equal, efficient, and economical use of the spectrum.

World Radiocommunication Conferences are held periodically to facilitate a continuous review and, if necessary, modify the contents of Radio Regulations. In addition, Regional Radiocommunication Conferences are also called by the ITU or a group of countries, in order to agree on a specific service or frequency band. The body within the ITU-R that guarantees the application of rules and dispositions contained by the Radio Regulations is the Radiocommunication Office, which also updates and manages the Master International Frequency Register that contains worldwide frequency assignments.

1.6.1 ITU-R Radio Regulations: Spectrum Parameters and Definitions

The basic concepts and definitions related to generic radiocommunication systems are included in the first volume of the Radio Regulations. This section compiles the most relevant concepts and definitions related to FWS links, which will be later used in further chapters of this book. The list of definitions that follow regard first spectrum-related terms and second equipment-related nomenclature. The frequency spectrum-related definitions are as follows:

Frequency Band Allocation: The allocation of a given frequency band is the definition of the purpose of its use by one or more terrestrial or space radiocommunication services or the radio astronomy service under specified conditions.

Frequency Allotment: A frequency allotment is the designation of a frequency or a channel in an agreed plan, adopted by a competent conference, for use by one or more administrations for a terrestrial or space radiocommunication service in one or more identified countries or geographical areas and under specified conditions.

Frequency Assignment: A frequency assignment is an authorization given by an administration for a radio station to use a RF or RF channel under specified conditions.

Emission characteristics and equipment-related nomenclature are as follows:

Emission: An emission is a radiation produced, or the production of radiation, by a radio transmitting station. Radiation is the outward flow of energy from any source in the form of radio waves.

Unwanted Emissions: Unwanted emissions consist of spurious emissions and out-of-band emissions.

Out-of-Band Emissions: Out-of-band emissions are those on a frequency or frequencies immediately outside the necessary bandwidth, which results from the modulation process, but excluding spurious emissions.

Spurious Emission: Spurious emission is the one produced on a frequency or frequencies that are outside the necessary bandwidth and the level of which may be reduced without affecting the corresponding transmission of information. Spurious emissions include harmonic emissions, parasitic emissions, intermodulation products, and frequency conversion products, but exclude out-of-band emissions.

Assigned Frequency Band: The assigned frequency band is the frequency range within which the emission of a station is authorized; the width of the band equals the necessary bandwidth plus twice the absolute value of the frequency tolerance.

Assigned Frequency: The assigned frequency is the center of the frequency band assigned to a station.

Characteristic Frequency: The characteristic frequency is a frequency that can be easily identified and measured in a given emission. A carrier frequency may, for example, be designated as the characteristic frequency.

Reference Frequency: The reference frequency is a frequency having a fixed and specified position with respect to the assigned frequency. The displacement of this frequency with respect to the assigned frequency has the same absolute value and sign that the displacement of the characteristic frequency has with respect to the center of the frequency band occupied by the emission.

Frequency Tolerance: The frequency tolerance is the maximum permissible departure by the center frequency of the frequency band occupied by an emission from the assigned frequency or by the characteristic frequency of an emission from the reference frequency.

Necessary Bandwidth: The necessary bandwidth is, for a given class of emission, the width of the frequency band, which is just sufficient to ensure the transmission of information at the rate and with the quality required under specified conditions.

Occupied Bandwidth: The occupied bandwidth is the width of a frequency band such that, below the lower and above the upper frequency limits, the mean powers emitted are each equal to a specified percentage $\beta/2$ of the total mean power of a given emission. Unless otherwise specified in an ITU-R Recommendation for the appropriate class of emission, the value of $\beta/2$ should be taken as 0.5%.

Peak Envelope Power: The peak envelope power of a radio transmitter is the average power supplied to the antenna transmission line by a transmitter during one RF cycle at the crest of the modulation envelope taken under normal operating conditions.

Mean Power: The mean power of a radio transmitter is the average power supplied to the antenna transmission line by a transmitter during an interval of time sufficiently long compared with the lowest frequency encountered in the modulation taken under normal operating conditions.

Carrier Power: The carrier power of a radio transmitter is the average power supplied to the antenna transmission line by a transmitter during one RF cycle taken under the condition of no modulation.

Antenna Gain: The gain of an antenna is the ratio, usually expressed in decibels, of the power required at the input of a loss-free reference antenna to the power supplied to the input of the given antenna to produce, in a given direction, the same field strength or the same power flux density at the same distance. When not specified otherwise, the gain refers to the direction of maximum radiation. The gain may be considered for a specified polarization. Gain is usually considered relative to an ideal isotropic radiator isolated in free space conditions and it is referred as isotropic gain or absolute gain (Gi).

Equivalent Isotropically Radiated Power: The EIRP is the product of the power supplied to the antenna and the antenna gain in a given direction relative to an isotropic antenna (absolute or isotropic gain).

Effective Radiated Power (ERP): The ERP in a given direction is the product of the power supplied to the antenna and its gain relative to a half-wave dipole in a given direction.

1.6.2 Frequency Allocations

The ITU-R has divided the world into three regions for regulatory purposes related to frequency allocation and associated procedures. These regions are shown in Figure 1.14 from ITU-R Radio Regulations.

The allocations of the frequency bands to radiocommunication systems in each one of the three ITU-R regions are compiled on the Table of Frequency Allocations of ITU-R. Currently, there are close to 40 different service types worldwide categorized by the ITU-R that have an entry on the Table of Frequency Allocations.

Service allocations are classified into primary and secondary. Stations associated with a secondary service should not cause harmful interference to stations of primary services to which frequencies are already assigned or to which frequencies may be assigned at a later date. Secondary services cannot claim protection from harmful

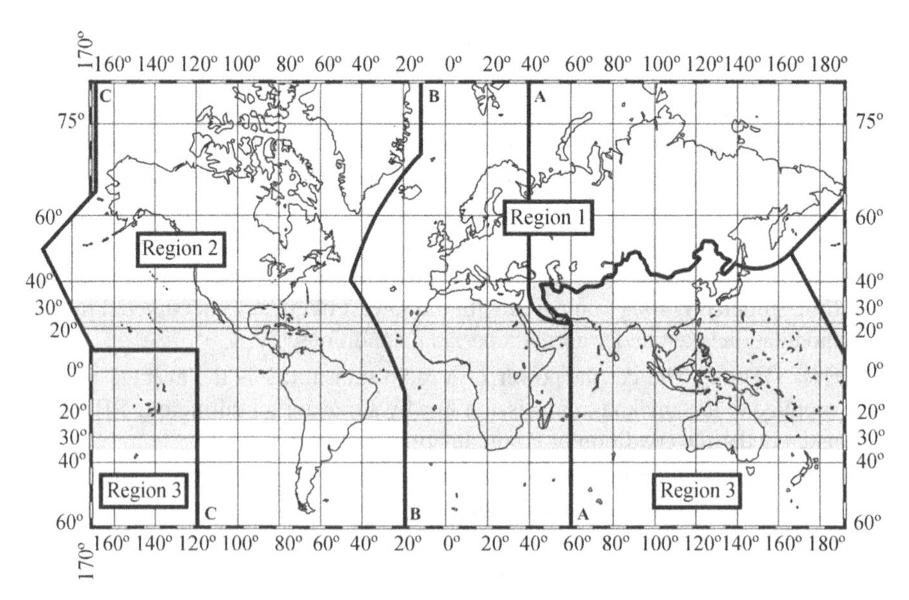

FIGURE 1.14 ITU-R regions. (*Figure Courtesy of ITU-R.*)

TABLE 1.1 ITU Frequency Bands and Global Allocations of the Fixed Service

Band Number	Band Symbol	Frequency Range (Lower Limit Excluded but Including Upper Limit)	Metric Subdivision	Fixed Service Allocation
4	VLF	3–30 kHz	Myriameter band	
5	LF	30–300 kHz	Kilometer band	
6	MF	300–3000 kHz	Hectometer band	
7	HF	3–30 MHz	Decameter band	
8	VHF	30–300 MHz	Meter band	
9	UHF	300–3000 MHz	Decimeter band	x
10	SHF	3–30 GHz	Centimeter band	x
11	EHF	30–300 GHz	Millimeter band	x
12		300–3000 GHz	Decimillimeter band	

VLF, very low frequency; LF, low frequency; MF, medium frequency; VHF, very high frequency; UHF, ultra high frequency; SHF, super high frequency; EHF, extremely high frequency

interference from stations of a primary service to which frequencies are already assigned or may be assigned at a later date. However, secondary stations can claim protection from harmful interference from stations of the same or other secondary services to which frequencies may be assigned at a later date.

Table 1.1 contains the frequency band division of the radio spectrum adopted by the ITU-R. This table includes, in relation to this book, those bands that contain global allocations for the Fixed Service.

Each one of the frequency bands allocated to the FS for use in microwave LOS links is divided according to different channel widths and a variety of possibilities for number of channels. These arrangements are described in Recommendations of the ITU-R F Series. Specifically, ITU-R Recommendation F.746 contains a useful guide of the ITU-R arrangements for microwave LOS links in different bands, with references to the documents that describe the specific arrangements of each band. The specifications for upper frequency bands have been in line with the technological developments in equipment and devices. Figure 1.15 shows the evolution in the use of HF bands during the last decades of the present and the past century.

The regulatory body that manages the use of the spectrum in the United States is the Federal Communications Commission (FCC). The FCC is a federal agency that establishes the technical specifications of radiocommunication systems and elaborates the Frequency Allocation Table for national use, based on the ITU-R's Table of Frequency Allocations.

In Europe, the Conference of European Post and Telecommunication Administrations (CEPT), through the Electronic Communications Committee (ECC), develops a common policy for regulating electronic communications, harmonizing the use of the spectrum resources at European level. All the documentation and regulatory documents produced by the ECC is distributed by the European Radiocommunications Office, which also provides detailed information about the work carried out by the ECC.

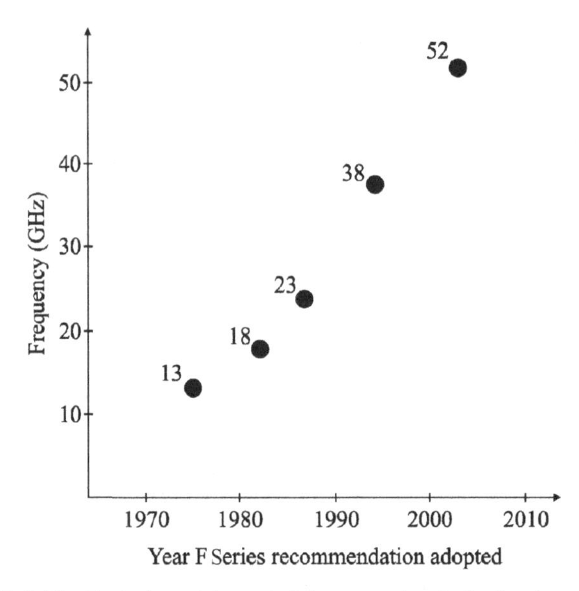

FIGURE 1.15 Evolution of the use of frequency bands for fixed service links.

1.7 FIRST APPROACH TO THE DESIGN OF A MICROWAVE LOS LINK

The design of a microwave LOS link system is a complex task that involves different calculation steps and design procedures that are (almost all of them) interrelated. Even though this topic will be described in detail in 3, 5, further chapters, a first approach to the design of a microwave LOS link should include at least the following phases and tasks:

Phase 1 Preliminary Studies. This phase involves the following generic tasks.

1. Analysis of the specifications and study of the application for which the link will be designed. This task involves an evaluation of the transport technologies upon which the link will be designed (TDM, PDH/SDH, ATM, IP, Ethernet) in relation to the capacity requirements of the application and the possible restrictions arriving from the network that the link is going to belong.
2. Study of the appropriate frequency band (in many cases, this is a specification that cannot be modified).
3. Equipment selection and equipment specification studies. Analysis of the capacities provided by different manufacturers and models, BB and multiplexation options, system upgrade and extension possibilities, diversity and redundancy schemes allowed, etc.
4. Study of the availability and error performance objectives and the allocation of a portion to the MW link in relation to the network where the system will be

installed. This task will be based on reference values found in ITU-T and ITU-R Recommendations for availability and error performance objectives unless the operator of the radio link specifies its own values from experience in previous designs on the specific geographic area.

5. First analysis of the link radio route and terrain profile. This first path analysis will identify the number of hops and the candidate sites for intermediate repeater stations, if those were required.

Phase 2 Detailed Link Design.

1. Design of an initial frequency plan. This task will propose the radio channel arrangements in each one of the hops of the link.
2. Detailed study of the radioelectric route. Intermediate repeater station choice and calculations associated with terrain profiles (antenna heights, clearance criteria, etc.).
3. Assignment of error performance objectives to the different sections (hops) of the radio link and analysis of the system threshold values.
4. Link budget design in each one of the link hops. Evaluation of system margins and preliminary decision about the use of diversity and redundancy techniques.
5. Interference analysis. Study of intrasystem interferences and optimization of the radio channel plan. Decision about the need for special antennas that might mitigate interference problems in complex frequency reuse scenarios.

Phase 3 Installation, Tests, Operation, and Maintenance.

1. Inspection of path obstacles and relevant spots in the field. Site redesign and antenna height recalculation if necessary.
2. Equipment setup and installation. System tests to evaluate background bit error rates (BBERs), system threshold checks, identification of unexpected interference problems, etc.
3. Link operation and maintenance.

The different design process blocks are depicted in Figure 1.16. The figure also contains a reference to the chapters where the different design procedures and calculations will be described in this book. The block diagram also contains an indication of the processes that are more closely related. Nevertheless, it is important to note that there are design dependencies that have not been drawn in the picture, for the sake of visual simplicity of the diagram. In fact, a microwave LOS link design should be regarded as a group of interrelated methods whose partial outputs influence practically the rest of design modules and steps.

The remaining sections of this chapter cover the basic calculations associated with the link budget evaluation, system threshold definitions, noise calculations, and preliminary introduction to interference calculations. These sections have been deliberately included in this introductory chapter due to their basic nature. Any radiocommunication engineer should be familiar to these basic concepts, as they are quite common practice in practically all radiocommunication systems.

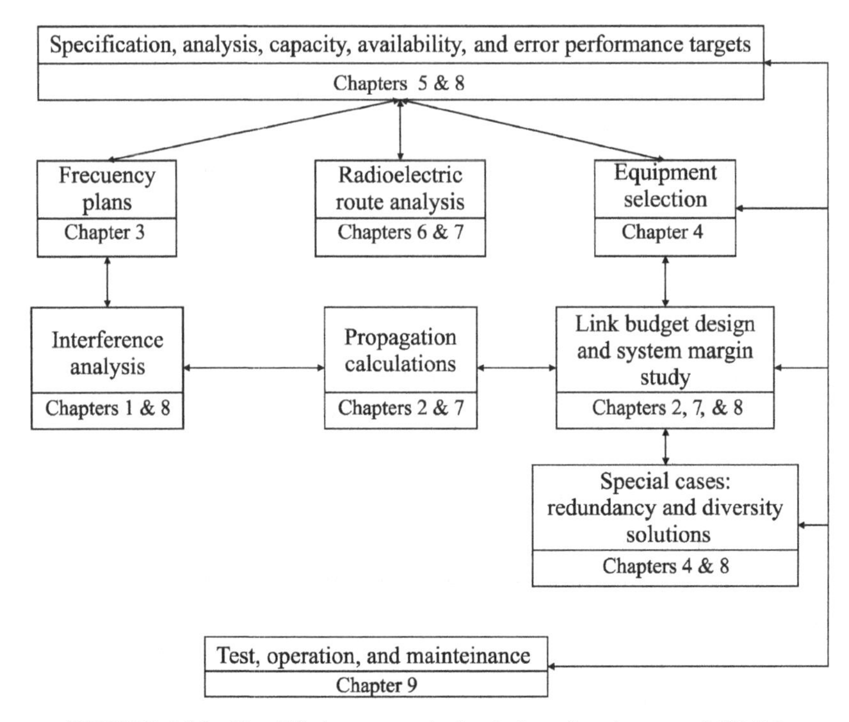

FIGURE 1.16 Simplified processes in the design of a microwave LOS link.

1.8 LINK BUDGET BASICS

1.8.1 Link Budget

The link budget of a microwave LOS link is the set of calculations that relate the available power at the receiver input with the transmitted power, the losses in transmission lines, losses in antenna distribution networks, antenna gain values, and attenuation suffered by the link signals along the propagation path.

Figure 1.17 shows a block diagram that includes all the generic elements of a radiocommunications system, from the transmitter to the receiver. This diagram is the model commonly used to illustrate the calculations involved in a link budget as well as the parameters that should be taken into account.

The block diagram contains the following elements associated with the transmission equipment:

1. Transmitter (TX).
2. Antenna distribution and coupling circuits: antenna feeder, multiplexors, etc. The interface T (physical interface) is defined between the transmitter and the antenna coupling elements.
3. Antenna circuit, which accounts for all the elements associated with losses of the antenna.

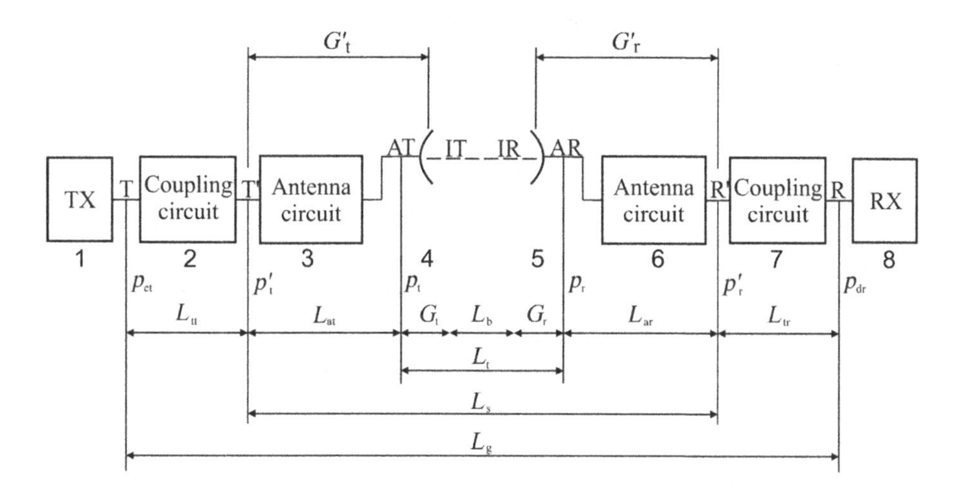

FIGURE 1.17 Model of a radiocommunication system for link budget calculation purposes.

4. Ideal directional antenna. This is a virtual block representing a lossless antenna which in combination with the antenna circuit is equivalent to the real physical antenna. The interface AT (virtual interface) is defined between the antenna circuit and the ideal antenna.

On the receiver side, the following blocks are defined:

1. Ideal lossless antenna for receiving purposes.
2. Reception antenna circuit. Equivalent to the antenna circuit in transmission. This block and the previous lossless antenna element compose the real receiving antenna.
3. Coupling circuits from the antenna to the receiver. These circuits are connected to the receiver at the physical interface R′ and are composed of duplexors, filters, transmission line sections, etc.
4. Receiver (RX).

1.8.1.1 Power Levels All the power magnitudes in the link calculation formulae are always expressed in decibels and associated logarithmic units. A compilation of the different parameters associated with power levels in a link budget is shown on Table 1.2. Absolute values are usually expressed in decibel-milliwatt units (power level relative to 1 mW).

The whole model assumes perfect impedance matching condition for all interfaces.

1.8.1.2 Gain and Losses Following the block diagram shown in Figure 1.17, Table 1.3 contains the definition of all the losses associated with the different elements of the system. All magnitudes shown are in decibels.

TABLE 1.2 Power Levels and Related Magnitudes in a Link Budget (as in Figure 1.17)

P_{tx} (dBm)	Power delivered by the transmitter to the antenna circuit
P'_t (dBm)	Power delivered to the real antenna at the input of the antenna circuit
P_t (dBm)	Virtual power delivered to the ideal lossless antenna and equivalent to the total radiated power
EIRP (dBm)	Equivalent isotropic radiated power in the direction to the receiver
P_r (dBm)	Available power (virtual) in the lossless ideal receiving antenna
P'_r (dBm)	Available power (real) at the antenna circuit output (coupling network input)
P_{rx} (dBm)	Available power at the receiver input

The losses associated with the antennas (antenna circuits), L_{at} and L_{ar}, are usually given in the form of efficiency values (η_{at}, η_{ar}) either in linear or logarithm units, as shown in equation (1.1):

$$\begin{aligned} l_{at} &= \frac{100}{\eta_{at}(\%)} \\ l_{ar} &= \frac{100}{\eta_{ar}(\%)} \end{aligned} \quad \text{or in logarithmic units} \quad \begin{aligned} L_{at} &= 10 \log(100/\eta_{at}) \\ L_{ar} &= 10 \log(100/\eta_{ar}) \end{aligned} \tag{1.1}$$

Additionally, a difference should be made between the power gain and directive gain of the antenna systems of the link. These gain values are represented by G'_t/G'_r for power gain and G_t/G_r for directive gain. Power gain values take into account the gain of the ideal antenna and reduce this value according to the antenna losses (efficiency). The basic loss, the transmission loss, and total loss are then:

$$\begin{aligned} L_t &= L_b - G_t - G_r \\ L_s &= L_t + L_{at} + L_{ar} = L_b - G'_t - G'_r \\ L_g &= L_s + L_{tt} + L_{tr} \end{aligned} \tag{1.2}$$

where L_t, L_s, and L_g have been defined in Table 1.3. In general, the losses associated with antenna circuits are usually low and power gain values are usually considered

TABLE 1.3 Losses in the Link Budget Calculation (as in Figure 1.17)

L_{tt}	Losses on the antenna distribution and coupling circuits, between T and T′ interfaces
L_{at}	Losses on the antenna circuits on the transmission side, between T′ and AT interfaces
L_{tr}	Losses on the distribution and coupling circuits connecting the antenna and the receiver, between R′ and R interfaces
L_{ar}	Losses on the antenna circuits on the receiver side, between AR and R′ interfaces
L_b	Basic propagation loss, function of the distance, frequency and propagation mechanism
L_t	Transmission loss. It is calculated as the basic propagation loss minus the ideal antenna gains (transmitter plus receiver)
L_s	System loss. It represents the difference in levels between the input to the real antenna in transmission and the output of the real antenna at the receiver side
L_g	Total loss, defined between the transmitter output and the receiver input

the same as directive gains. In this case, equation (1.2) can be rewritten as:

$$L_t = L_s = L_b - G_t - G_r \tag{1.3}$$

1.8.1.3 Link Budget Expression The link budget equation relates the available power at the receiver input with the transmitted power and all the loss sources reviewed in the previous section, including the ones associated with equipment and also those related to propagation:

$$P_{rx} = P_{tx} - L_{tt} + G_t - L_b + G_r - L_{tr} \tag{1.4}$$

where

P_{rx} (dBm) = available power level on the receiver input
P_{tx} (dBm) = power delivered by the transmitter to the antenna distribution network
L_{tt} (dB) = losses associated with the antenna distribution network (transmitter to antenna)
L_{tr} (dB) = losses associated with the antenna distribution network (antenna to receiver)
L_b (dB) = basic propagation loss
G_t (dB$_i$) = transmitter antenna gain (directive gain, assuming negligible antenna losses)
G_r (dB$_i$) = transmitter antenna gain (directive gain, assuming negligible antenna losses)

1.8.2 Propagation Losses

Under standard propagation conditions, in absence of hydrometeors or anomalous refraction behavior, and provided that the system is adequately designed to ensure adequate clearance above terrain obstacles, the basic propagation loss should equal the free space propagation loss value. Nevertheless, both rain, nonstandard refraction conditions, and other phenomena such as multipath produce fading. In each hop of a generic microwave LOS link route, the propagation loss can be divided into two components:

L_{fs} Basic free space loss
L_{bexc} Basic losses in excess to the free space loss

The free space loss formula is given by equation (1.5)

$$L_{FS} = 92.45 + 20\log f + 20\log d \tag{1.5}$$

where

f (GHz) = frequency
d (km) = propagation distance

The basic losses in excess to the free space loss include all the additional losses suffered by the signal along the propagation path that cannot be associated with the free space loss. Strictly speaking, all the different excess loss components will be represented by statistic variables that will depend on how the different physical phenomena involved in propagation (refraction, diffraction, reflection, scattering, etc.) affect the signal at each moment.

From a practical standpoint, not all the propagation losses have a variation high enough to be worth modeling as variables. Consequently, for system design purposes, the excess losses are usually divided into constant or fixed excess losses and variable excess losses. The loss sources commonly considered constant are as follows:

L_{gases} Gas (O_2) and water vapor (H_2O) absorption

$L_{vegetation}$ Vegetation attenuation values. If necessary, this term is calculated using empirical attenuation values for different vegetation and polarization situations. Under standard system deployment conditions, if the link is designed correctly, this term should be zero.

The excess losses that are variable over time, L_{bexcv}, are those ones that show a variability that is relevant enough to be included in the link design and calculation processes. Those variable losses are usually referred as *fading*. Fading sources can be classified as follows:

$L_{diffraction}$ Diffraction in obstacles (caused by anomalous refraction, see Chapter 2).

$L_{scintillation}$ Tropospheric scintillation fading

$L_{hydrometeors}$ Fading caused by rain and other hydrometeors

$L_{multipath}$ Multipath fading, including reflection effects on earth surface and multipath originated due to various refraction phenomena in higher troposphere layers.

L_{XPD} Depolarization losses (associated with hydrometeors and anomalous refraction)

$L_{misaligment}/L_{beam\ spreading}$ These losses are caused by anomalous refraction conditions. The refraction creates in both cases a variation of the angles of the signal at different sections of path (angle variation at the transmitter/receiver antenna, changes along the path defocusing the signal beam). This variation is independent of frequency and occurs specially on coastal regions and humid climates, where refraction phenomena are more relevant.

Figure 1.18 shows a diagram representing the power level values at different points of the link model. This type of graph is called hypsogram. The figure introduces the concept of link margin, an important concept in the system design process. In this specific case, the figure represents the difference between fixed losses (associated with free space conditions plus absorption from gases) and variable losses.

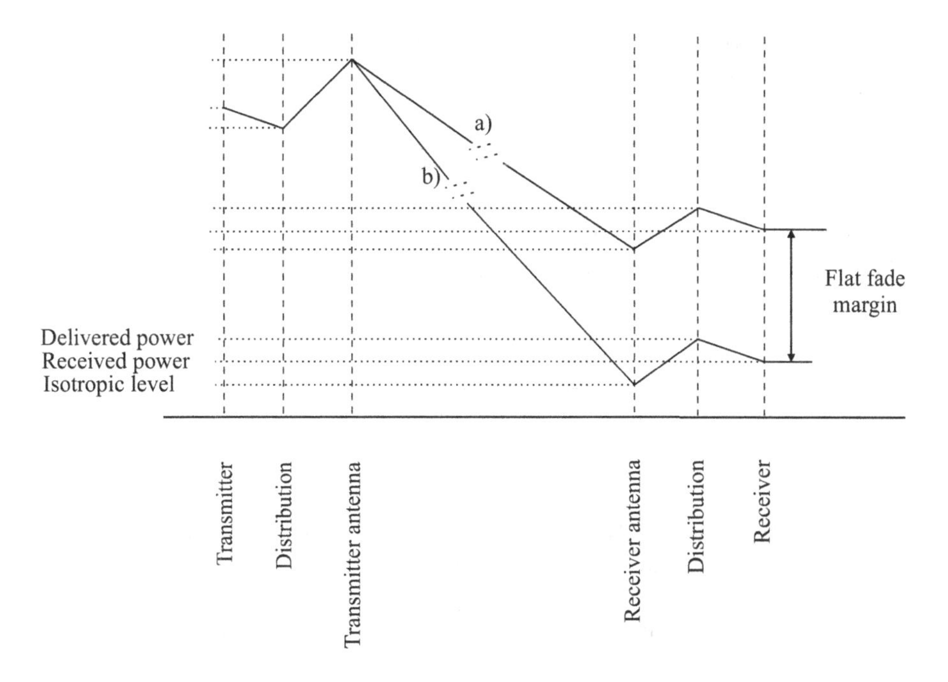

a) Fixed basic losses
- L_{fs} : Propagation basic losses (free space)
- L_{gases} : Gases and water vapor absorption
- $L_{vegetation}$: Vegetation attenuation

b) Time-varying losses in excess
- $L_{diffraction}$: Diffraction losses (anomalous refraction)
- $L_{scintillation}$: Atmospheric scintillation losses
- L_{rain}: Losses due to rain and other hydrometeors
- $L_{mutipath}$: Losses due to multipath
- $L_{beam\ spreading}$: Fading due to beam spreading

FIGURE 1.18 Hypsogram (power level diagram) in a microwave LOS link.

The equipments in real links have a certain capacity to cope with flat fading using automatic gain control (AGC) equipment that could compensate, at least partially, fading occurrences of the received signal. The benefit obtained with AGC circuits will depend on each manufacturer's compromise choice between complexity, gain, bandwidth, and associated noise and nonlinear degradations that any active circuit will introduce in a receiving system.

1.8.3 Threshold Values and Gross Fade Margin

The gross fade margin or fade margin in a link budget is the difference between the power level received in nominal conditions, that is, absence of fading, and the system threshold level associated with a specified error performance condition. The fade

TABLE 1.4 Theoretical Thresholds ($w = E_b/N_0$) Associated with Different Modulation Schemes

	E_b/N_0 (dB) at (BER = 10^{-3})	E_b/N_0 (dB) at (BER = 10^{-6})
BPSK	6.8	10.5
QPSK	6.8	10.5
4-DPSK	9.1	12.8
8-PSK	10.0	13.8
16-QAM	10.4	14.4
64-QAM	14.7	18.8
256-QAM	19.3	23.5

QPSK, quadrature phase shift keying; QAM, quadrature amplitude modulation; BPSK, binary phase shift keying; DPSK, differential phase shift keying

margin is expressed in decibels. It is obvious that the study of the receiver thresholds is a requirement for obtaining the fade margin.

The receiver threshold is the minimum power level at the input receiver that is associated with a specific BER. This parameter is usually referred as T_h and it is commonly expressed in decibel-milliwatt. Under thermal noise perturbation conditions, the threshold will depend on the modulation scheme, the bitrate, the channel coding algorithm, and the equivalent thermal noise at the receiver input. The theoretical calculation of the threshold value is based on the E_b/N_0 value associated with the BER objective. The E_b/N_0 is the relationship between the energy per bit and the spectra density of the noise. Table 1.4 shows a list of commonly used modulation schemes and associated E_b/N_0 threshold values for usual error performance objectives (10^{-3} and 10^{-6}).

Once the value of w is set, the power level threshold value is immediate, provided the bitrate R and the receiver noise figure F are known. The calculation in its linear version is:

$$w = \frac{e_b}{n_o} = \frac{T_h}{kT_o f R} \tag{1.6}$$

or in decibel units:

$$T_h(\text{dBm}) = W(\text{dB}) + F(\text{dB}) + 10 \log R(\text{bps}) - 174 \tag{1.7}$$

where

T_h = power level threshold associated with a specific BER
K = Boltzman constant
T_0 = standard temperature (usually 270°)
f = receiver noise factor (F, noise figure in decibels)
R = bitrate

TABLE 1.5 BER and ITU-R Recommendations for Threshold Calculation in Microwave LOS Links

Criteria		ITU Recommendations	BER
SESR	Availability	ITU-T: G.821, G.827 // ITU-R F.1703	10^{-3}
SESR	Error performance	ITU-T: G.826, G.828, G.829 // ITU-R F.1668	10^{-6}, $1.7 \cdot 10^{-5}$
BBER	Error performance	ITU-T: G.826, G.828, G.829 // ITU-R F.1668	10^{-12}

In practice, the threshold values are slightly higher than the ones based on theoretical expressions. Theoretical values, such as the ones shown in Table 1.4, are usually calculated for a Gaussian channel transmission model. The real propagation channel will be affected by different perturbation sources in addition to the thermal noise (interferences, phase noise in carrier recovery circuits, jitter in clock recovery modules, distortion, nonlinearities, etc.). Exception made for interference, it is assumed that the threshold values that manufacturers provide, already account for any degradation arriving from the equipment design or manufacturing tolerances.

There is a quite numerous group of ITU-R and ITU-T recommendations that provide guidance on the study of system thresholds and performance objectives. As it will be developed in further chapters, in the case of microwave LOS links, this analysis will be based on a twofold process in order to take into account unavailability and error performance. Unavailability will be associated with periods where the link is completely unusable due to equipment failures or severe propagation impairments. Error performance degradation will relate to periods where the link is operating with different degrees of degraded performance. Table 1.5 contains a first summary of the relevant recommendations for microwave LOS link design.

A new set of parameters have been deliberately introduced in Table 1.5. These parameters are used to define the criteria for evaluating availability and error performance. These parameters are the severely errored second rate and the background block error rate (or BBER) and they will be described in depth in Chapter 5. The reason for defining these parameters arises from the difficulty of a single BER value to provide accurate information about the system behavior. In the event of a certain BER value, questions would arise such as the following: How long has the BER value measured? How is the statistical distribution of the errors across the measured bit group? How accurate are bit error measurements if the system is on service? These questions make BER specifications inadequate, especially for monitoring the system quality once it is under normal operation. In order to provide a solution to these uncertainties associated with BER in operating links, the ITU has defined severely errored second rate and BBER parameters that are based on evaluating the existence of errors in blocks of transmitted bits. The advantage of this approach is the fact that most digital systems already group bites into blocks for channel coding and error correction and, at the same time, already provide mechanisms to identify erroneous blocks (where all errors have not been corrected). Thus, specifying the quality performance objectives is much more convenient for the operator of the microwave LOS link. The disadvantage is the fact that SESR and BBER are not related directly to E_b/N_0, and thus, in any case, there should be a reference BER value that is associated with the system performance objectives in order to evaluate the system power thresholds. Table 1.5 summarizes a long, complex, and controversial discussions over the past years in order to associate ITU objective parameters with BER values.

The SESR parameter from Recommendation UIT-T G.821 is 10^{-3}, which is mostly used for unavailability calculation purposes. Error performance objectives after Recommendation ITU-T G.826 (including ITU-R F.1668) are usually associated with a BER value of 10^{-6} (sometimes $1.7{\cdot}10^{-5}$). Usually, BBER is associated with a BER 10^{-12}.

The threshold parameter W will be indexed with the appropriate exponential of the BER objective; thus, T_{h3} and W_3 will be associated with a BER of 10^{-3}, T_{h6} and W_6 to 10^{-6} , T_{h5} and W_5 to $1.7{\cdot}10^{-5}$, and finally T_{h12} and W_{12} to 10^{-12}.

The manufacturer will provide the equipment thresholds under defined conditions. Should this information not be available, a common practice is the use of the reference theoretical value associated with the modulation and coding schemes and increase this threshold in 5 dB. Once the thresholds are established, the link margin calculation will be straight ahead. The gross fade margin also called thermal margin or flat fade margin for different BER thresholds (10^{-6}, $1.7{\cdot}10^{-5}$, and 10^{-12}) will be:

$$\begin{aligned} M_3 &= C - T_{h3} \\ M_5 &= C - T_{h5} \\ M_6 &= C - T_{h6} \\ M_{12} &= C - T_{h12} \end{aligned} \tag{1.8}$$

where C (dB) is the power level at the receiver input. In relation to the threshold values and system margin, the system gain can be expressed as:

$$\begin{aligned} GS_3 &= P_t - T_{h3} \\ GS_5 &= P_t - T_{h5} \\ GS_6 &= P_t - T_{h6} \\ GS_{12} &= P_t - T_{h12} \end{aligned} \tag{1.9}$$

The system gain is a parameter usually provided by the manufacturer of microwave LOS link equipment that enables a fast estimation of the link length and antenna gain requirements.

1.9 NOISE

When analyzing the design or the performance of a radiocommunication system, in addition to the study of the fading associated with propagation phenomena in the atmosphere, it is necessary to extend the study for taking into account additional performance degradation caused by thermal noise and interference. Special countermeasures might be necessary to cope with these additional sources of impairment and reduce their impact the on system performance.

The radio noise received by a radiocommunication system is a random process, which is associated with RF radio noise signals that do not convey any information;

although, in certain situations, it might have associated information about its source, nature, and location, and there is a possibility that might be superimposed on or combined with the wanted signal. The ITU-R Recommendation P.372 provides radio noise reference data that are classified depending upon the noise source type:

- Noise from natural sources, such as space radioelectric sources, earth surface and other obstacles intersecting the antenna beam, noise originated by lightning, and noise emissions from hydrometeors and atmospheric gases.
- Man-made noise or artificial noise, which is the aggregation of different unwanted emissions from electric machines, electric and electronic equipment, energy transmission lines, internal combustion engine switching, etc. As opposed to noise arriving from natural sources, that has a practically flat frequency response, man-made noise is lower as the frequency increases. In consequence, it is only considered for practical designs below 1 GHz.

The noise becomes an impairment that imposes a limit to the performance to any radiocommunications system. The evaluation of its influence is carried out through the normalized total noise power level, which includes the noise received by the antenna, the noise generated by the antenna and the distribution network, and the internal noise of the receiver. If the network net gain is equivalent to zero decibel, the normalized power can be expressed by the general equation (1.10). This equation is linear so the net gain is considered to be 1.

$$p_{\mathrm{n}} = kT_0bf \tag{1.10}$$

where

k = Boltzmann's constant $1.38 \cdot 10^{-23}$ J/K
T_0 = reference (standard) temperature (°K), usually 290°K
b = receiver noise equivalent bandwidth (Hz)
f = receiver system noise factor

Equation (1.10) can be expressed in logarithmic units as follows:

$$P_{\mathrm{n}} = F + B - 204 \tag{1.11}$$

where

P_{n} = noise available power (dBW)
$B = 10 \log b$ (dBHz)
$-204 = 10 \log k\, t_0$ (dBW/Hz)
F = noise figure of the receiving system (dB)

The noise factor of the receiving system, f, is composed of a number of noise sources in the receiver chain. This parameter should include both the noise coming from outside and internal noise of the system as mentioned in previous paragraphs. The generic calculation method of the system noise factor follows.

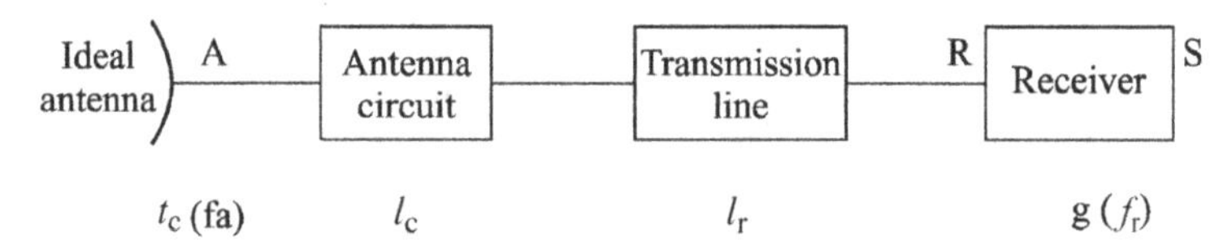

FIGURE 1.19 General radiocommunications receiver model for noise calculations.

The general model for noise calculations in a radiocommunications receiver is illustrated in Figure 1.19.

The model has three interfaces that are relevant for the calculation process:

- Interface "A": The ideal antenna output.
- Interface "R": RF receiver input.
- Interface "S": Receiver output.

The usual reference point to calculate noise is the input of the ideal equivalent receiver lossless antenna. This is a virtual interface due to the loss-free nature of the device because the terminals of this losseless antenna do not exist physically. For receivers free from spurious responses, the system noise factor is given by equation (1.12):

$$f = f_a + (f_c - 1) + l_c (f_t - 1) + l_c l_t (f_r - 1) \tag{1.12}$$

where f_a is the external noise factor defined by equation (1.13):

$$f_a = \frac{p_{nA}}{k t_0 b} \tag{1.13}$$

and the rest of involved parameters are

p_{nA} = available noise power from an equivalent lossless antenna
l_c = antenna circuit loss (available input power/available output power)
l_r = transmission line loss (available input power/available output power)
f_r = receiver noise factor
f_c = noise factor associated with the antenna circuit losses
f_t = noise factor associated with the transmission line losses
t_c = actual temperature of the antenna and nearby ground (°K)
t_t = actual temperature of the transmission line (°K)

$$f_c = 1 + (l_c - 1)\left(\frac{t_c}{t_0}\right) \tag{1.14}$$

$$f_t = 1 + (l_t - 1)\left(\frac{t_t}{t_0}\right) \tag{1.15}$$

if $t_c = t_t = t_0$, equation (1.12) simplifies to:

$$f = f_a - 1 + f_c f_t f_r \tag{1.16}$$

1.10 INTERFERENCES

Interference is defined by the ITU-R as the effect of unwanted energy due to one or a combination of emissions, radiations, or inductions upon reception in a radiocommunication system (microwave LOS link), manifested by any performance degradation, misinterpretation, or loss of information that could be extracted in the absence of such unwanted energy.

There have been defined three levels of interference for administrative purposes: permissible, accepted, and harmful interference. Permissible interference describes a level of disturbance, which in the given conditions involves degradation of reception quality to an extent considered insignificant, but which must be taken into account in the planning of systems. The level of permissible interference is usually laid down in ITU-R Recommendations and other international agreements. Acceptable interference describes a higher level of interference involving a moderate degradation of reception quality which in given conditions is deemed to be acceptable by the administrations concerned. The third term describes a level of interference that seriously degrades, obstructs, or repeatedly interrupts a radiocommunication service. Additional interference classifications can be found in ITU-R literature:

- Intrasystem, when the interference source and the interfered system are the same.
- Intersystem, where interfered and interfering are different systems.
- Simple interference, when there is a single interference source.
- Multiple interferences, when the interference is produced by multiple sources.
- Co-channel. The interference signal is produced in the same carrier frequency as the one used by the wanted signal, when the same radio channel is used by two or more emissions.
- Adjacent channel interference. The frequency of the interfering signal corresponds to adjacent contiguous channels. An adjacent channel is, in a given set of radio channels, the RF channel which characteristic frequency is situated next above or next below that of a given channel.

The degradation produced in the system performance caused by interference implies a higher required C/N ratio to maintain the system error performance objectives. The parameter usually employed to characterize interferences is the carrier to interference (C/I) ratio and is defined as the ratio between the level of the wanted signal and the interference power level (aggregate power in the case of multiple interference sources).

Interference analysis will be based on evaluating the C/I ratios between the interfering and interfered systems at the receiver side and ensure that those ratios are not below the specified minimum C/I threshold. The C/I calculation will be associated with the evaluation of possible radio channel frequency arrangements. If a selected radio channel scheme provides C/I ratios below the threshold, different countermeasures are possible: changing the channel scheme, using special antennas, etc. The minimum C/I ratio will be provided by the equipment manufacturer.

Interference calculations play a relevant role in applications with high link density, especially in urban areas, metro networks, mobile cellular networks, etc. The detailed study of interference calculation procedures will be covered in Chapter 8.

BIBLIOGRAPHY

An overview of the ETSI-BRAN HIPERACCESS physical layer air interface specification. K. Fazel, C. Decanis, J. Klein, G. Licitra, L. Lindh and Y.Y. Lebret. 13th Personal, Indoor, Mobile, Radiocommunications Conference. Lisbon. 2002.

Broadband Optical Access Networks. L.G. Kazovsky, N. Cheng, W. Shaw, D. Gutierrez, and S. Wong. Wiley-Blackwell, Hoboken, NJ. 2011.

Fundamentals of Telecommunications. R.L. Freeman. Wiley-Interscience, Hoboken, NJ. 4th Edition. 2004.

ITU-R Handbook: Digital Radio-Relay Systems. International Telecommunication Union. Radiocommunication Sector. ITU-R. Geneva. 1996.

ITU-R Rec. F.1399: Vocabulary of terms for wireless access. International Telecommunication Union. Radiocommunication Sector. ITU-R. Geneva. 2001.

ITU-R Rec. F.1668: Error performance objectives for real digital fixed wireless links used in 27 500 km hypothetical reference paths and connections. International Telecommunication Union. Radiocommunication Sector. ITU-R. Geneva. 2007.

ITU-R Rec. F.1703: Availability objectives for real digital fixed wireless links used in 27 500 km hypothetical reference paths and connections. International Telecommunication Union. Radiocommunication Sector. ITU-R. Geneva. 2005.

ITU-R Rec. F.746: Radio-frequency arrangements for fixed service systems. International Telecommunication Union. Radiocommunication Sector. ITU-R. Geneva. 2007.

ITU-R Rec. P.372: Radio Noise. International Telecommunication Union. Radiocommunication Sector. ITU-R. Geneva. 2009.

ITU-T Rec. G.821: Error performance of an international digital connection operating at a bit rate below the primary rate and forming part of an Integrated Services Digital Network. International Telecommunication Union. Telecommunication Standardization Sector. ITU-T. Geneva. 2002.

ITU-T Rec. G.826: End-to-end error performance parameters and objectives for international, constant bit-rate digital paths and connections. International Telecommunication Union. Telecommunication Standardization Sector. ITU-T. Geneva. 2002.

ITU-T Rec. G.827: Availability performance parameters and objectives for end-to-end international constant bit-rate digital paths. International Telecommunication Union. Telecommunication Standardization Sector. ITU-T. Geneva. 2003.

ITU-T Rec. G.828: Error performance parameters and objectives for international, constant bit-rate synchronous digital paths. International Telecommunication Union. Telecommunication Standardization Sector. ITU-T. Geneva. 2000.

ITU-T Rec. G.829: Error performance events for SDH multiplex and regenerator sections. International Telecommunication Union. Radiocommunication Standardization Sector. ITU-T. Geneva. 2002.

Mobile Backhaul. J. Salmelin, E. Metsälä. Wiley-Blackwell, Hoboken, NJ. 2012.

Radio Regulations. International Telecommunication Union. Radiocommunication Sector. ITU-R. Geneva. 2008.

Satellite Communications Systems Engineering: Atmospheric Effects, Satellite Link Design and System Performance (Wireless Communications and Mobile Computing). L.J. Ippolito Jr. Wiley, Hoboken, NJ. 2008.

TS 101 999: Broadband Radio Access Networks (BRAN); HIPERACCESS; Physical (PHY) layer protocol specification. European Telecommunications Standards Institute. ETSI. Geneva. 2002.

TS 102 177: Broadband Radio Access Networks (BRAN); HiperMAN; Physical (PHY) layer protocol specification. European Telecommunications Standards Institute. ETSI. Geneva. 2010.

WiMAX Technology and Network Evolution (The ComSoc Guides to Communications Technologies). K. Etemad and M. La. Wiley-IEEE Press, Hoboken, NJ. 2010.

CHAPTER 2

LOSS AND FADING ASSOCIATED WITH TROPOSPHERE PROPAGATION PHENOMENA

2.1 INTRODUCTION

This chapter provides a description of the physical phenomena associated with the propagation of the signals involved in fixed wireless service (FWS) systems. Microwave LOS link systems fall into this radiocommunication category. The chapter will describe the parameters that are commonly used to characterize the propagation phenomena and their influence on the perturbations suffered by the microwave LOS link signals along the transmitter–receiver path. This chapter also covers the description of the phenomena, while the calculation procedures and algorithms to quantify fading, signal depolarization, and other effects of propagation have been compiled in Chapter 7 with the aim of including them in the context of the system design. These algorithms are mostly part of the International Telecommunication Union, Radiocommunication Sector (ITU-R) P Series recommendations.

The propagation in the troposphere is a variable process that is influenced by different factors. Some of them are associated with the composition and structure of the troposphere itself, others are linked to meteorological phenomena, and a third group is caused by the influence of the Earth's surface and elements situated on it.

The troposphere is not a homogeneous medium. Its structure depends on relative humidity (RH), atmospheric pressure and temperature. All these factors are also variable, and this variability can be considered from different points of view. The pressure, temperature and water vapor density depend on the height at which they are measured, and they are usually characterized by height models. Additionally, the variation as a function of the height is not always the same in all places and it strongly

Microwave Line of Sight Link Engineering, First Edition. Pablo Angueira and Juan Antonio Romo.

depends on the latitude and longitude coordinates. The third axis of variability is time. The time variation is associated with short-term meteorological variations or midterm seasonal variations over a period of time.

When designing a radiocommunication system in the troposphere, the heterogeneous character of the troposphere is modeled through the refractive index of the troposphere or other parameters derived from the basic refraction measure. The inherent variability of the refractive index causes the curvature of the signal path along the troposphere, creates duct propagation, defocusing effects and multipath propagation.

Hydrometeors are additional relevant phenomena from the propagation perspective. Specifically, rain is present in many geographical areas with different intensity and occurrence probability statistics. Rain and other hydrometeors (such as snow) are crucial for the choices taken and availability calculations required for designing a FWS link at frequencies above 10 GHz.

Additionally, when radio signals travel along paths close to the Earth's surface, there is a possibility that the wave front is partially obstructed by terrain irregularities such as hills or mountains, or even by artificial structures (buildings). This phenomenon is called diffraction and is a key factor in the design of a microwave LOS link. As described in detail in Chapter 6, path analysis is one of the most important steps as well as measures taken in the system design process to ensure that the link is free of obstructions from terrain and man-made structures.

Some wireless communication systems in the troposphere also require an evaluation of the attenuation, depolarization and scattering effects caused by vegetation areas, or isolated trees. In the case of the microwave LOS link systems, measures will be taken such that the signal paths are not obstructed by vegetation volumes. In the extraordinary case that this is not possible, the calculation models are complex and not very accurate as the effects will be highly dependent on the exact features of the vegetation area.

Finally, the molecular composition of the troposphere is relevant at certain frequencies. Oxygen (O_2) and water vapor (H_2O) absorb the energy of an electromagnetic wave in a degree that will depend on the frequency of the signal. Absorption is not usually considered below 10 GHz for practical purposes, but it might be a key design factor in frequencies such as 60 GHz. This dependency is closely related to the resonant frequencies of both molecules and also on the concentration of these gases in the lower layers of the atmosphere.

In conclusion, the link design will have quite a long list of influencing factors associated with propagation that in most cases will be variable depending on the specific latitude–longitude position, the frequency and the link length. Table 2.1 shows the first compilation of propagation phenomena in the troposphere and the impact on the FWS link design.

2.2 INFLUENCE OF REFRACTION ON PROPAGATION IN THE TROPOSPHERE

The troposphere is not a homogeneous medium and tends to a layered structure. Troposphere layers' composition, distribution, height, and duration are variable.

TABLE 2.1 Effects Associated with Propagation Through the Troposphere and Influence on Fixed Wireless Service (FWS)

Cause	Effect	Description	Statistics	Degradation
Refraction	Ray bending	Curvature in the propagation path followed by the signal between transmitter and receiver	Depends on the statistics of the refractive index gradient along the transmitter receiver path	Fading by diffraction in terrain obstacles during extreme bending periods
	Ducting	Propagation confined between troposphere layers (ducts) or between layers and the Earth's surface	Depends on the probability of super-refractive troposphere occurrences	Flat fading; selective fading; long distance interference;
	Troposphere multipath	Propagation through multiple components that refract over different troposphere layers. The reflected component in the Earth's surface is sometimes also considered in the multipath ensemble	Depends on the statistics of the refractive index gradient	Selective fading (distortion) in broadband systems (medium and high capacity links). Flat fading in narrowband systems (low capacity links)
	Beam spreading	Divergence (defocusing) of the radio signal. Caused because different components of the beam suffer different bending effects.	Depends on the statistics of the refractive index gradient	Enhacements, flat fading
Refraction	Launch and arrival angle variation. (Beam misalignment effect)	Variation of the launch and arrival angle due to anomalous refraction values. The effect is a virtual pointing error between transmitting and receiving antennas.	Depends on the statistics of the refractive index gradient	Flat fading

TABLE 2.1 ***(Continued)***

Cause	Effect	Description	Statistics	Degradation
Terrain diffraction	Attenuation	Diffraction on one or several terrain obstacles or artificial structures	Associated with the same statistics of the refractive index gradient. Under standard conditions diffraction should not occur in a FWS	Flat fading
Vegetation effects[a]	Attenuation depolarization	Absorption associated with the leaves and branches of vegetation, usually trees.	Variable as a function of the foliage cycles and wind	Flat fading, scintillation and depolarization
Troposphere absorption	Attenuation	Energy absorption of molecules forming the lower troposphere: Oxygen and water vapor	Depends mostly on water vapor density statistics, as oxygen concentration is assumed to be constant.	Attenuation
Hydrometeors	Absorption	Energy absorption by rain drops, mist/clouds, water particles, and snow flakes	Hydrometeor statistics. Specific models provided by the ITU-R	Attenuation
	Depolarization	Change in the polarization of the wave front caused by the nonspherical shape of rain drops, snow flakes, or ice particles	Hydrometeor statistics (ITU-R)	Fading due to coupling losses in the antenna (gain) associated with depolarization. Intra-system interference in dense reuse scenarios, either in the same or in different links.

[a]A correct design of a microwave LOS link should ensure that vegetation does not influence the system performance

Intuitively, having in mind Snell's law, a signal propagating along a stratified medium will likely suffer refraction, so that the wave front will not follow a straight path, as it occurs in vacuum or in homogeneous media. The curvature of the path followed by the wave front will depend on the variation of the refraction along the transmitter–receiver path. The parameter that is used to model the effect of refraction is the Refractive Index.

2.2.1 Refractive Index and Radio Refractivity *N*

The Refractive Index relates the wave propagation velocity in vacuum to the one in the medium under consideration, in our case the troposphere. The refractive index has a variable distribution through the atmosphere. In any case, this variation is very small and the values are always very close to unity. The troposphere is not an exception and the values are usually slightly higher than one. The proximity of the values to one supports the supposition that the electromagnetic waves propagate through the troposphere at the speed of light.

The refraction phenomena do not depend on the value of the refractive index but on its variation. The absolute value being very close to one, the difference variations in the index range from $2\ 10^{-4}$ to $3\ 10^{-4}$. For this reason, a simple change of variable is defined to provide a more convenient range of values for calculations related to refractive index differences and their statistics. The new parameter is called radio refractivity or radio refractive index, N. The relationship between the refractive index and the radio refractivity is given by equation (2.1):

$$n = 1 + N 10^{-6} \tag{2.1}$$

$$N = (n - 1)\, 10^{6} \tag{2.2}$$

In the same way as the refractive index, the radio refractivity is an adimensional parameter and its values are usually referred to as "N units." The mean value of the refractive index in the troposphere is 1.00315, so the equivalent value is 315 N units. The radio refractive index can be expressed as a function of the atmospheric conditions, specifically as a function of the pressure, the temperature and the relative water vapor concentration. The expression, which is valid for frequencies below 100 GHz provided by ITU-R Recommendation P.453:

$$N = N_{\text{dry}} + N_{\text{wet}} = \frac{77.6}{T}\left(P + 4\,810\frac{e}{T}\right) \tag{2.3}$$

where

N_{dry} = "dry term" of radio refractivity
N_{wet} = "wet term" of radio refractivity
P = atmospheric pressure in milibars
T = temperature (°K)
e = water vapor pressure in milibars

If mean values of temperature, pressure and water vapor pressure are considered, the radio refractivity is 315 as expected. This value is considered the reference for mean values at sea level. The radio refractive has a significant variability with height, and as depicted in Section 2.2.2, its values will be commonly referred to the mean valuc at sea level.

2.2.2 Radio Refractivity Gradient

The exact value of the radio refractivity in a specific location is not a relevant datum for designing a microwave loss link on that area. The consequences of the refraction on the propagation of radio signals depend on the radio refraction gradient (height dependency). The vertical gradient of N is defined by equation (2.4):

$$\Delta N = \frac{\partial N(h)}{\partial h} \tag{2.4}$$

From the radio design perspective it will be interesting to know the value of the gradient in the area where the link is being designed and also the variation of this gradient along the path from the transmitter to the receiver. Additionally, in order to build a design with stable performance characteristics over time, it will also be required to have a model (or an estimation) of the variation of the gradient under different conditions. The models used for describing the vertical gradient of N include dependencies with the specific geographic location, month of the year, specific meteorological conditions and obviously the height above median sea level.

2.2.2.1 Variation with Height The radio refractivity gradient varies with height in line with the equivalent vertical variation of pressure, temperature and water vapor pressure. ITU-R Recommendation P.385 provides a model to describe the variation of the three atmospheric magnitudes with height. Based on that model, Recommendation ITU-R P.453 contains an expression to calculate directly the variation of the radio refraction with height, using the value at sea level as the reference:

$$N(h) = N_0 \exp\left(\frac{-h}{h_0}\right) \tag{2.5}$$

where

N_0 = average value of atmospheric refractivity extrapolated to sea level
h_0 = scale height for normalization purposes
h = height over the average sea level

N_0 and h_0 depend on the climatic characteristics of each region. The mean values are 315 N units and 7.35 km respectively. There are cases where it is useful to have the radio refraction value at the Earth's surface at locations above sea level. In those cases, the following expression can be used:

$$N_s = N_0 \exp\left(\frac{-h_s}{h_0}\right) \tag{2.6}$$

where N_s represents the radio refractivity value at h_s meters above sea level, and both N_0 and h_0 are the same parameters as in equation (2.5).

For radiocommunication system design purposes, the statistics of N_0 and the radio refractivity gradient ΔN are given by ITU-R Recommendation P.453. Figure 2.1 shows an example of the data map contained by the mentioned recommendation. Depending upon the value of ΔN a classification of the conditions of the troposphere can be made. Thus, troposphere conditions are referred to as normal troposphere, subrefractive troposphere and super-refractive troposphere. Normal troposphere is associated with a gradient of -39 N units/km. This value is the median in most locations on the Earth, and it is commonly used by most radiocommunication system design procedures.

2.2.3 Subrefraction Conditions

Subrefractive troposphere conditions are associated with radio refractivity values close to zero or even positive. According to the statistics provided by the ITU-R, subrefractive conditions are possible in practically any place or geographic area for a percentage lower than 10% of the average year. In fact, the value not exceeded for 99% of the average year is 30 N units/km in most countries.

The atmospheric conditions that cause subrefraction are either negative steep temperature gradients or water vapor pressure positive gradients, or both conditions at the same time. There are several meteorological conditions and phenomena that are associated temperature negative gradients and water vapor pressure positive gradients. Subrefraction is more probable in areas where the temperature is high and the humidity is low ($T > 30°$C, RH $< 40\%$). In these cases, subrefraction is associated with the conduction physical effect, when the surface temperature is heated significantly by the sun, creating significant temperature gradient conditions as height increases. Similar conditions occur due to auto-convection in dry terrains (typically sandy terrains and dessert), where there is a heat transfer from the surface to the troposphere that creates important temperature gradients.

Subrefraction zones are also created when cool moist air volume moves over a hot dry ground (advection) and creates positive humidity gradients. Subrefraction also occurs when frontal weather processes (storm fronts) pushes warm moist air over cool dry air on ground areas (see Figure 2.2).

Some areas of the globe are well known for having usual subrefraction conditions. These are the polar areas, west coast of Japan in winter, Mediterranean Sea also in winter, and the Grand Banks in the coast of the Labrador Peninsula.

2.2.4 Super-Refraction Conditions

Super-refraction conditions are considered if the radio refractivity gradient is lower than -100 N units/km. Super-refraction conditions are usually related to temperature

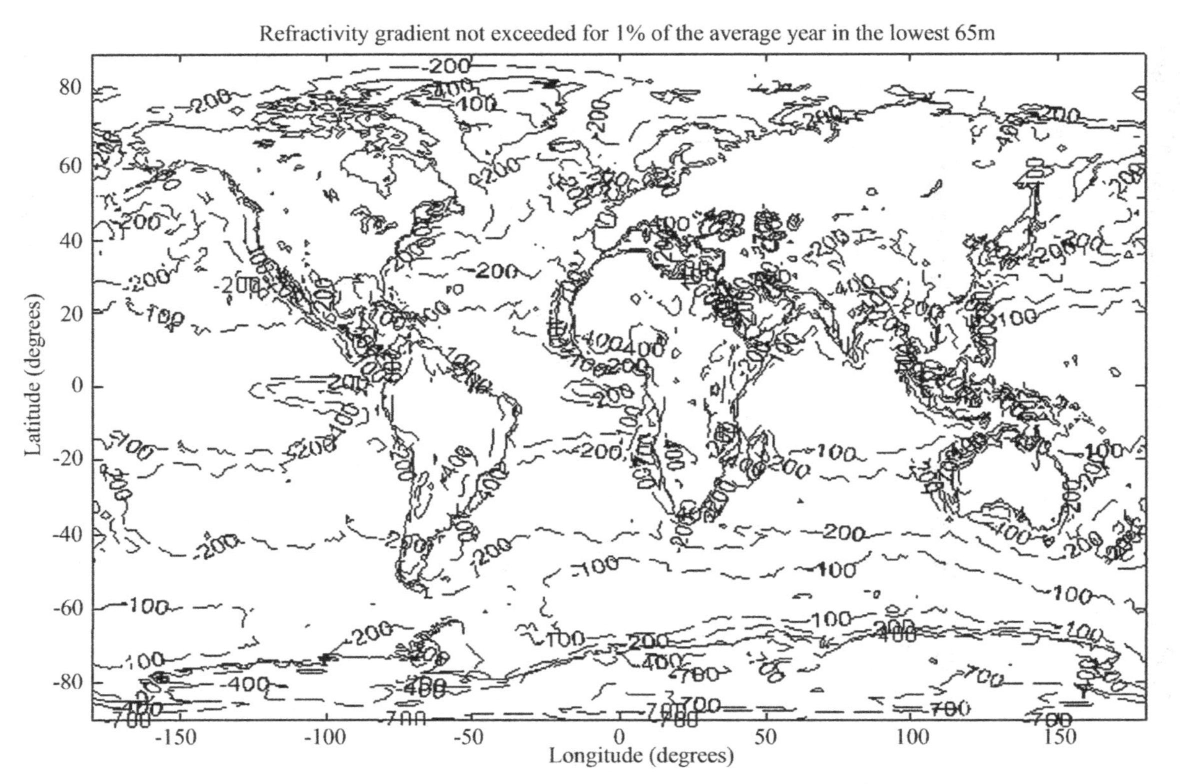

FIGURE 2.1 Refractivity gradient not exceeded for 1% of the average year in the lowest 65 m of the atmosphere. (*Figure Courtesy of ITU-R.*)

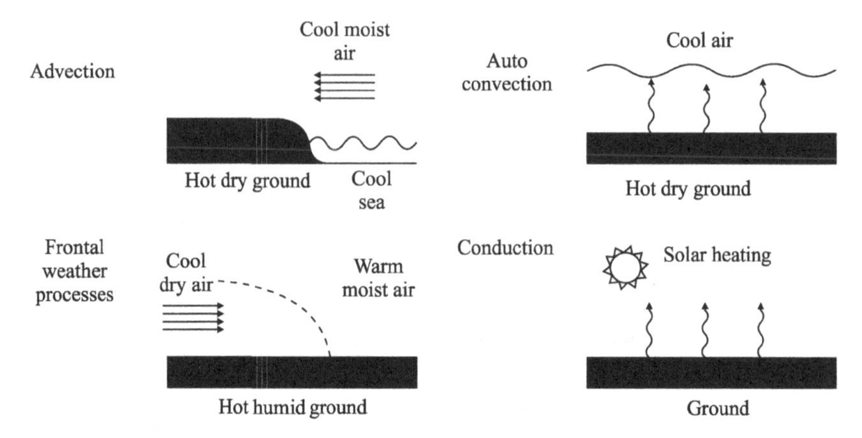

FIGURE 2.2 Meteorological phenomena that cause subrefractive troposphere conditions.

inversion phenomena and very negative water vapor gradients, which among others, are associated with advection, subsidence and radiation phenomena, as well as to the formation of wide layers of evaporation over large water volumes (see Figure 2.3).

Advection is the transfer of a property of the atmosphere, such as heat, cold, or humidity, by the horizontal movement of an air mass. In the case that these air mass movements generate negative temperature gradients, super-refraction might occur. Examples of these phenomena can be found when a cold air mass moves over warm air volumes. Another related phenomena is quasi-advection, which occurs when a strong cool wind blows over a warm wet surface, a warm sea for example, can also lead to a steep negative water vapor pressure gradient proportional to the strength

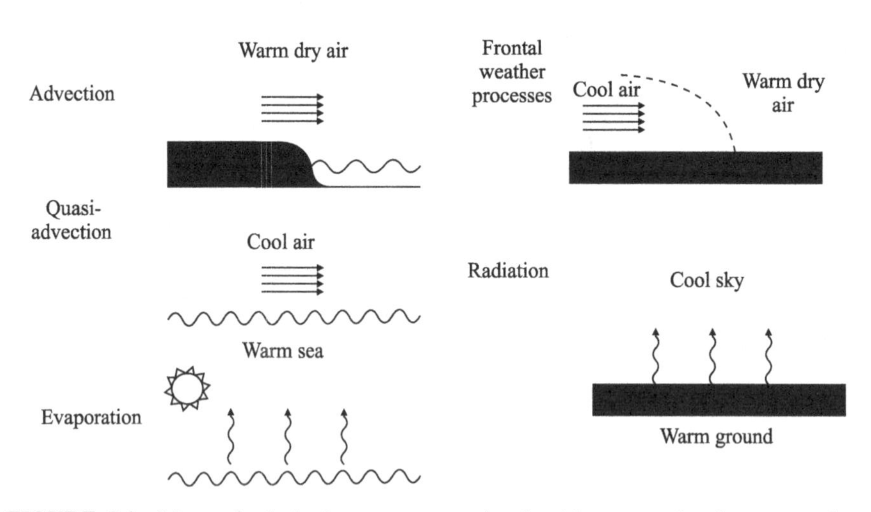

FIGURE 2.3 Meteorological phenomena associated with super-refraction troposphere conditions.

of the wind. Additional super-refraction examples are movements of warm air mass towards the sea in coastal areas.

Subsidence creates an inversion of the temperature gradient. This situation is originated when stratocumulus are situated under an anticyclone inversion. In this case, the water vapor gradient creates very negative refraction gradients. In regions where stratocumuli are probable, subsidence is the dominant cause of super-refraction.

The evaporation over the surface of the sea and over moist ground areas creates a water vapor concentration layer that, in turn, creates a steep temperature gradient.

Additionally, the radiation (cooling of the ground) during the night under clear air conditions creates a temperature inversion that is associated with super-refraction. Associated with radiation, if dew is deposited on the ground, sunrise will cause dew evaporation creating a steep negative water vapor gradient.

Super-refractive conditions can be found over wide areas of the globe. Super-refraction is common in tropics, the Red Sea, Arabian Gulf, or the Mediterranean Sea in summer. Additionally, during small percentages of time, in many additional areas, negative radio refraction gradients will create duct propagation conditions. Ducts will be described more in detail in a further section of this chapter.

2.2.5 Ray Bending: Effective Earth Radius

The propagation path followed by the wave front between the transmitter and the receiver of a microwave LOS link is a bent trajectory. If we would analyze the radio horizon (the distance when the wave front will intercept the Earth's curvature) it would be confirmed that this distance can be either higher or lower than the geometrical distance to the optical horizon (tangent to the Earth passing by the transmitter) depending on the conditions of the troposphere. Figure 2.4 illustrates the phenomenon.

The path followed by the wave front depends on the radio refractivity value. In order to understand the qualitative effect, we can imagine the propagating signal as a flat wave front. The radio refractivity will depend upon the height therefore the relative propagation velocity of different points of the same front will not be constant, and thus the propagation path will tend to bend, either upward or downward, depending

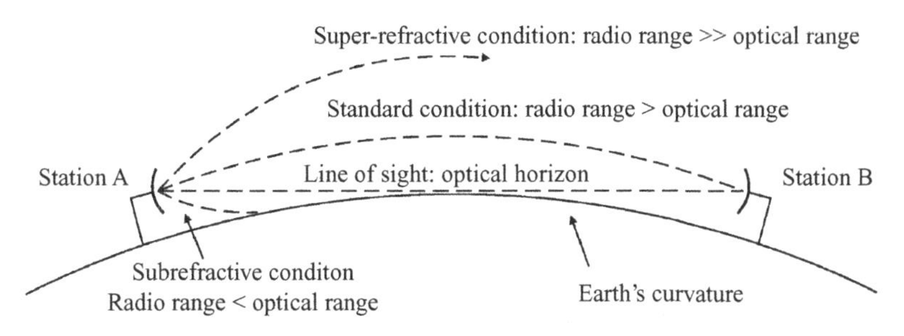

FIGURE 2.4 Optical horizon and radio horizon.

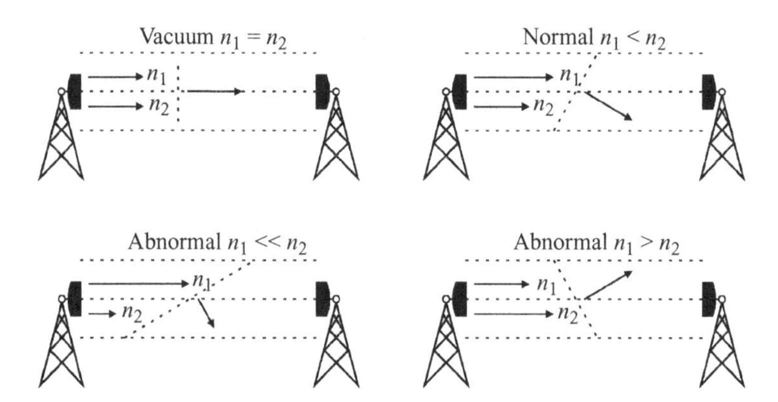

FIGURE 2.5 Impact of radio refractive gradient values on the wave front bending.

on the sign and value of the radio refractivity gradient. Figure 2.5 shows a simplified summary of the possible situations.

Under standard atmospheric conditions, if the radio refractivity gradient is close to −39 N units/km, the bending radius of the wave front path is higher than the optical path. If subrefractive conditions occur, the wave front path is bent severely and the radius of the trajectory followed by the propagated signals is shorter than the one associated with the optical path and in this case, the radio horizon is shorter than the optical horizon.

Should super-refraction conditions occur, the trajectory of the wave front diverges with respect to the Earth's curvature. In this super-refractive conditions, for negative values of the radio refractivity gradient lower than −150 N units/km, ducts are usually formed, and severe perturbations usually appear in the radiocommunication systems involved, either due to flat fading or unexpected long distance interferences.

For practical purposes, a simplification is usually made, leaving aside the fact that the gradient is not constant along the transmitter and receiver path. In real designs it is assumed that the signal follows a curve trajectory with constant radius that will depend on the average value of the gradient along the transmitter–receiver path. The path radius ρ can be obtained applying Snell's law:

$$\frac{1}{\rho} = -\frac{\cos\varphi}{n}\frac{\mathrm{d}n}{\mathrm{d}h} \tag{2.7}$$

where

ρ = bending radius of the wave trajectory
n = refraction index
$\mathrm{d}n/\mathrm{d}h$ = vertical gradient of the refraction index
h = height above ground level
φ = angle measured with respect to the horizontal path

Assuming that in most cases, the launch angle from the transmitting antenna is almost zero (grazing incidence) and the refraction index is close to one:

$$\rho = \frac{1}{R} = \frac{\partial n(h)}{\partial h} \tag{2.8}$$

If now the relation between N and n is included in equation (2.8), the curvature can be expressed in terms of the radio refractivity:

$$\Delta N = \frac{\partial n(h)}{\partial h} = -10^6 \frac{1}{R} = -10^6 \rho \tag{2.9}$$

So far the path-bending radius has been expressed as a function of the radio refractivity gradient. In fact, the relevant analysis is related to the distance between the signal path and the height of the terrain, usually referred to as clearance. There are different geometrical models used to represent the wave path and the Earth's surface heights. The model widely used in microwave LOS link design is based on the so-called effective Earth radius. The model represents the propagation of the radio wave as a straight line between the transmitter and the receiver and compensates the Earth's curvature with the real propagation trajectory followed by the radio signal. The modified Earth radius is called Effective Earth radius. This model defines a factor, *k factor*, that is the ratio between the real and effective Earth radius. Figure 2.6 illustrates the model.

The value of the k factor is calculated as a function of the radio refraction gradient and the real Earth radius R_0:

$$k = \frac{1}{1 + R_0 \Delta N \cdot 10^{-6}} = \frac{157}{157 + \Delta N} \tag{2.10}$$

According to the different specific refraction conditions explained in previous sections, Table 2.2 shows the relationship between the k factor and the radio refraction

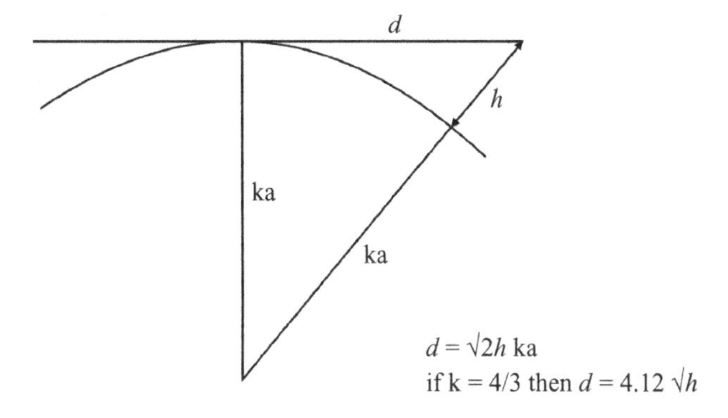

FIGURE 2.6 Effective earth radius.

TABLE 2.2 K Factor Values for Different Refraction Conditions

Troposphere condition	ΔN (N units/km)	k
Normal	$0 \leq \Delta N < -39$	$1 \leq k < 4/3$
Subrefractive	$\Delta N > 0$	$0 \leq k < 1$
Super-refractive	$\Delta N < -39$	$k > 4/3$
Ducting Conditions	$\Delta N \leq -157$	$k \infty$ [a]

[a] k has a function discontinuity for trajectories parallel to the Earth's surface (bending radius is equal to Earth's radius)

values. Under standard propagation conditions, k is 4/3. This value is obtained by averaging the values over a long period of time (usually, these calculations are done over a year). The deviations from the average value are usually expressed by different percentile values. It is expected that the variation of the ΔN and consequently k, will depend on the link path length, because of the fact that the value that is used is averaged along the whole path.

2.2.6 Ducts

A duct is a region of the troposphere that because of anomalous radio refractivity gradient values, keeps the propagation of a certain signal confined between two layers of the troposphere or between a layer of the troposphere and the Earth's surface.

Under the presence of a duct, if certain launch and frequency conditions are fulfilled, the signal will propagate along the duct and the trajectory followed by the wave front will be quite different from the one that would be expected under standard refraction conditions. Duct propagation usually cause longer ranges than standard propagation, and in consequence, it is usually a problem associated with unexpected long distance interferences. Countermeasures will be taken in the design process in order to avoid as much as possible the influence of ducting in microwave LOS links.

Ducts are created in super-refraction conditions when the radio refractivity gradients are lower than -157 N units/km. Duct propagation is difficult to characterize quantitatively. A possible approach might be to use ray-tracing techniques that account for the different trajectories within the duct. Nevertheless, this is more a theoretical exercise than a realistic design practice in microwave LOS links.

In order to provide a simple model of refraction in cases where the gradient is lower than -157 N units/km, a new parameter, referred to as modified radio refractivity (M) is defined by ITU-R Recommendation P.453. The ducts are then described as a function of M in equation (2.11):

$$M(h) = N(h) + 157h \tag{2.11}$$

where M is adimensional (M units) and h is the height above ground level in kilometers.

There are three types of ducts: surface, elevated-surface, and elevated ducts. Surface ducts create a region that confines the propagation between a lower layer of the troposphere and the Earth's surface. Under such condition and as a consequence of the extreme value of the gradient, the signal path bends down to the Earth's surface, where it is reflected toward the troposphere. The signal is then propagated along the duct through successive reflection and bending.

The second type of duct is known as elevated duct and it is created between two layers of the troposphere. This duct is usually located in the first few kilometers of the troposphere. Finally, if the elevated duct is produced close to the Earth's surface, it will be called elevated surface duct. Ducts are described by means of their strengths S_s or E_s (M units) and their thickness S_t or E_t (meters). Two additional parameters are used to describe elevated ducts: namely, the base height of the duct E_b (m), and E_m (m), the height within the duct of maximum M.

Because of rather few cases of elevated-surface ducts in comparison with surface ducts, the statistics have been derived by combining these two types into one group called surface ducts. Figure 2.7 illustrates the three duct types and associated parameters M and S.

Surface and surface-elevated ducts cause deep fading and even unexpected signal enhancements due to constructive combination of all the signal paths propagating across the duct at the receiving antenna. Elevated ducts are usually associated with

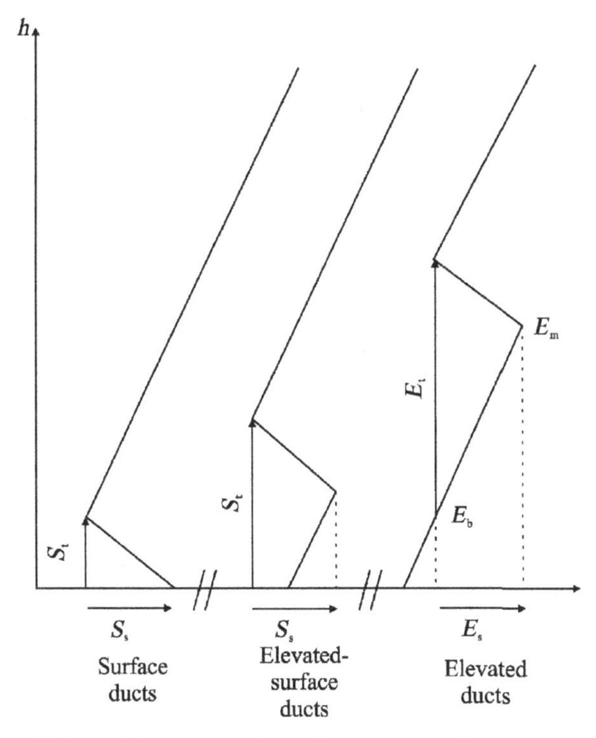

FIGURE 2.7 Surface, elevated, and surface-elevated ducts.

TABLE 2.3 Some Areas Where Ducts are Probable

Country/Region	Area	Season
North America	North East Coast	Summer
North America	Florida	Winter
North America	West Coast	Indifferent
East Europe	British Islands and North Sea	Summer
Europe	Mediterranean Sea	Summer
East Africa	Arabic Peninsula and Indian Ocean	Dry Season (October–May)
Bay of Bengal	India/Sri Lanka/Myanmar–Birmania/ Bangladesh	Dry Season (October–May)
Pacific Ocean	Korea	Indifferent
Pacific Ocean	West Coast of Australia	Indifferent
Pacific Ocean	Japan Sea	Summer
Pacific Ocean	South China Sea	Winter

Note: As it comes out from a quick inspection of the table, the information available in literature usually concerns maritime zones, due to the relevance of ducting for maritime navigation and maritime radar systems.

multipath effects that will create fast flat fading to narrow band signals or distortion (selective fading) to signals with wider bandwidths. Recommendation ITU-R P.453 provides statistical data of ducting for practical designs. Table 2.3 summarizes some of the most relevant ducting areas of the globe.

The available statistics are the probability of duct occurrence, the intensity of the duct (M units) as well as the thickness and height of the different duct types. The statistics of ducting will depend on the climate of the region under study. In areas where the climate is affected by monsoon variations, duct statistics also follow the monsoon seasonal variation. On the contrary, in places where subrefraction is associated with low-pressure conditions, the duct conditions vary daily.

2.2.7 Beam Spreading

Beam spreading is a phenomenon that appears under severe super-refraction conditions. Beam spreading appears as a succession of different regions of convergence and divergence of the beam that composes the radio signal along the transmitter–receiver path. The beam spreading is caused by different curvatures suffered by different launch angles at the transmitting antenna on the vertical plane. The angle difference between the launch direction with maximum elevation and the launch direction with minimum elevation will be very small, a few degrees or a few tens of a degree. If the propagation path is long (long hops) even a small difference can create significant spreading if the super-refraction condition is severe.

Beam spreading does not depend on frequency and from a practical standpoint can be modeled as a decrease of the antenna gain. The effect on the signal transmission is a fading occurrence that usually is slower than multipath variation. As illustrated by Figure 2.8, if the receiving antenna is located on the region called "radio hole," the level at the receiver input will suffer a severe fade. In the same way, it would

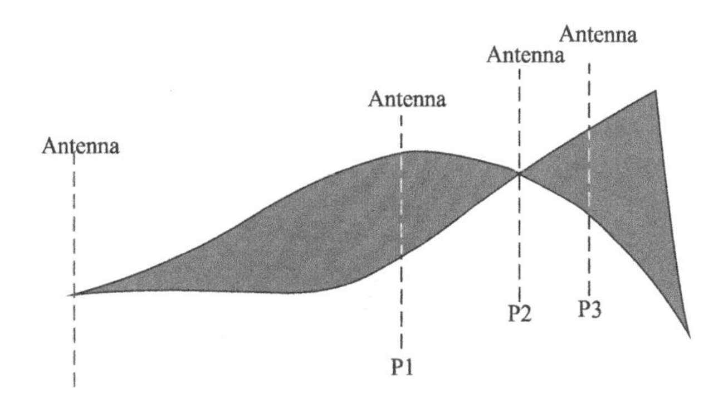

FIGURE 2.8 Beam spreading due to refraction on the transmitter–receiver path.

be possible that beam spreading would create a signal enhancement if the antenna is located at the convergence point.

Beam spreading effects can be significant in case of grazing angle paths (similar transmitter and receiver antenna heights), where the duct is located immediately below the line that links transmitter and receiver.

Beam spreading is usually a phenomenon that concerns satellite links, where the path is significantly long. In the case of microwave LOS links, it will be only considered if the hop is very long and the super-refraction conditions are extreme.

2.2.8 Variation in the Angles of Launch and Arrival

Similar to the case of beam spreading, extreme refraction under clear air conditions can create variations in the angles of launch and arrival from and to the transmitting and receiving antenna respectively. This variation is frequency independent and it is limited to the vertical plane. The range of variation angles is wider in coastal moist areas than in dry continental ones.

As it could be expected, the narrower the antenna beams are, the higher the impact on the system performance. Additionally in long hops two effects will contribute to system degradation. First, the angle variation will also increase the probability of anomalous refraction in higher layers of the troposphere originating multipath distortion. Second, a longer path implies severe misalignment problems even in variations of tens of a degree.

Launch and arrival variations are relevant in the process of antenna alignment, which if carried during anomalous refraction periods might lead to a suboptimal link installation. The Recommendation ITU-R P.530 suggests performing alignment procedures during a period of a few days in the case of system installations in critical conditions (long paths in coastal areas).

2.2.9 Troposphere Multipath Propagation

Multipath propagation in the troposphere is the main degradation source in microwave LOS links that operate in frequencies below 10 GHz. This phenomenon

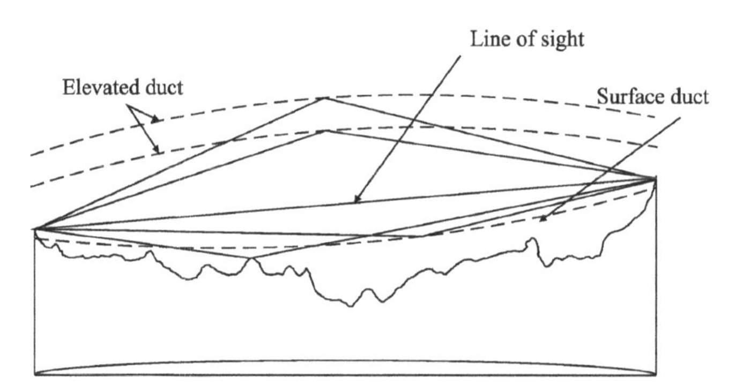

FIGURE 2.9 Troposphere multipath sources in a microwave LOS link.

involves the reception of multiple components of the signal, which have arrived following different paths through refraction in layers of the troposphere with steep negative radio refractivity gradients. Troposphere multipath is caused by either elevated or surface ducts. In the case of surface ducting, the associated gradient condition is subrefractive, and multiple paths include reflections on the Earth's surface. Figure 2.9 illustrates both effects.

In addition to multipath components originated by refraction, depending on the terrain path profile, the height of the antennas, and the antenna radiation pattern, reflected components on the Earth's surface and diffraction could also create multipath components. In these cases, a careful design of the antenna heights and radiation pattern choices should be enough to minimize its influence.

Multipath produces variable fading on the radio channels of the microwave link. The fading depth will depend on the amplitude and phase ratios amongst all the components that contribute to the overall received signal.

In the case of low capacity links, if the bandwidth is narrow enough (usually systems with a single radio channel), all the spectral components are equally faded. The consequence is a flat or nonselective fading, because, even though the multipath creates a frequency response that is inherently selective, the effect of that fading in a narrow frequency range can be considered flat with reasonable accuracy. In the case of medium and high capacity systems, the multipath will usually be a selective effect and will create signal distortion and inter-symbol interference. Flat fading and selective fading will be treated separately when designing a microwave LOS link. Their impact on the system performance parameters is completely different in the case of flat and selective fading, as well as the countermeasures usually taken to prevent the system being depredated by both effects.

The fading produced by atmospheric multipath can be characterized by different statistic methods. Recommendation ITU-R P.530 contains various methods to be applied in the specific case of microwave LOS links. Actually the models include an algorithm that includes the combined effect of atmospheric multipath, which usually originates selective and fast fading, and beam spreading, which usually will create flat and slow fading. These methods will be described in detail in Chapters 7 and 8 of this book.

2.2.10 Scintillation

Scintillation is characterized by a random variation of the received signal level around the median value. These are fast and low intensity variations below 40 GHz. Scintillation is associated with small-scale irregularities in the troposphere, which are changing their features very fast. The effect of scintillation inn a microwave LOS link can be regarded as negligible.

2.3 TERRAIN DIFFRACTION LOSSES: FRESNEL ZONES

Microwave LOS links are designed, by definition, to be free of any diffraction caused by terrain irregularities. Under standard conditions, the line that links transmitter and receiver should be well above the terrain obstacles and thus the diffraction losses should be zero. Nevertheless, as described in previous sections, when severe subrefraction conditions occur, the path followed by the wave front can be bent down and intersect with terrain obstacles. These obstacles will block partially (or even completely) the propagation of the signal energy. Diffraction is studied through the analysis of the degree of obstruction suffered by the so-called First Fresnel Ellipsoid or First Fresnel Zone.

The energy at the receiver location is the contribution of the infinite paths that are contained by the plane that is perpendicular to the Pointing vector of a flat wave front. This concept is based on Huygens' principle that states that any of the points of a flat wave front originates a spherical wave, that in turn, with all the spherical waves associated with other points of the flat wave front, will create an envelope that forms again a flat wave front at a distance r. Figure 2.10, illustrates the infinite number of propagation paths followed by the energy from the transmitter to the receiver. The contribution of each path to the total received power will depend on the phase of each component associated with the flat wave front.

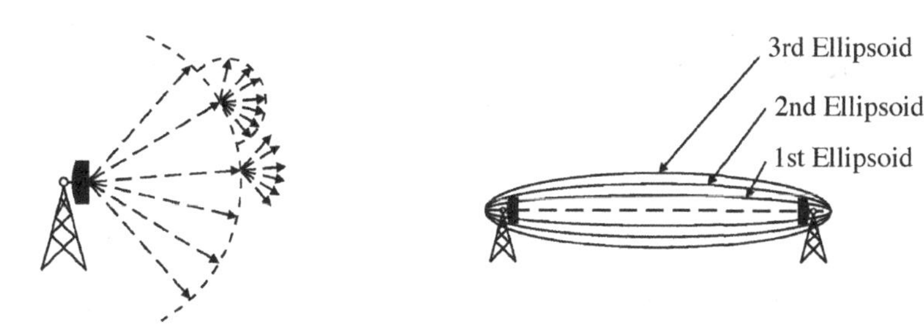

FIGURE 2.10 Huygens principle and Fresnel Zones[1]

[1]Actually, the antennas are located at the focal points of the Fresnel ellipsoids. Nevertheless, the reader will find that most references, for simplicity reasons, draw the ellipsoids tangent to the antenna at the end of the major semiaxes. This simplified approach will be also followed in the rest of figures in this book.

Under these premises, Fresnel demonstrated that the energy between transmitter and receiver is distributed into an infinite number of ellipsoids whose foci are transmitter and receiver locations. Following Fresnel's model, some of the ellipsoids will contribute to the total received power and a second group of ellipsoids will create detraction. Obviously the total received power will be the balance of constructive and destructive contributions from all the ellipsoids. This fact is shown in Figure 2.10, where odd order ellipsoids will contribute positively and even order ellipsoids negatively. The radius of each ellipsoid for each location within the transmitter–receiver path profile depends on the frequency and the order of the ellipsoid. Equation (2.12) shows the expression to calculate these radii:

$$R_{\mathrm{n}} = 548\sqrt{\frac{n \cdot d_1 \cdot d_2}{f \cdot d}}(m) \tag{2.12}$$

where

N = order of the ellipsoid
R_{n} = radius of the nth order ellipsoid
d_1 = distance to the transmitter (km)
d_2 = distance to the receiver (km)
d = transmitter–receiver distance (d_1+d_2) (km)
f = frequency (MHz)

The energy contribution of the different ellipsoids to the total received power is a decreasing function of the order of ellipsoids. In consequence, for practically any radiocommunication application design the diffraction losses are evaluated considering the first ellipsoid only and neglecting any effect due to the rest of ellipsoids.

As mentioned before, in microwave LOS links, any diffraction losses will be associated with anomalous refraction conditions. The first Fresnel ellipsoid will be free of obstacles (at least a 60% of its radius should be above any relevant terrain obstacle of the path profile). In exceptional cases where this condition cannot be fulfilled and for calculation purposes of diffraction associated with the subrefractive condition, there are different methods to estimate the diffraction loss value. Chapter 7 will describe these calculation algorithms. The method to be applied in each case will depend on the number of obstacles, the shape of the obstacles and the position of those obstacles in the transmitter–receiver path.

2.4 VEGETATION ATTENUATION

Vegetation produces a variety of perturbations on the propagation of an incident wave that have been exhaustively described by scientific literature. The associated phenomena worth a mention in this book are fading, depolarization and signal scattering. The mentioned high number of studies have been carried out at different frequencies and with different vegetation types and density conditions. Nevertheless, the variety of situations that exist in reality, impose severe difficulties for providing a universally applicable mathematic model with a decent accuracy. This section describes the possible effects that might be relevant for a microwave LOS link system.

First, it should be clearly stated that any perturbation effect from vegetation is a degradation that should be easily avoided by an adequate design in most microwave LOS links. A careful choice of transmitting and receiving antenna choices should be enough for considering the influence of vegetation negligible. Some sources recommend calculating the tower heights with an additional clearance of 15 meters over any terrain obstacle to avoid interaction of the signal with any possible tree or vegetation mass on that specific spot. This recommendation is a practical design advice that does not arise from a mathematical model from any model.

Recommendation ITU-R P. 833 provides methods to calculate the attenuation by vegetation in frequencies between 30 MHz and 60 GHz. In addition, this recommendation provides depolarization data as well as dynamic effects (caused by wind moving the vegetation) and scattering. The methods proposed in ITU-R Recommendation P.833 vary depending on the frequency band and the link path and vegetation geometry.

Finally, it should be stressed once again that due to the poor accuracy of the calculation methods in a generic situation, it is not realistic to include them in the design process of a microwave LOS link.

2.5 ATMOSPHERIC GAS AND VAPOR ABSORPTION

This section describes the attenuation caused by the elements that compose the atmosphere. The elements that compose the atmosphere in the first few kilometers of the atmosphere are oxygen (O_2), nitrogen (N_2), carbon dioxide (CO_2), and water vapor (H_2O). The molecules of each gas interact in diverse ways with a radio signal. The key factor is the ratio of the signal frequency and the resonant frequencies of each gas. For practical purposes, in frequency bands allocated to the fixed service (FS), the gases that have a relevant effect in system performance are oxygen and water vapor.

In order to study the influence of gas absorption in radiocommunication system design, the "specific attenuation" parameter is defined as the loss associated with the absorption by each one of these two elements in dB/km. The specific attenuation of the oxygen molecule is usually referred to as specific attenuation in dry air. Figure 2.11 shows the specific attenuation by oxygen and water vapor between 1 and 1000 GHz. Specific attenuation increases with frequency, and the functions that describe its behavior present several relative maxima that correlate with individual absorption lines associated with specific resonant frequencies of the elements involved. Between those maxima, there are relative minima that enable transmission windows for general use in radiocommunication systems.

Only in very special cases, as in short range communications (body area networks) frequencies close to attenuation relative maxima will be used. Microwave LOS links

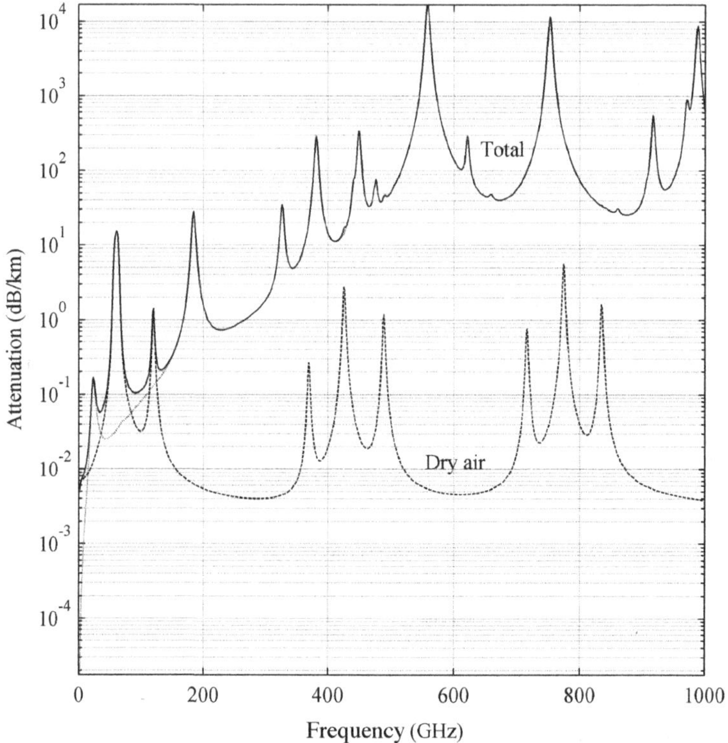

FIGURE 2.11 Specific attenuation values associated with oxygen (dry air) and water vapor as provided by the model available in Recommendation ITU-R P.676.

use FS frequency allocations associated with transmission windows around relative minima.

The specific attenuation values will depend on the pressure, the temperature and the water vapor density. At each geographic location on the globe, it would be necessary to obtain these parameters with the associated annual statistics. Usually, the calculations are carried out with a pressure of 1013 hPa, a temperature of 15°C and a water vapor density of 7.5 g/m^3.

If more accurate calculations would be required, Recommendation ITU-R P.835 provides reference values of a standard atmosphere for different latitudes.

2.6 HYDROMETEORS

Hydrometeors are one of the most influential factors on a microwave LOS link performance. Rain, clouds, fog, snow, or hail fall into the hydrometeor category that might have influence on a radiocommunication system performance. Among all of these phenomena, rain is the most relevant one, because of its occurrence statistics and also for its impact on the link budget and system margin calculations.

The perturbations that are associated with hydrometeors are two. First, the energy of the radio signal is absorbed and scattered by raindrops or ice particles that rain, snow and hail are composed of. Second, the ice particles and raindrops, because of their nonspherical shape, create a polarization rotation effect that will be associated with the shape, size and distribution of raindrops (or ice crystals). The models that describe rain include the size and distribution of rain volumes as well as the shape, size and distribution of raindrops. As expected, in order to account for all those variations, both absorption and depolarization caused by hydrometeors are described by empirical statistical models.

2.6.1 Absorption Due to Hydrometeors

The classical model used to describe the attenuation created by a rain cell (rain volume), assumes that the energy is faded exponentially along distance travelled within the rain volume. Additionally, the raindrops are supposed to be slightly nonspherical and it is assumed that the total absorption will be the aggregate consequence of absorption and scattering. Under those conditions, the specific attenuation associated with rain is:

$$\alpha = 4.34 \int_{0}^{L} \left[N_0 \int Q_t(r, \lambda, m) e^{-\Lambda r} dr \right] dl \tag{2.13}$$

In equation (2.13), N_0 and Λ are factors that depend upon the size and flattening of rain drops. The function Q_0 (r, λ, m), represents the dispersion and absorption coefficient of rain drops and it depends on the rain drop radius, the wavelength and the complex refraction index of those rain drops. If equation (2.13) is observed carefully, it can be seen that the calculation depends on specific attenuation as a function of a drop with radius r, and later is extended to the rain cell L.

Nevertheless these parameters and their statistics are not easy to obtain. In consequence, are more usual to work with empirical models as a function of the rain rate R (mm/h) which statistics are well known. Equation (2.14) provides the specific attenuation of a rain with intensity R (mm/h) as a function of two empirical factors α and k:

$$A = (kR^{\alpha}) \tag{2.14}$$

where

A = specific attenuation of the rain (dB/km)
k and α = empirical values that depend on the frequency and polarization
R = rain rate (mm/h)

After obtaining the specific attenuation per length unit, it is necessary to know the length of the rain cell, in order to calculate the total attenuation in the link. This parameter is also statistical in nature and, in the models used for microwave LOS links, it is also calculated using an empirical law that depends also on the rain rate R (mm/h).

At this point it is thus clear that the relevance of having an accurate model for the statistics of R, because to a significant extent, the availability figures in fixed links will depend on this accuracy. Rain is the main cause for severe fading in links that work in frequencies higher than 10 GHz and in consequence rain is the most common cause behind unavailability. The availability objectives in modern systems being 99.9% (and even higher), it is noteworthy that the need for accurate R data will always be accurate only if they are based upon long-term accurate meteorological data. The ITU-R Recommendation P.837 provides reference values for those cases where local values are not available.

The calculation steps of attenuation caused by rain and associated statistics in a microwave LOS link will be described in detail in Chapter 8, in relation to methods and models for quantifying availability and error performance objectives.

2.6.2 Cross-Polarization: Coupling Losses

Cross-polarization plays a crucial role in the study of the interferences in a microwave LOS link design. The cross-polarization will influence the choice of radio channels, because it might be a source of interference between the systems using the same or adjacent frequencies. The parameter that is commonly used to quantify cross-polarization is the cross-polarization discrimination (XPD). This parameter is graphically described in Figure 2.12. This figure shows the transmission of two signals linearly polarized, one vertically and the other horizontally, and the corresponding components received with associated cross-polarization components.

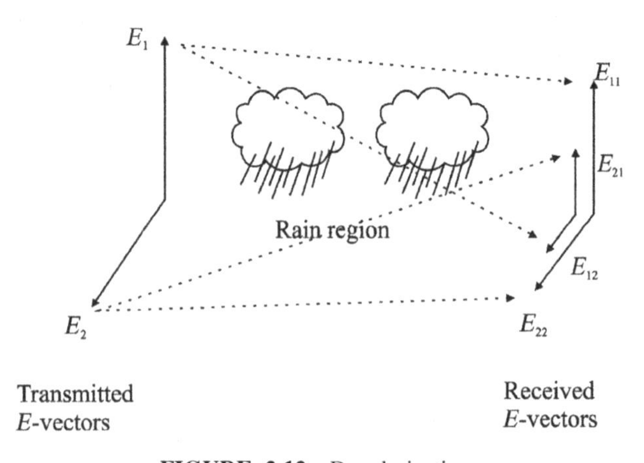

FIGURE 2.12 Depolarization.

The cross-polarization discrimination parameter is calculated using the following equation:

$$\mathrm{XPD} = 20 \log \frac{|E_{11}|}{|E_{12}|} \tag{2.15}$$

where

E_{11} = component received with the original transmitted polarization
E_{12} = component received with the orthogonal polarization caused by cross-polarization.

Cross-polarization (decrease in XPD) is originated by multiple factors, including the cross-polar antenna patterns, atmospheric multipath and hydrometeors. In absence of rain and other hydrometeors, cross-polarization is referred to as clear air cross-polarization and it is based on the same statistics as fading.

Cross-polarization caused by hydrometeors is usually more severe than clear air XPD reduction. In this case, the XPD reduction is associated with the shape and position of raindrops or ice crystals. The flattening of raindrops, which increases with the rain drop radius, and the raindrop orientation angle, are some of the factors that will affect the rotation of the signal polarization. Once again, the complex nature of this phenomenon is overcome by statistic empirical models.

The Recommendation ITU-R P. 530 provides an expression to estimate XPD as a function of the fading associated with a cell rain. It is assumed that both the parameters follow the same statistical behavior and the relationship between them is:

$$\mathrm{XPD} = U - V(f) \log(\mathrm{CPA}) \tag{2.16}$$

where

U, V = empirical parameters function of the frequency
CPA = co-polar attenuation produced by rain (equation (2.14))

2.7 REFLECTION

In frequencies higher than 1 GHz, the space wave is composed of a direct path and a reflected path on the Earth's surface. Figure 2.13 shows a simplified model describing the effect of reflection. The signal received at each terminal station of the microwave link will be the vector sum of the direct and reflected field strength components.

Depending upon the relative phases of the direct and reflected components, the vector sum might lead to attenuation or enhancement. In the first case, it is possible to observe severe fading, thus reducing the link margin significantly. In order to avoid this link margin reduction, the reflection can be evaluated by a calculation of the relative phase difference, assuming that the reflection itself will also change the phase of the reflected path (usually assumed π radians: perfect reflection).

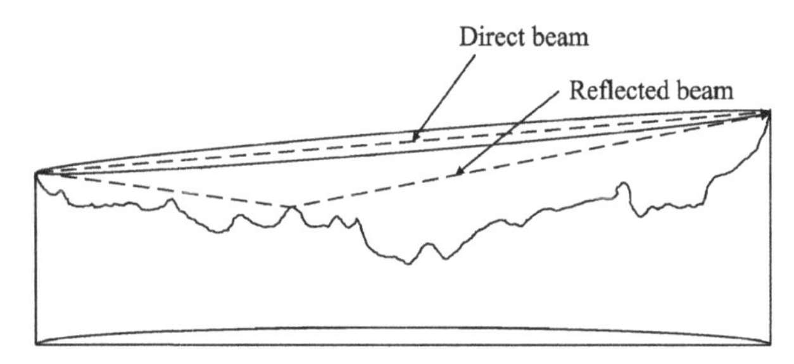

FIGURE 2.13 Reflection on the Earth's surface.

The influence of reflection can be significantly avoided if antennas with high directivities are used, if the antenna height is chosen such that the reflection path is blocked intentionally taking advantage of possible terrain obstructions on the link path, or if vertical polarization is used. If any of those techniques are not applicable and in cases where the reflection point is flat and the terrain has high conductivity and permittivity, the reflected component might be significant and usually disturbs the link performance. A clear example of these critical conditions occurs in links over wide water surfaces.

Additionally the higher the conductivity of the surface, the higher the reflected component. The worst-case condition can be found not only in links over seas and lakes, but also flat grounds such as steppes, and to a lower extent in areas with vegetation.

The reflection coefficient of the terrain is the ratio between the amplitudes of the incident and reflecting waves on the Earth's surface. This parameter is usually represented by R and depends on the conductivity, dielectric constant of the soil and also on the frequency, polarization, and angle of incidence. This coefficient will also depend effectively on the Earth's curvature.

If the Earth's surface is a water mass or a soil ground with good conductivity, the reflected component will have an amplitude that will be similar to the one of the incident wave and opposite phase ($R = -1$ or 1_{π}) almost independent of the incidence angle. Table 2.4 shows some standard values of the reflection coefficient in different soil types.

TABLE 2.4 Values of the Reflection Coefficient on Earth's Surface for Different Soils

Terrain	Reflection Coefficient (module)
Sea	1
Steppes without vegetation	0.9
Cultivated lands	0.8–0.9
Hills with bushes	0.6–0.7
Hills with trees	0.1–0.2

If vertical polarization is used, there is an incidence angle, close to 20° where the reflection is minimal. This angle is known as Brewster's angle. In fact, vertical polarization, especially in lower frequencies, does not have significant reflection coefficients independent of the incidence angle.

The variation of the reflection coefficient with frequency is because of two factors. First, the reflection coefficient tends to decrease for lower frequencies. Second, the section of the Fresnel ellipsoid at the reflection point will be wide enough in low frequencies to consider terrain roughness and irregularities. In those cases, a correction factor is applied to the reflection coefficient in order to account for the scattering that creates an additional energy loss at the reflection point.

Sometimes, in microwave LOS link design, in order to minimize the influence of reflection on the transmitter–receiver path, it might be possible to take advantage of the terrain profile characteristics and adjust the antenna heights so that the reflected path is obstructed by any of the obstacles of the terrain. In addition, systems installed over conductive soils, lakes and seas, usually include space diversity configurations, with (at least) two vertically separated antennas. An optimal value of the separation distance would provide a maximum value of received signal when there is a fade in the other one and vice versa.

A detailed calculation of reflections will be described in Chapter 6, which covers the path link engineering methods.

2.8 DISTORTION DUE TO PROPAGATION EFFECTS

Distortion is defined as the perturbation in amplitude and phase of spectral components within the frequency range associated with the radio channel plan of a microwave LOS link. Distortion may affect one or more radio channels of the link plan.

The major cause of distortion in microwave LOS links is multipath propagation. The conditions that create multipath have been described in the previous sections of this chapter, with emphasis on the effects of refraction under nonstandard conditions. Generally two multipath sources are considered: multipath in the troposphere and multipath generated by different reflections on the Earth's surface. Troposphere multipath will happen for anomalous radio refractivity gradient values where as multipath on Earth's surface will be relevant in links over soils with reflection coefficients close to -1.

2.8.1 Channel Impulse Response

During the periods of anomalous propagation, because of super-refractive or subrefractive conditions, the transmission channel will be subject to distortion and other perturbations that are variable in time.

Nevertheless, the phenomenon that creates those fluctuations is usually considered to vary slowly enough to allow the evaluation of the wideband signal distortion using a complex multipath transfer function (MTF). This function is referred to as $M(x,f)$

and its equivalent impulse response is *m(x,t)*, where x is the position vector of the receiver with respect to the transmitter. The dependency on the x variable, accounts for the path geometry and it is especially relevant for special diversity calculations.

When multipath propagation occurs, it is possible that two or more paths are configured between the transmitter and the receiver. Each one of the paths being different, the relative amplitude, the delays and the phases of components arriving from different paths are obviously different. Some of the anomalous paths will arrive from reflections on the Earth's surface, with propagation delays that range from fractions of 1ns to 10 ns.

The influence of all the possible paths on the propagation channel response may vary significantly from one link path to another. Moreover, the influence of the aggregated signal resulting from multipath combination at the receiver will depend very much on the channel bit rate and modulation scheme. There are different models that try to describe the process at the receiver input. The simplest one is the two-ray model.

2.8.2 Two-Ray Model: Frequency Response

The Two-ray model was proposed by Bell Labs in the 80's. This model assumes that as a consequence of the different multipath propagation phenomena, two paths will prevail—one of them being the LOS path between the transmitter and the receiver.

The second path might be caused by a reflection on the Earth's surface or refraction on upper layers of the troposphere. The model assumes that the second path (reflected component) will usually have a relative amplitude b ($b < 1$) lower than the LOS component, and it will be delayed τ time units from the direct component arrival time. The relative phase between both components, φ, will usually have a random variation between 0 and 360°. The transfer function of the model can be then expressed as:

$$H(\omega) = a(1 - b\exp(\pm j(\omega - \omega_0)\tau) \tag{2.17}$$

where a represents the channel transfer function amplitude. Usually, it is considered that a is made equal to one in equation (2.17) to make attenuation calculations independent of multipath analysis and work with normalized transfer functions.

The frequency ω_0 is the reference frequency where the problem is being analyzed (usually the center of the radio channel or the radio channel plan in case of multiple spectra). If only one of the signs of the exponential is taken (i.e., the negative) the real part of the transfer function $H(\omega)$ is:

$$H(\omega) = a(1 - b\cos((\omega - \omega_0)\tau) \tag{2.18}$$

and the imaginary part is:

$$X(\omega) = a\, b\sin((\omega - \omega_0)\tau) \tag{2.19}$$

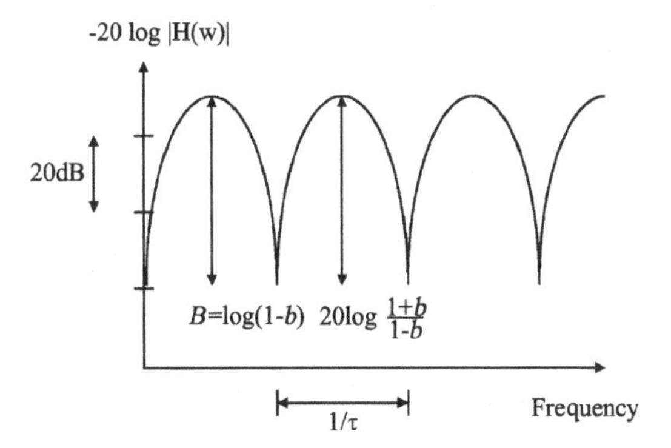

FIGURE 2.14 Channel amplitude response in a two-ray model for multipath distortion analysis.

Thus, the amplitude response can be expressed as:

$$P(\omega) = a \cdot \sqrt{(1 + b^2) - 2b \cdot \cos\left((\omega - \omega_o)\,\tau\right)} \tag{2.20}$$

and the relative phase φ can be obtained by:

$$\varphi = arctg\left(\frac{X(\omega)}{R(\omega)}\right) \tag{2.21}$$

that is the basis to calculate the group delay (GD):

$$\mathrm{GD} = -\frac{\partial \varphi}{\partial \omega} \tag{2.22}$$

Figures 2.14 and 2.15 shows the frequency response and group delay of the two-ray model as a function of the frequency ω. Some conclusions may be highlighted in the figure:

- The resulting curves have crests and valleys associated with constructive and destructive interference situations between the two components of the model. Crests will have $(1 + b)$ amplitude and valleys $(1 - b)$.
- Crests (and valleys) are periodic with a period equivalent to $1/\tau$.
- The whole diagram shifts to left or right depending on the value of the relative phase $\varphi = (\omega - \omega_0)\tau$.

The GD follows a similar diagram, even though in this case the relative maxima are higher, and depend on $(1 - b)$. It is also worth mentioning that there are two situations depending on the relative amplitudes and delays of the two components of the model, which as propagated signals, are of statistical nature. In situations where

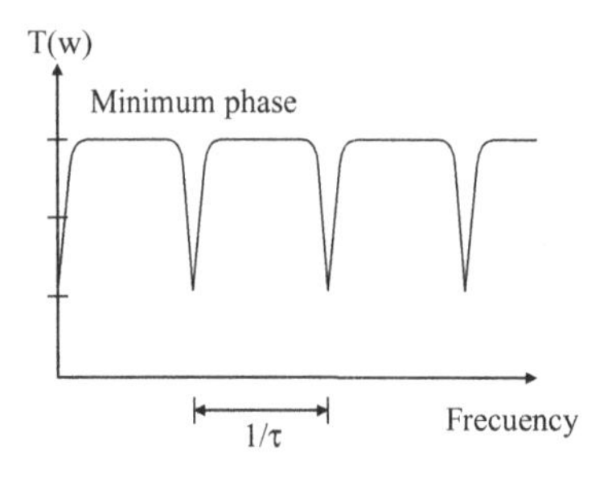

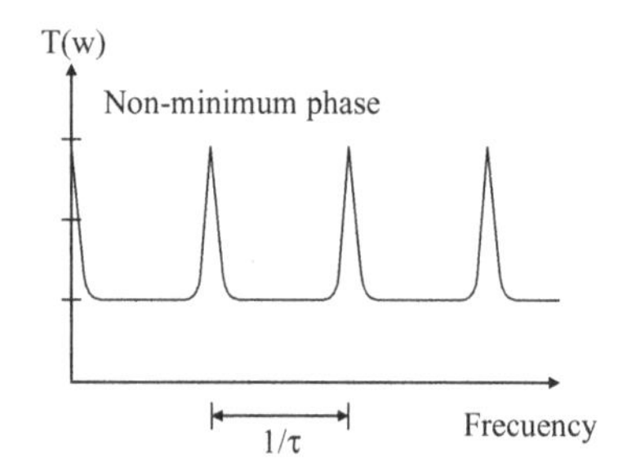

FIGURE 2.15 Delays for minimum phase and non-minimum phase conditions in a two-ray model for multipath distortion analysis.

the relative amplitude of the reflected component is less than the main path and the delayed signal occurs after the main signal or when the relative amplitude of the echo is greater than the main beam and the delayed signal occurs before the main beam, the condition is known as a minimum phase (MP) condition. The opposite situations are regarded as non-minimum phase conditions.

In this model, b, τ, and φ are parameters that vary with time. The phase φ varies extremely fast, because it will change 360° with any variation in the path equivalent to a wavelength. Consider for example the wavelength of 5 cm at 6 GHz in a signal propagating through a usual link length of 40–50 km. As a consequence, it is usually assumed that due to the variations in these parameters, there will be fast variations of the frequency response amplitude and group delay. It is assumed that the RF channel arrangements, if located randomly with respect to the two-ray model frequency response, will suffer the fast variations of amplitude and frequency that finally will be quite an accurate representation of flat and selective fading associated with multipath.

BIBLIOGRAPHY

Digital radio outage due to selective fading—observation vs. prediction from laboratory simulation. C.W. Lundgren and W.D. Rummler. *Bell System Technical Journal*. 1979.

ITU-R Rec. P. 453: The radio refractive index: its formula and refractivity data. International Telecommunication Union. Radiocommunication Sector. ITU-R. Geneva. 2003.

ITU-R Rec. P. 530: Propagation data and prediction methods required for the design of terrestrial line-of-sight systems. International Telecommunication Union. Radiocommunication Sector. ITU-R. Geneva. 2009.

ITU-R Rec. P. 676: Attenuation by atmospheric gases. International Telecommunication Union. Radiocommunication Sector. ITU-R. Geneva. 2010.

ITU-R Rec. P. 833: Attenuation in vegetation. International Telecommunication Union. Radiocommunication Sector. ITU-R. Geneva. 2007.

ITU-R Rec. P. 834: Effects of tropospheric refraction on radiowave propagation. International Telecommunication Union. Radiocommunication Sector. ITU-R. Geneva. 2007.

ITU-R Rec. P. 835: Reference Standard Atmospheres. International Telecommunication Union. Radiocommunication Sector. ITU-R. Geneva. 2005.

ITU-R Rec. P. 836: Water vapour: surface density and total columnar content. International Telecommunication Union. Radiocommunication Sector. ITU-R. Geneva. 2009.

ITU-R Rec. P. 837: Characteristics of precipitation for propagation modeling. International Telecommunication Union. Radiocommunication Sector. ITU-R. Geneva. 2007.

ITU-R Rec. P. 838: Specific attenuation model for rain for use in prediction methods. International Telecommunication Union. Radiocommunication Sector. ITU-R. Geneva. 2005.

Propagation of radiowaves. Les Barclay. The Institution of Engineering and Technology. IET. Hertfordshire. 2003.

PUB1310 Radar Navigation and Maneuvering Board Manual . National Geospatial-Intelligence Agency (Author)and M. Nicholson (Editor). ProStar Pubs. Inc. 7th Edition. 2008.

Radar and ARPA Manual: Radar and Target Tracking for Professional Mariners, Yatcshmen and users of Marine Radar. A. Bole, B. Dineley and A. Wall. Elsevier. 2005.

Radio Regulations. International Telecommunication Union. Radiocommunication Sector. ITU-R. Geneva. 2008.

CHAPTER 3

FREQUENCY PLAN FOR A FIXED SERVICE MICROWAVE LINK

3.1 FREQUENCY PLANNING OVERVIEW

Planning, management and efficient use of radio frequency (RF) spectrum are of vital importance to all network operators in order to be able to deploy microwave LOS links. The use of the spectral resources will be determined by the limited nature of frequency bands, compliance with the offered service and specified traffic quality, as well as maintaining minimum radioelectric interference.

On a global basis, the use of the spectrum is managed by diverse international bodies, with special mention of International Telecommunication Union, Radiocommunication Sector (ITU-R). The Radio Regulations from ITU-R are regarded as the main international directive reference to the allocation of the spectrum resources to radiocommunication services.

Following ITU-R terminology, the spectrum is allocated to the following service categories: fixed terrestrial, fixed satellite, mobile (terrestrial, aeronautical and maritime), broadcast (sound, television), meteorological, spatial (space operations, space research, earth exploration, intersatellite), radioastronomy, amateur and radiodetermination.

The spectrum allocation is approved by ITU administrations at the World Radiocommunication Conferences that are held by ITU periodically. The use of the spectrum on specific areas of the globe is decided at Regional Radiocommunication Conferences.

Microwave Line of Sight Link Engineering, First Edition. Pablo Angueira and Juan Antonio Romo.

A further step after frequencies have been allocated to services, is the process of allotments. An allotment is an entry of a designated frequency channel in an agreed plan, adopted by a competent conference, for use by one or more administrations for a terrestrial or space radiocommunication service in one or more identified countries or geographical areas and under specified conditions

Allocations and allotments to each service in each ITU-R region are compiled by the Radio Regulations in the Table of Frequency Allocations.

Most national administrations worldwide use the Table of Frequency Allocations as the basis to define their own National Table of Frequency Allocations. Each country will in addition specify the availability of frequencies and bands, any prioritization of their use and (usually) a differentiation between the frequencies and bands reserved for exclusive governmental use and those for general public use.

In any case, from the point of view of a specific microwave link design, national tables of frequency allocations constitute only a guideline. Actually, the use of frequency resources implies the design of the radio channel plan within the frequency band where the link will operate. The radio channel plan consists of the specification of the frequency parameters that describe the number, distribution, polarization and limit of each one of the radio signals implied in the interchange of information, usually bidirectional, between stations of a microwave.

The ITU-R prepares and publishes a long list of Recommendations within the F Series throughout the Radiocommunication Study Group 9, where radio channel arrangements of each band allocated to terrestrial fixed services (FSs) are studied. It should be taken into consideration that ITU-R Recommendations do not always show the latest radio frequency (RF) arrangements that are actually used in different countries. Initiatives to use new channels frequently come from a specific country or International Regulatory bodies and are only recognized some time later in ITU-R Recommendations.

Thorough arrangement of spectrum in Europe is a matter for the European Communications Committee (ECC) that is part of the European Conference of Postal and Telecommunications (CEPT). The ECC produces recommendations and decisions to provide the guidelines for frequency harmonization in Europe.

Regulation and assignment of radio frequencies is one of the major tasks of the governmental agency, Federal Communications Commission (FCC) in the United States. The Office of Communications (Ofcom) does the same in the United Kingdom.

Therefore, it is possible that certain countries promote the use of radio channel arrangements that differ from the ones standardized by ITU-R. As a result, from the manufacturers' perspective, when new microwave LOS link plans are going to be introduced in the international market it is advisable to know in advance the specific development in the channel plan for each country.

Finally, when the link is being designed, a specific frequency plan needs to be defined. This plan will contain a detailed channel arrangement for all the stations that form part of the link. The term "assignment" refers to the authorization given by an administration for using a radio-frequency (RF) plan by a specific station.

This chapter describes and develops examples of channel arrangements in different frequency bands. Standard radio channel arrangement plans for use in most cases will be described in detail. The designs will depend on the kind of application the microwave LOS link is used for, its capacity and the interferences that are present at the operation scenario.

3.2 BANDWIDTH AND CAPACITY OF A MICROWAVE LOS LINK

The total bandwidth used by a radio link is the whole spectrum used by its radio channels. The number of channels will depend on the operational characteristics microwave LOS link and will include the service, reserve and/or diversity.

Thus, the simplest arrangement of a link will consist of a single radio channel (one signal for each transmission direction). This scheme will be referred to as a $1 + 0$ configuration. The general case will consist of providing M service radio channels and N reserve radio channels, and will be referred to as $M + N$ configuration. Provided that single carrier modulation is used, the necessary transmission bandwidth of the modulated signal associated with each radio channel is calculated through the same equation applied to any digital transmission system:

$$B_{\mathrm{RF}} = (1 + \alpha)V' \tag{3.1}$$

where

α = roll-off factor of the channel filter
V' = modulation rate, in symbols per second

The roll-off factor depends on the filter used in each system. The objective of this filter consists of decreasing the out-of-band emissions that consequently will reduce the impact of the adjacent-channels interference. Moreover, these filter specifications should keep intersymbol interference within acceptable limits in the time domain. The α factor is also referred to as excess bandwidth factor, because it indicates the excess bandwidth over Nyquist bandwidth (equal to symbol rate). Current systems are very close to Nyquist spectrum efficiency, operating with roll-off factors down to 0.1.

Equation (3.1) can be rewritten as a function of the gross bit rate of the baseband (BB) signal before modulation process (bit/s). The variable change from symbol rate to bitrate is performed through the number of modulation states (number of bits per symbol):

$$B_{\mathrm{RF}} = (1 + \alpha)V_{\mathrm{b}}\frac{1}{\log_2 M} \tag{3.2}$$

where

α = roll-off factor of the channel filter ($0 < \alpha < 1$)
V_{b} = gross bit rate of the modulated signal in bit/s
M = number of modulation states

3.2.1 Gross and Net Bit Rate

The gross bit rate of the BB-modulating signal is the total bitrate transmitted through a specific radio channel. On the other hand, the net bit rate is that portion of the gross bit rate that is available to the transport of useful information.

The net bitrate associated with the digital information that will be transmitted is thus increased by a certain percentage (roughly about 5% in most cases) to form the gross bitrate. This increase is produced in information processing stages and BB data aggregation, and will be associated with the following functional blocks:

1. Forward error correction.
2. Additional overhead bits for maintenance, supervision or internal service of the system.
3. Internal multiplexing of several standardized bit streams.

Net bit rates used by BB time division multiplexing (TDM) systems, or asynchronous transfer mode (ATM), are standardized. These specifications are described by ITU-T Recommendations G.702, G.703 and G.704 for plesiochronous digital hierarchy (PDH) systems. In the case of synchronous digital hierarchy (SDH) and synchronous optical network (SONET) systems, the reference recommendations are G.707, G.708 and G.709. Bit rates of BB streams in these multiplexing schemes usually include multiples (n) of the following modules:

1. DS1 or T1 (1.544 Mbit/s) and DS3 or T3 (44.736 Mbit/s) (nDS1 and nDS3).
2. E1 (2048 Mbit/s) and E3 (34.368 Mbit/s) (nE1 and nE3).
3. STS-1 or sub-STM-1 (51.84 Mbit/s) and STM-1 (155.52 Mbit/s)(nSTS-1 and nSTM-1).

Multiplexing techniques that are based on PDH or SDH (SONET) in conjunction with ATM are now being replaced with IP/Ethernet based technologies systems. IP/Ethernet based systems give more flexibility and facilitate the integration of transport network into the communications network they serve to. The communications network will be generally based on Internet protocol (IP) technology. There is no predefined rate hierarchy in such cases. Currently, products that exceed several Gbit/s for each radio channel can be found on the market. The flexibility provided by high bitrate IP technology implies extra requirements: a relevant increase in management quality of service and tight synchronization requirements.

The application paradigm of the mentioned IP technology can be found in radio access networks for broadband wireless mobile communications service, including high-speed packet access, evolution-data optimized, long-term evolution, and heterogeneous networks, where transport links use a technology based on IP/Ethernet packet switching. Radio-relay links used for local area network, interconnection are an additional noteworthy example.

As transmission capacity is concerned, digital fixed wireless systems can be subdivided into the following categories:

1. Low-capacity systems for digital signal transmission at gross bit rates up to 10 Mbit/s inclusive.
2. Medium-capacity systems for digital signal transmission at gross bit rates from 10 Mbits/s to about 100 Mbits/s.
3. High-capacity systems for digital signal transmission at gross bit rates above 100 Mbits/s.

3.2.2 Spectrum Efficiency

One of the key objectives when planning FS radio-relay link systems is the efficiency in the use of RF spectrum. Spectrum use can be evaluated through the spectrum efficiency that is defined as the total bit rate of information transmitted between the two ends of the radio-relay link divided by the total bandwidth used. It is expressed in bits per second per occupied hertz of bandwidth.

For phase and amplitude modulation schemes (e.g., phase shift keying and quadrature amplitude modulation (QAM)), the key factor for calculating the necessary bandwidth is the gross bit rate of the multiplex that will be conveyed by the link. Table 3.1 provides a list of spectrum efficiencies obtained for different modulation schemes and radio channel bandwidths.

In the case of coded modulation, necessary bandwidth depends, furthermore, on the number of modulation levels and redundancy added by coding. For a modulation with 2^n states, redundant bits make bit rate increase by the f factor calculated according to the equation:

$$f = \frac{n+1}{n+z} \tag{3.3}$$

TABLE 3.1 Efficiency and Bit Rate for Each Radio-Frequency Channel Using Uncoded Modulations

	Efficiency (b/Hz)		Gross Bit Rate Depending on Radio-Frequency Channel Width (Mbps)					
			3.5 MHz		7 MHz		28 MHz	
Roll-off (α)	0.5	0.1	0.5	0.1	0.5	0.1	0.5	0.1
QPSK	1.33	1.82	4.67	6.36	9.33	12.73	37.33	50.91
8-PSK	2.00	2.73	7.00	9.55	14.00	19.09	56.00	76.36
16-QAM	2.67	3.64	9.33	12.73	18.67	25.45	74.67	101.82
64-QAM	4.00	5.45	14.00	19.09	28.00	38.18	112.00	152.73
128-QAM	4.67	6.36	16.33	22.27	32.67	44.55	130.67	178.18
256-QAM	5.33	7.27	18.67	25.45	37.33	50.91	149.33	203.64
1024-QAM	6.67	9.09	23.33	31.82	46.67	63.64	186.67	254.55

QPSK, quadrature phase shift keying; PSK, phase shift keying

TABLE 3.2 Efficiency and Bit Rate for Each Radio-Frequency Channel Using Coded Modulations

Modulation Scheme	Specific Cases	Bandwidth
BCM	96-BCM-4D (QAM one-step partition)	$V_b/6$
	88-BCM-6D (QAM one-step partition)	$V_b/6$
	16-BCM-8D (QAM one-step partition)	$V_b/3.75$
	80-BCM-8D (QAM one-step partition)	$V_b/6$
	128-BCM-8D (QAM two-step partition)	$V_b/6$
TCM	16-TCM-2D	$V_b/3$
	32-TCM-2D	$V_b/4$
	128-TCM-2D	$V_b/6$
	512-TCM-2D	$V_b/8$
	64-TCM-4D	$V_b/5.5$
	128-TCM-4D	$V_b/6.5$
	512-TCM-4D	$V_b/8.5$
MLCM	32-MLCM	$V_b/4.5$
	64-MLCM	$V_b/5.5$
	128-MLCM	$V_b/6.5$
QPR with AZD	9-QPR with coherent detection	$V_b/2$
	25-QPR with coherent detection	$V_b/3$
	49-QPR with coherent detection	$V_b/4$

BCM, block coded modulation; TCM, Trellis coded modulation; MLCM, multilevel coded modulation; QPR, quadrature partial response; AZD, ambiguity zone detection

Z is a parameter that ranges from 0 to 1 in equation (3.3) according to the coded modulation used in each case.

Symbol rate (baud) and bandwidth B_{RF} are related to each other through the expression:

$$V = \frac{B_{\mathrm{RF}}}{n + z} \tag{3.4}$$

Values for some coded modulation are shown in the Table 3.2, without taking into account roll-off correction factor, $(1 + \alpha)$.

Finally, it is worth reminding that instantaneous bit rate will change according to the optimized pattern for the propagation channel in systems with adaptive modulation described in Chapter 4.

3.2.3 Spectrum Resources

Some techniques based on spectrum optimization are available, as an answer to market demand on capacity increase and more efficient spectrum use. Highlighted among them are radio channel aggregation, polarization multiplexing (co-channel reuse) and multiple input multiple output (MIMO).

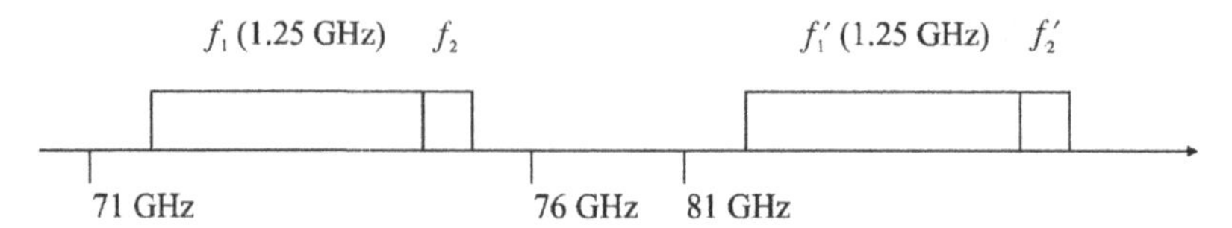

FIGURE 3.1 An example of radio channel aggregation suggested by the ECC05047 Recommendation for 71–76/81–86 GHz bands.

Channel aggregation consists of aggregating several carriers of a frequency plan, with the aim of using larger bandwidths for each single radio channel. A multicarrier system consists of n (being $n > 1$) modulated carrier signals, simultaneously transmitted or received by the same RF equipment. In fact, some recent regulations widen traditional channel bandwidths of 2.5–3.5, 7, 14, and 28 MHz up to 110 and 240 MHz radio channel bandwidths, for bands above 40 GHz. Figure 3.1 illustrates an example of a CEPT through ECC recommended radio channel plan in Europe (ECC Rec 0507). Radio channel aggregation from individual 250 MHz into 1.25 GHz channels in the 71–76/81–86 GHz bands is shown.

Polarization multiplexing or co-channel reuse consists of using the same frequencies by two radio channels with different crossed polarizations, vertical and horizontal respectively. Using co-channel systems can provide double capacity if compared with conventional radio-relay link systems for a specific bandwidth, Ideally, cross-polar discrimination (XPD) could be assumed to be infinite. However, on the one hand, polar and cross-polar patterns of the antennas will not show an ideal discrimination. On the other hand, propagation phenomena such as hydrometeors and multipath will modify the polarization of the transmitted signal, therefore, creating interference among channels with different polarization. The degree of discrimination between the two polarizations, called XPD, must be high enough in order to ensure interference free operation. The nominal value of XPD is called XPD_0 and it is given by the cross-polarization patterns of the antennas. The fact that propagation factors associated with variations in the polarization are time varying may produce an important degradation on the XPD level. As XPD decreases, interference level will increase in channel, and will cause a degradation of the system threshold, increasing the number of errors on the data traffic.

To mitigate this effect and make radio-relay link systems resistant to XPD variation, most manufacturers nowadays use cross-polarization interference cancellation circuits on receivers, with signal processing algorithms that eliminate the unwanted signal that has been filtered from the opposite polarization toward the desired one.

Finally, using MIMO techniques in LOS conditions provides an increase of the spectrum efficiency. Currently, MIMO techniques are being widely used on mobile and wireless communication systems. Basically, $n \, x \, n$ MIMO systems are composed of n transmitters and n receivers transmitting n different signals. At the receiver, it is possible to use all the received components to cancel the interfering signals. Ideally, in a 2×2 MIMO system, capacity could be twice of that in a conventional system. More details about MIMO techniques in microwave LOS links are provided in Chapter 4.

3.3 ITU-R FREQUENCY PLANS

This section provides a description and analysis of the most relevant characteristics of the channel arrangement plans in bands allocated to the FS. The guidelines for creating radio channel plans have been studied by ITU-R Study Groups and are available in F Series Recommendations.

Channel plans or "Channel Arrangements" aim at subdividing a specific allocated frequency band into smaller parts, so-called RF channels, so that the necessary spectrum of the modulated signal produced by a transmitter can be arranged within each RF channel.

Tables 1 and 2 of Recommendation ITU-R F.746 provide a comprehensive list of all the published Recommendations about channel arrangements for each of the allocated bands to the FS, with applicable frequency ranges and the different bandwidth possibilities for radio channels of each plan. Table 3.3 contains a simplified summary of this information.

The numbers in column "Band (GHz)" reflect the frequency GHz that is usually employed to designate the associated band. Frequency ranges of each band, where the specific channel plan is valid are indicated in the second column. Lower and upper frequency limits define every frequency range.

Associated with each one of the frequency bands, there will be one or more F Series ITU-R Recommendations, shown in the third column of the table. The same frequency range may have different channel arrangements that will be described in specific recommendations or annex numbers (third column). Finally, in column entitled "Separation between channels (MHz)" various channel arrangements are included. This column allows the calculation of a fast rough estimate of the capacity associated to each channel arrangement.

The ITU-R is not the only regulatory body developing frequency channel arrangements for radiocommunication services in general, and microwave LOS links in particular. Regional regulatory bodies such as, CEPT in Europe and FCC in the United States should be mentioned. In the case of CEPT, Electronic Communications Committee (ECC) regulates frequency bands. Table 3.4 summarizes frequency bands and their associated channel arrangements.

The reference regulatory document in the case of specific allocations and channel arrangements for fixed-service in the US is Code of Federal Regulations, where Title 47 focuses on Telecommunications. The specific technical regulations associated to each radiocommunication service are listed in Part 101. In this case, it is the only document that describes channel arrangements of any of the fixed-service bands. Table 3.5 summarizes bands and channel arrangements for microwave LOS links. It needs to be clearly understood that the FCC legislation is mainly based on establishing the maximum bandwidth of a signal for each band, and then offer different possibilities for channel width under that maximum bandwidth.

TABLE 3.3 Radio-Frequency Channel Arrangements for Fixed Service Systems

Band (GHz)	Frequency Range (GHz)	F Series ITU-R Recommendations	Separation Between Channels (MHz)
0.4	0.4061–0.430 0.41305–0.450	1567. Annex 1 1567. Annex 1	0.05, 0.1, 0.15, 0.2, 0.25, 0.6, 0.25, 0.3, 0.5, 0.6, 0.75, 1, 1.75, 3.5
1.4	1.35–1.53	1242	0.25, 0.5, 1, 2, 3.5
2	1.427–2.69	701	0.5
	1.7–2.1, 1.9–2.3	382	29
	1.9–2.3	1098	3.5, 2.5
	1.9–2.3	1098. Annexes 1, 2	14
	1.9–2.3	1098. Annex 3	10
	2.3–2.5	746. Annex 1	1, 2, 4, 14, 28
	2.29–2.67	1243	0.25, 0.5, 1, 1.75, 2, 3.5, 7, 14, 2.5
3.6	3.4–3.8	1488. Annex 1	25[a]
	3.4–3.8	1488. Annex 2	0.25[b]
4	3.8–4.2	382	29
	3.7–4.2	382. Annex 1	28
	3.6–4.2	635	10
	3.6–4.2	635. Annex 1	90, 80, 60, 40, 30
U4	4.4–5.0	1099	10
	4.4–5.0	1099. Annex 1	40, 60, 80
	4.4–5.0	1099. Annex 3	28
	4.54–4.9	1099. Annex 2	40, 20
L6	5.925–6.425	383	29.65
	5.85–6.425	383. Annex 1	90,
	5.925–6.425	383. Annex 1	60, 40
	5.925–6.425	383. Annex 2	28
	5.925–6.425	383. Annex 3	40, 20, 10, 5
U6	6.425–7.11	384	40, 30, 20, 10, 5
	6.425–7.11	384. Annex 1	80
7	7.25–7.55	385. Annex 5	3.5
	7.425–7.725 (7.125–7.425)[c] (7.250–7.550)[c] (7.550–7.850)[c]	385	7, 14, 28
	7.425–7.725	385. Annex 1	28
	7.435–7.75	385. Annex 2	5
	7.11–7.75	385. Annex 3	28
10	10.0–10.68	746. Annex 2	3.5, 7, 14, 28
	10.15–10.65	1568. Annex 1	28[a]
	10.15–10.65	1568. Annex 2	30[a]
	10.3–10.68	746. Annex 2	5, 2
	10.5–10.68	747. Annex 1	7, 3.5
	10.55–10.68	747. Annex 2	5, 2.5, 1.25

TABLE 3.3 ***(Conitinued)***

Band (GHz)	Frequency Range (GHz)	F Series ITU-R Recommendations	Separation Between Channels (MHz)
11	10.7–11.7	387	40
	10.7–11.7	387. Annex 1	67
	10.7–11.7	387. Annex 3	60
	10.7–11.7	387. Annex 2	80
	10.7–11.7	387. Annex 4	5, 10, 20
12	11.7–12.5	746. Annex 3, § 3	19, 18
	12.2–12.7	746. Annex 3, § 2	20
13	12.75–13.25	497	28, 14, 7, 3.5
	12.7–13.25	746. Annex 3, § 1	25, 12.5
14	14.25–14.5	746. Annex 4	28, 14, 7, 3.5
	14.25–14.5	746. Annex 5	7, 14, 28
15	14.4–15.35	636	28, 14, 7, 3.5
	14.5–15.35	636. Annex 1	2.5
	14.5–15.35	636. Annex 2	2.5
18	17.7–19.7	595	220, 110, 55, 27.5
	17.7–19.7	595. Annex 1	60 (block)
	17.7–19.7	595. Annex 2	50, 40, 30, 20, 10, 5, 2.5
	17.7–19.7	595. Annex 3	7, 3.5
	17.7–19.7	595. Annex 4	27.5, 13.75, 7.5, 5, 2.5,
	17.7–19.7	595. Annex 5	1.25
	17.7–19.7	595. Annex 6	7, 3.5, 1.75
	17.7–19.7	595. Annex 7	55, 110
	18.58–19.16	595. Annex 7	55, 27.5, 13.75
			60
23	21.2–23.6	637	3.5, 2.5
	21.2–23.6	637. Annex 1	112–3.5
	21.2–23.6	637. Annex 2	28, 3.5
	21.2–23.6	637. Annex 3	112 to 3.5
	21.2–23.6	637. Annex 4	50
	21.2–23.6	637. Annex 5	112–3.5
	22.0–23.6	637. Annex 1	112–3.5
27	24.25–25.25	748	3.5, 2.5
	24.25–25.25	748. Annex 3	40[a]
	25.25–27.5	748	3.5, 2.5
	25.27–26.98	748. Annex 3	60[a]
	24.5–26.5	748. Annex 1	112–3.5
	27.5–29.5	748	3.5, 2.5
	27.5–29.5	748. Annex 2	112–3.5
31	31.0–31.3	746. Annex 7	25, 50
	31.0–31.3	746. Annex 8	28, 14, 7, 3.5
32	31.8–33.4	1520. Annex 1	3.5, 7, 14, 28, 56
	31.8–33.4	1520. Annex 2	56[a]

(continued)

TABLE 3.3 (*Conitinued*)

Band (GHz)	Frequency Range (GHz)	F Series ITU-R Recommendations	Separation Between Channels (MHz)
38	36.0–40.5	749	3.5, 2.5
	36.0–37.0	749. Annex 2	112–3.5
	37.0–39.5	749. Annex 1	140, 56, 28, 14, 7, 3.5
	38.6–39.48	749. Annex 2	60[a]
	38.6–40.0	749. Annex 2	50[a]
	39.5–40.5	749. Annex 3	112–3.5
52	51.4–52.6	1496. Annex 1	56, 28, 14, 7, 3.5
57	55.78–57.0	1497. Annex 1	56, 28, 14, 7, 3.5
	57.0–59.0	1497. Annex 2	100, 50

[a] Bandwidth of the frequency block
[b] Basic frequency slot for aggregating wider frequency block bandwidth
[c] Alternating bands in brackets.

TABLE 3.4 Frequency Allocation to Fixed Service Radio-Relay Links and Associated Channel Arrangements (CEPT/ECC)

Frequency Band (GHz)	ECC Recommendation	Channel Arrangements (MHz)
1.0–2.3	T/R 13–01	2, 1, 0.5, 0.25, 0.025, 3.5 // 14, 7, 3.5, 1.75
3.4–3.6	ERC/REC 14–03	14, 7, 3.5, 1.75
3.6–4.2	ERC/REC 12–08	20, 40 // 30, 15 //14, 7, 3.5, 1.75
5.925–6.425	ERC/REC 14–01	29.65, 59.3
6.425–7.125	ERC/REC 14–02	40, 30, 20, 14, 7, 3.5
7.125–8.500	ECC/REC/(02)06	56, 28, 14, 7, 3.5, 1.75 // 59.3, 29.65
10.0–10.68	ERC/REC 12–05	0.5 // 56, 28, 14, 7, 3.5, 1.75
10.7–11.7	ERC/REC 12–06	80, 56, 40, 28
12.75–13.25	ERC/REC 12–02	56, 28, 14, 7, 3.5, 1.75
15.23–15.35	ERC/REC 12–07	56, 28, 14, 7, 3.5, 1.75
17.7–19.7	ERC/REC 12–03	110, 55, 27.5, 13.75
22.0–29.5	T/R 13–02	112, 56, 28, 14, 7, 3.5
24.5–26.5, 27.5–29.5, 31.8–33.4	ECC/REC/(11)01	As T/R 13–02
31.0–31.3	ECC/REC/(02)02	28, 14, 7, 3.5
31.8–33.4	ERC/REC/(01)02	112, 56, 28, 14, 7, 3.5
37.0–39.5	T/R 12–01	112, 56, 28, 14, 7, 3.5
48.5–50.2	ERC/REC 12–10	28, 14, 7, 3.5
51.4–52.6	ERC/REC 12–11	56, 28, 14, 7, 3.5
55.78–57.00	ERC/REC 12–12	56, 28, 14, 7, 3.5
57–64	ECC/REC/(09)01	Not specified
64–66	ECC/REC/(05)02	Not specified
71–76, 81–86	ECC/REC/(05)07	250. Aggregation possible up to 4.5 GHz

TABLE 3.5 Frequency Allocation to Fixed Service Radio-Relay Links and Associated Channel Arrangements (FCC)

Frequency Band (GHz)	Channel Arrangements (MHz)
1850–1990	5, 10
2110–2130	Max. 3.5
2130–2200	0.8
2450–2500	630
3700–4200	20
5925–6425	0.4, 0.8, 1.25, 2.5, 3.75, 5, 10, 30
6425–6525	1, 8, 25
6525–6875	0.4, 0.8, 1.25, 2.5, 3.75, 5, 10
10550–10680	0.4, 0.8, 1.25, 2.5, 3.75, 5
10700–11700	1.25, 2.5, 3.75, 5, 10, 30, 40
17700–19700	1.25, 2, 2.5, 5, 10, 20, 30, 40, 50, 80
21200–23600	5, 10, 20, 20, 30, 40, 50
37000–40000	Max. 50
71000–76000	1.25 GHz blocks that can be aggregated
81000–86000	
92000–94000	
94100–95000	

3.3.1 General Description of a Radio-Frequency Channel Arrangement Plan

Homogeneous RF channel arrangement plans divide into two halves the frequency range of the band. Duplex separation is done through frequency division multiplex, using RF channels of each subband for each transmission direction. A simplified diagram of an arrangement plan is shown in Figure 3.2.

RF channels are characterized by a center frequency and sequential numbering within each subband. It should be reminded here that a radio channel is defined by the combination of two subchannels or spectra in the frequency plan. Each one will

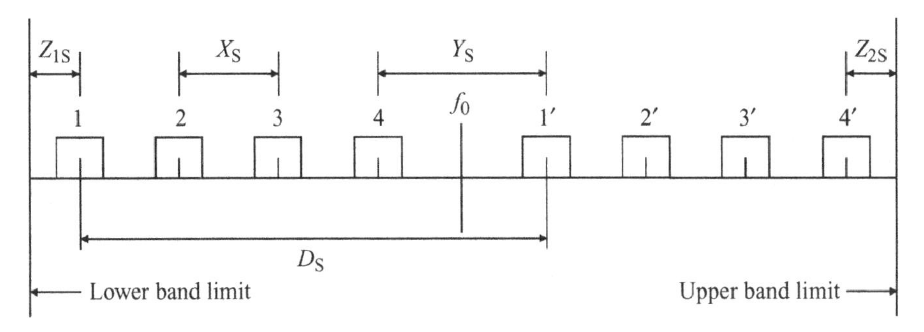

FIGURE 3.2 Radio-frequency channel arrangement parameters with contiguous blocks of spectrum.

be associated with the forward and return directions respectively. Forward and return spectra of the same radio channel are associated with the same number, and they are differentiated using the prime suffix. The couple of center frequencies of RF channels of order n are called f_n and f'_n respectively. In most RF channel arrangements defined by ITU-R forward and return channels are contained in contiguous blocks of spectrum. The main spectrum parameters that characterize a radio channel arrangement plan, for a specific band and specified capacity are the following:

1. Center frequency of the occupied frequency band, f_0.
2. Separation between center frequencies of adjacent RF channels, defined as the separation between the center frequencies of adjacent channels on the same polarization and in the same direction of transmission, X_S. In practice, the channel bandwidth is considered to be approximately equal to separation between radio channels.
3. Y_S, separation between the center frequencies of forward and return radio channels that are closest between each other, also called innermost channels. In the case of forward and return frequency subbands not to be in contiguous blocks, Y_S includes real separation between both spectrum frequencies.
4. Guard band, Z_S, defined as the RF separation between the center frequencies of the outermost RF channels and the edge of the frequency. In the case of different lower and upper separations, Z_{1S} specifies lower separation and Z_{2S} upper separation. In the case of noncontiguous go and return frequency subbands, Z_{Si} is defined for the innermost edges of both subbands and it is included in Y_S.
5. D_S, Tx/Rx duplex spacing. It is the separation between center frequencies of forward and return channels of the same radio channel. These frequencies will have the same order number in the arrangement plan.

Horizontal or vertical polarization is usually represented by plotting the spectrums above and below the frequency axis. Figure 3.3 shows the main parameters of a generic RF channel plan with duplex frequencies in contiguous spectrum blocks.

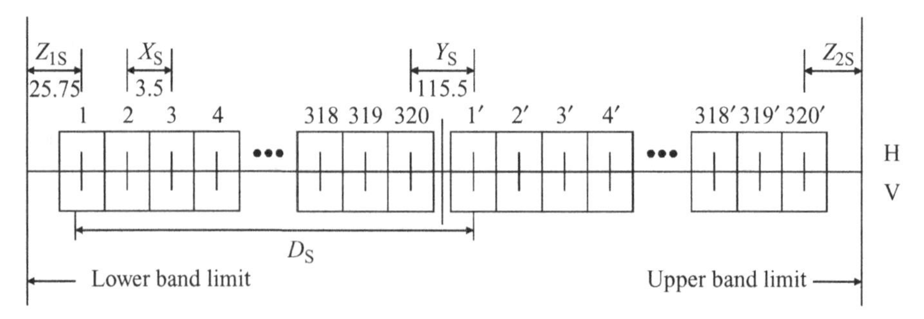

FIGURE 3.3 Example of radio-frequency channel arrangement for the 23 GHz band in the United Kingdom, as defined in Recommendation ITU-R F.637.

3.3.2 Radio Channel Arrangement Options

ITU-R Recommendation F.746 describes the general guidelines that should be followed to create RF channel arrangements for fixed wireless systems. The most appropriate homogeneous patterns that are recommended as a basis for choosing RF channel arrangements are also indicated. Basically, three homogeneous patterns are considered:

1. Alternated RF channel arrangements.
2. Cochannel band reuse RF channel arrangements.
3. Interleaved band reuse RF channel arrangements.

Features such as type of equipment, considerations about interferences and necessary spectrum efficiency should be taken into account in order to choose one or another pattern and the most appropriate channeling parameters. The particular type of pattern will be chosen depending on XPD and net filter discrimination (NFD) values that contribute to the value of carrier/interference ratio. Both parameters are defined by equations (3.5) and (3.6) respectively:

$$\mathrm{XPD}_{\mathrm{H(V)}} = \frac{\text{Power received on polarization H(V) transmitted on polarization H(V)}}{\text{Power received on opposite polarization V(H) transmitted on polarization H(V)}} \tag{3.5}$$

$$\mathrm{NFD} = \frac{\text{Adjacent channel received power}}{\text{Adjacent channel received power after RF, IF, and BB filters}} \tag{3.6}$$

3.3.2.1 Alternated Radio Channel Arrangements When using this arrangement (Figure 3.4), adjacent channels operate on opposite polarization, while the same polarization is used every two channels.

This configuration can be used (co-polar adjacent-channel contribution to the interference level neglected) if requirement for the total amount of interfering power is met:

$$\mathrm{XPD_{min}} + (\mathrm{NFD} - 3) \geq (\mathrm{C/I})_{\mathrm{min}} \tag{3.7}$$

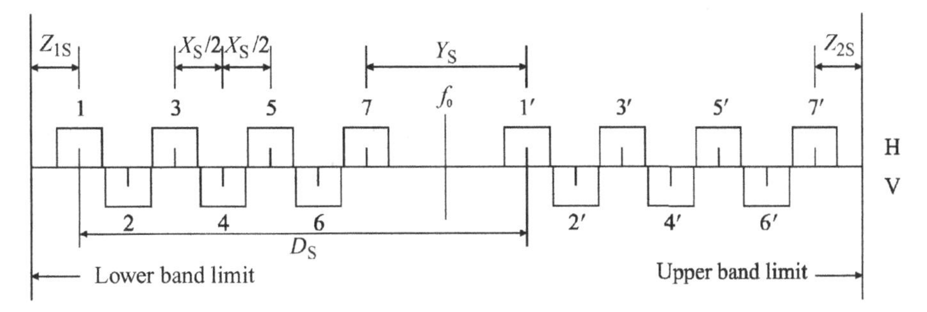

FIGURE 3.4 Alternated channel arrangement pattern.

where

$(C/I)_{min}$ = minimum carrier/interference acceptable ratio for the modulation scheme adopted, and for a given BER limit

XPD_{min} = minimum XPD during the percentage of time required

Alternated channel arrangement is the traditional arrangement used in most radio-relay link system applications. This RF pattern is described in the main text of all ITU recommendations for channel arrangements in radio-relay links.

Each RF channel has only one polarization assigned, so it is easy to implement and economically attractive. Commonly-used antennas and standard radio equipment comply with the required performance for expected interferences in this scenario. Nevertheless, as cross-polar frequency reuse is not being taken into account, this arrangement cannot be considered efficient in terms of spectrum, when compared with other arrangements.

3.3.2.2 Radio Channel Arrangements for Co-Channel Band Reuse

When this arrangement is implemented, every RF channel is reused two times in the same hop, on opposite polarization, and each one conveys different traffic. The resulting homogeneous arrangement is schematically represented in Figure 3.5.

This arrangement is used in scenarios with polarization multiplexing when spectrum is intensively used within the band under consideration. This way, we can double the capacity by reusing the same channel and changing polarization.

As a counterpart, this arrangement is more demanding on antenna and radio equipment features. Equation (3.8) is given as a reference in ITU-R Recommendation 746, that is more restrictive than the one associated with alternated arrangements:

$$10\log \frac{1}{\frac{1}{10^{\frac{XPD+XIF}{10}}} + \frac{1}{10^{\frac{NFD_a-3}{10}}}} \geq (C/I)_{min} \tag{3.8}$$

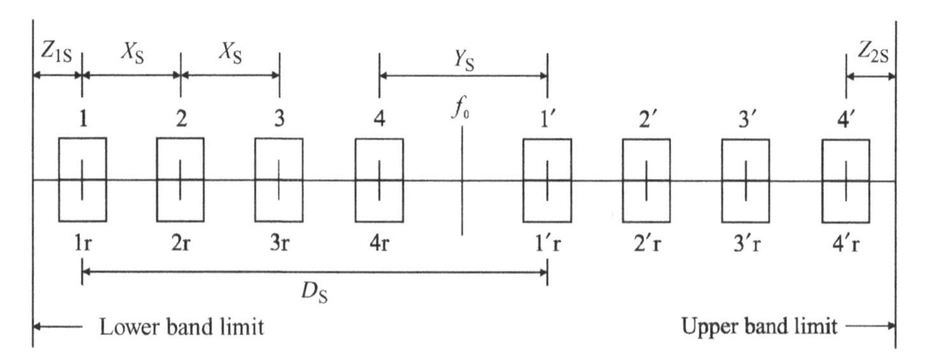

FIGURE 3.5 Co-channel band reuse radio-frequency channel arrangements.

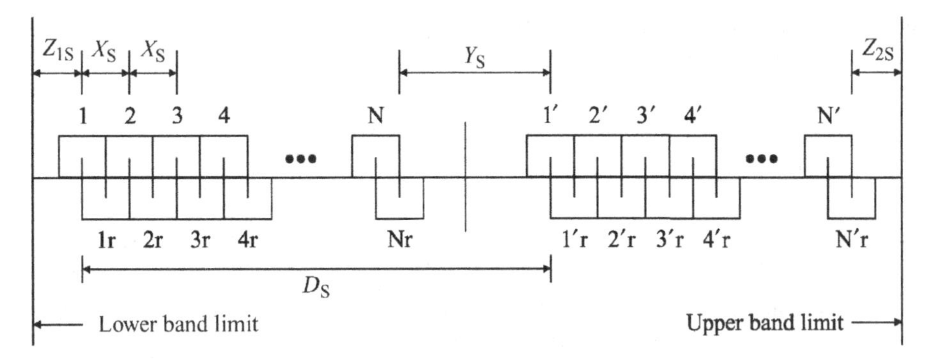

FIGURE 3.6 Channel arrangement for interleaved band reuse.

where

$(C/I)_{min}$ = minimum carrier/interference acceptable ratio for the modulation scheme adopted, and for a given BER limit

XPD_{min} = minimum XPD during the percentage of time required

NFD_a = NFD calculated with *XS* frequency spacing

XIF = XPD improvement factor of any cross-polar interference countermeasure, when implemented in the interfered receiver

3.3.2.3 Radio Channel Arrangements for Interleaved Band Reuse

Channel arrangement for interleaved band reuse is an intermediate solution between alternated and co-channel arrangements. It optimizes the bandwidth use by inserting new channels in between the main ones. Center frequency of the channel on opposite polarization is interleaved between the adjacent-channel frequencies. Figure 3.6 shows the resulting homogeneous arrangement.

This RF channel arrangement can be typically applied to low capacity systems where signal bandwidth can be smaller than separation between channels. ITU-R Recommendation F.746 provides an inequation that must be complied for this arrangement to be applied, like in the above-mentioned case. It depends on polar discrimination characteristics and filtering requirements:

$$10\log\frac{1}{\dfrac{1}{10^{\frac{XPD+(NFD_b-3)}{10}}}+\dfrac{1}{10^{\frac{NFD_a-3}{10}}}}\geq (C/I)_{min} \tag{3.9}$$

where

$(C/I)_{min}$ = minimum carrier/interference acceptable ratio for the modulation scheme adopted, and for a given BER limit

XPD_{min} = minimum XPD during the percentage of time required

NFD_a = NFD calculated with XS frequency spacing

NFD_b = NFD calculated with *XS*/2 frequency spacing

XIF = XPD improvement factor of any cross-polar interference countermeasure, when implemented in the interfered receiver

3.3.3 Radio-Frequency Channel Arrangement Plan: A Detailed Example

This section details and analyzes the plan for the upper 6 GHz band (6.425–7.125 MHz), in order to clarify concepts associated with RF channel arrangement plans. The following arrangements are set out for the frequency band:

1. Alternated RF channel arrangement based on 40 MHz channel spacing, using up to eight forward and eight return RF channels.
2. Alternated RF channel arrangement based on 20 MHz channel spacing, using up to 16 forward and 16 return RF channels.
3. Co-polar channel arrangements with multicarrier transmission.
4. RF channel arrangement based on 30 MHz channel spacing, using up to 10 forward and 10 return RF channels.
5. RF channel arrangement based on 10 MHz channel spacing, using up to 32 forward and 32 return RF channels.
6. RF channel arrangement based on 5 MHz channel spacing, using up to 64 forward and 64 return RF channels.

3.3.3.1 40 MHz Alternated Radio Channel Arrangement This arrangement is planned for up to eight RF channels, with eight forward channels and eight return channels, with a separation of 40 MHz between the carriers. This is the main pattern of the recommendation and it is suited to RF channels of net bit rate in the order of 140 Mbit/s for PDH or STM-1 of SDH.

Following the frequency assignment steps defined in ITU-R recommendations, first, the band is divided into two halves in order to assign the center frequencies to each one of the subband channels. Frequencies of each RF channel are expressed in terms of the following relationships:

Lower half of the band: $f_n = f_0 - 350 + 40\,n$	MHz
Upper half of the band: $f'_n = f_0 - 10 + 40\,n$	MHz

where

f_0 = center frequency of the occupied frequency band. It is recommended to be 6770 MHz

f_n = center frequency of one of the RF channels in the lower half of that band (MHz)

f'_n = center frequency of one of the RF channels in the upper half of that band (MHz)

$N = 1, 2, 3, 4, 5, 6, 7, 8$

The RF channel arrangement is alternated. Adjacent radio-channels in the same half of the band use different polarizations alternately. Depending on the number

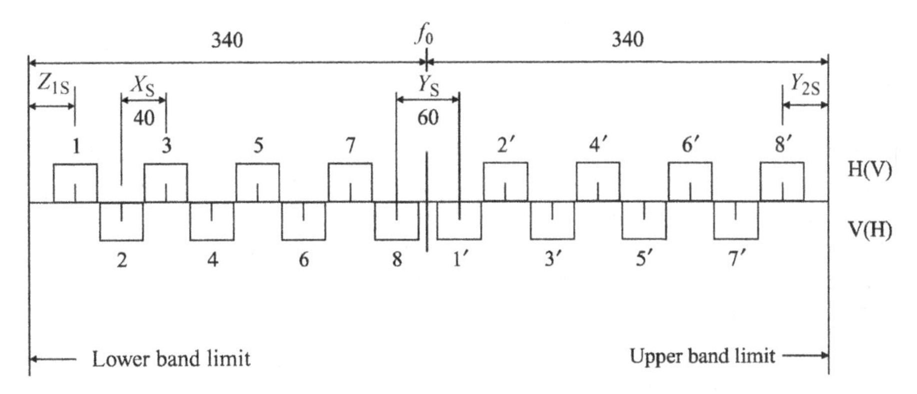

FIGURE 3.7 Radio-frequency channel arrangement in the case of dual polarized antennas (40 MHz channel arrangements).

(one or two) and polarization of the antennas used, the recommendation offers two possibilities for this pattern implementation.

Associated channel pairs (forward and return) will have different polarization in systems with dual polarized antennas. This case is illustrated in Figure 3.7.

The arrangement in Figure 3.8, where associated channel pairs will have the same polarization, is used in systems furnished with antennas with single polarization or by common transmission/reception dual polarized antennas.

If common transmission/reception antennas are used and no more than four RF channels are transmitted with the same antenna, it is better to choose RF channel frequencies according to one of the following combinations:

$n = 1, 3, 5$, and 7 in both halves of the band.

$n = 2, 4, 6$, and 8 in both halves of the band.

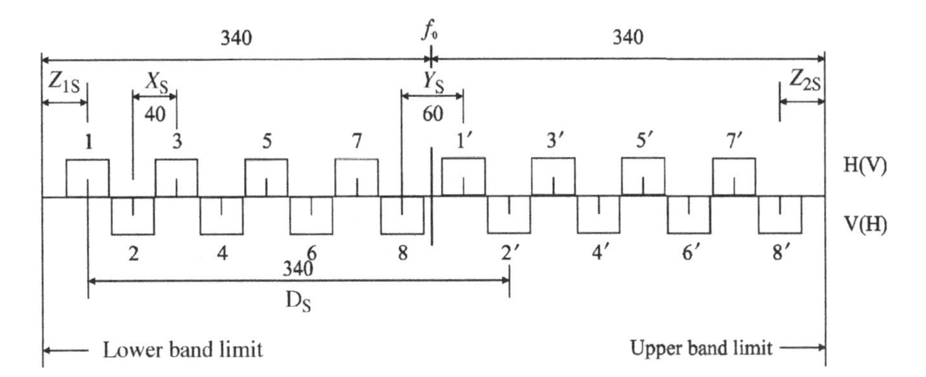

FIGURE 3.8 Radio-frequency channel arrangement in the case of antennas with single polarization (40 MHz channel arrangements).

For smaller bandwidths, recommendation leaves an option for RF channel number to increase by using separation between carriers of 20 MHz, 10 MHz, and 5 MHz subdividing the 40 MHz channels of this arrangement.

3.3.3.2 20 MHz Alternated Radio Channel Arrangement

This arrangement is planned for up to 16 RF channels with a separation of 20 MHz between carriers. This arrangement is suited to medium capacity PDH or SDH systems. The arrangement is built by alternating additional RF channels in addition to the ones in the main channel plan. Frequencies of each RF channel are expressed in terms of the following relationships:

Lower half of the band: $f_n = f_0 - 350 + 20\,n$	MHz
Upper half of the band: $f'_n = f_0 - 10 + 20\,n$	MHz

where

f_0 = center frequency of the band, 6770 MHz
f_n = center frequency of one of the RF channels in the lower half of that band (MHz)
f'_n = center frequency of one of the RF channels in the upper half of that band (MHz)
$n = 1, 2, 3, \ldots, 14, 15, 16$

The resulting RF channel arrangement is alternated. Adjacent radio-channels in the same half of the band use different polarizations alternately. In accordance with the same idea that in 40 MHz arrangement, depending on the number (one or two) and polarization of the antennas used, recommendation offers two possibilities for the implementation of this pattern. Associated channel pairs will have different polarization for the implementation in systems based on dual polarized antennas. Figure 3.9 illustrates this case.

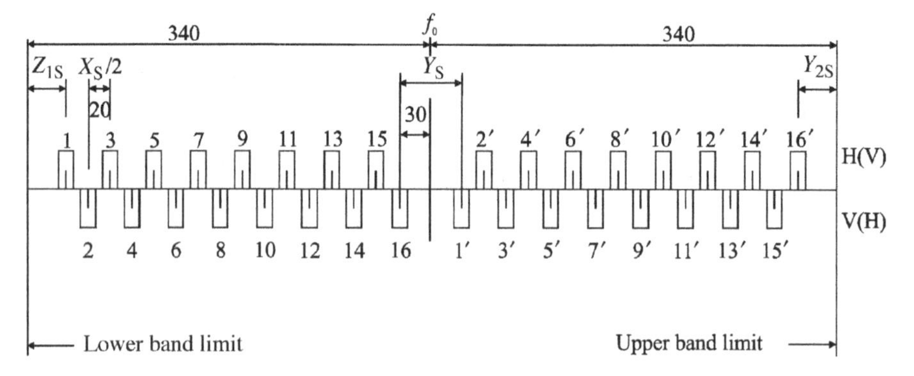

FIGURE 3.9 Radio-frequency channel arrangement in the case of dual polarized antennas (20 MHz channel arrangements).

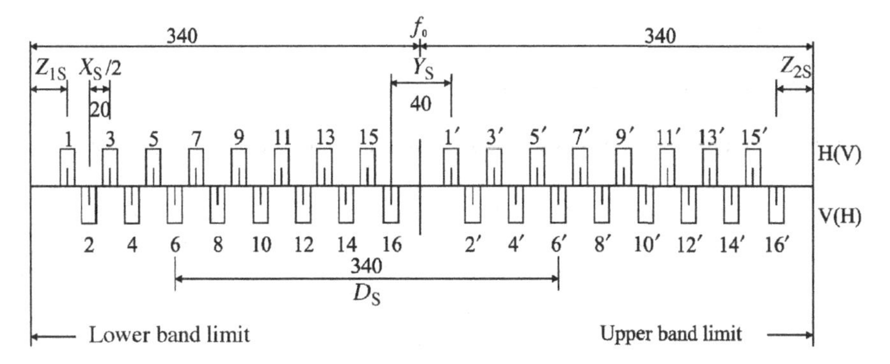

FIGURE 3.10 Radio-frequency channel arrangement in the case of antennas with single polarization (20 MHz channel arrangements).

For the implementation in systems based on antennas with single polarization the arrangement shown in Figure 3.10 is used, were associated channel pairs will have the same polarization. If common transmission/reception antennas are used and no more than four RF channels are transmitted with the same antenna, it is better to choose RF channel frequencies using one of the following combinations in both halves of the band:

$n = 1, 5, 9, 13$, or
$n = 2, 6, 10, 14$, or
$n = 3, 7, 11, 15$, or
$n = 4, 8, 12, 16$

3.3.3.3 30 MHz Radio Channel Arrangement This arrangement is planned for up to 10 RF channels with a separation of 30 MHz between carriers. This arrangement is suited to RF channels with a net bit rate of about 155 Mbit/s (PDH) or any SDH bit rate. Frequencies of each RF channel are expressed in terms of the following relationships:

Lower half of the band: $f_n = f_0 - 340 + 30\,n$	MHz
Upper half of the band: $f'_n = f_0 + 30\,n$	MHz

where

f_0 = center frequency of the band, 6 770 MHz
f_n = center frequency of one of the RF channels in the lower half of that band (MHz)
f'_n = center frequency of one of the RF channels in the upper half of that band (MHz)
$n = 1, 2, 3, 4, 5, 6, 7, 8, 9, 10$

In order to improve spectrum efficiency, this RF channel arrangement may use co-channel frequency reuse when equipment and network features allow it, and as long as interested administrations previously agree to do so.

3.3.3.4 10 MHz Radio Channel Arrangement This arrangement is planned for up to 32 RF channels with a separation of 10 MHz between the carriers. This arrangement is suited to medium-capacity RF channels (PDH and SDH). Frequencies of each RF channel are expressed in terms of the following relationships:

Lower half of the band: $f_n = f_0 - 340 + 10\,n$	MHz
Upper half of the band: $f'_n = f_0 + 10\,n$	MHz

where

$$n = 1, 2, 3, \ldots 31, 32$$

3.3.3.5 5 MHz Radio Channel Arrangement This arrangement is planned for up to 64 RF channels with a separation of 5 MHz between carriers. Frequencies of each RF channel are expressed in terms of the following relationships:

Lower half of the band: $f_n = f_0 - 340 + 5\,n$	MHz
Upper half of the band: $f'_n = f_0 + 5 + 5\,n$	MHz

where

$$n = 1, 2, 3, \ldots 63, 64$$

3.3.3.6 Co-Polar Channel Arrangements with Multicarrier Transmission Based on the previous channel arrangements with separation between channels of 40 and 20 MHz, multicarrier RF channel arrangements can be designed. This configuration arranges wider high capacity radio-channels, in the spectrum resulting of the aggregation of several basic channels. Figure 3.11 shows a co-channel

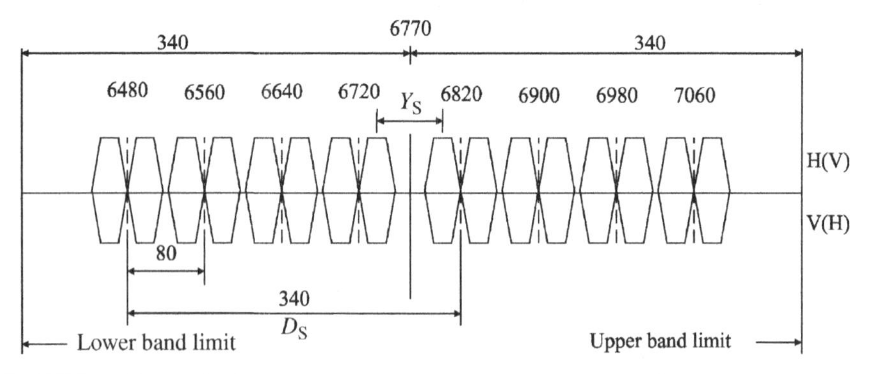

FIGURE 3.11 Radio-frequency channel arrangement for a wireless fixed system at 2 × 2 × 155.52 Mbit/s (4 × STM-1) (All frequencies in MHz).

arrangement of aggregated channels, where center frequencies of aggregated channels are obtained from 20 MHz channel dispositions, making n = 3, 7, 11, 15. Separation between adjacent channels is 80 MHz. This channel arrangement enables co-polar transmission of 2 × 2 × 155.52 Mbit/s signals, with 64-QAM modulation scheme.

3.4 ASSIGNMENT OF RADIO-FREQUENCY CHANNELS

The main purpose of the frequency plan for a radio-relay link is to assign the specific transmission and reception frequencies to each station of the radio-relay link, together with the polarization of the radio wave. Frequencies to be assigned are chosen according to a RF channel arrangement plan of the selected frequency band.

The general purpose of any frequency assignment should be an efficient use of the available spectrum. This means reusing available frequencies as much as possible, as well as ensuring protection against intra and inter system interference. It will therefore be necessary to take into account the coordination with the rest of equipment installations over the potential interference area radius and at the same time minimize the equipment costs. As an additional means of protection against interferences, cross polarization discrimination could be considered.

The frequency plan must be designed on a long-term basis, considering not only the situation at the time of implementation but also taking into account further potential needs and installations that will be required over the service life of the radio-relay link. At each station, frequencies of one half of the band will be used for transmission while the ones of the other half for reception. Lower subband is usually called A Band, and upper subband B Band. Similarly, stations that transmit in the lower half of the band and receive in the upper half are usually called Type A stations, and stations that transmit in the upper half of the band and receive in the upper half are called Type B stations. For a specific hop, transmitter and receiver RFs of a bidirectional channel will be paired. This means they will have the same channel order number, with the selected index within A/B subbands. Channels in the lower half of the band, A, will be named without the prime suffix (channel number $3, f_3$), and e.g., channels in the upper half of the band, B, are numbered and identified with the prime suffix (e.g., channel number $3', f'_3$).

There are several methods of assigning frequencies to radio-relay link emissions. Studies on interference and coordination of installations should be conducted for a specific RF channel assignment plan to be validated.

3.4.1 Assignment of Half Bands to Stations

A or B stations are assigned in a different manner depending on the radio-relay link topology: ring or linear (tandem).

3.4.1.1 Linear Topology Figure 3.12 shows the typical case of a radio-relay link of a backbone network using linear topology. This radio-relay link has 5 hops with two terminal stations (Stations 1 and 6), and 4 repeater stations (2,3,4,5).

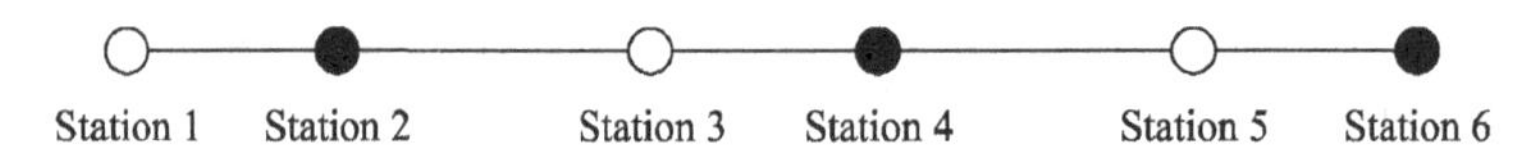

FIGURE 3.12 Diagram of a radio-relay link with six stations using a linear topology.

Once the type of one of the stations is chosen, Station 1 as Type A, for example, the type of the rest of the stations is automatically determined. If one station of a hop is a determined type (A or B), the contiguous station will be the opposite type (B or A). Possible passive repeaters along the link route are disregarded in frequency assignment for A and B stations, due to they do not convert frequencies. In the simplest case of a radio-relay link with an only hop, one of the stations will be Type A and the other one Type B.

3.4.1.2 Ring Topology

If radio-relay link uses a ring topology and there is an even number of stations (Figure 3.13), previous assignment procedure can be followed.

Assignment turns out to be more complex when there are an odd number of stations. Two kinds of solutions can be considered for this assignment: utilization of the same band on the whole radio-relay link system or utilization of a different frequency band for each hop of the radio-relay link.

Use of a Single Frequency Band: The solution is to use a single frequency band for the whole radio-relay link. One of the stations must be chosen, station P in Figure 3.14, for dual operation of A/B station type, thus transmitting and receiving in both halves of the band, upper and lower. The usual assignment procedure is followed for the rest of the stations.

This setting increases the possibility of creating interferences at the critical location, P. Certain frequency spacing between transmitter and receiver must be achieved at that location in order to solve this problem. At the same time, these interference issues can limit the number of radio channels to be transmitted within this scenario.

A further problem is over-shoot interference, because P neighboring stations, (3 and 4) in the figure, will disturb each other.

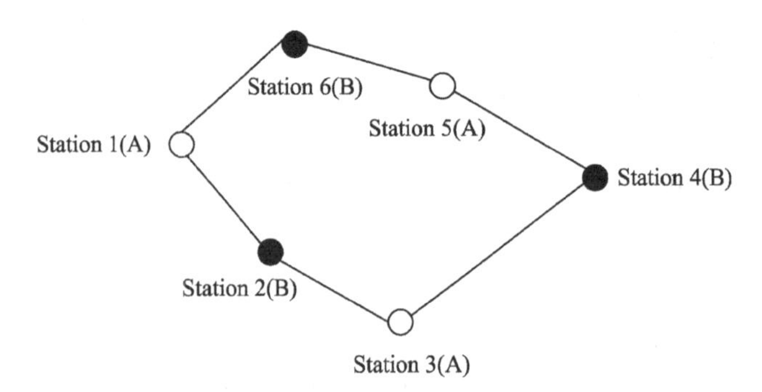

FIGURE 3.13 Diagram of a ring topology and an even number of stations.

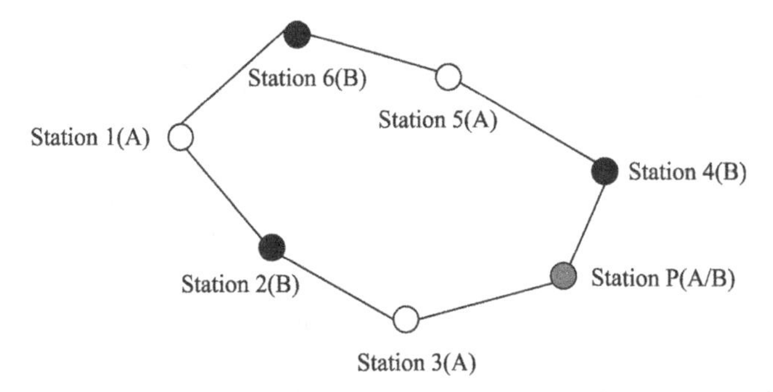

FIGURE 3.14 Diagram with an odd number of stations in a ring topology using a single frequency band.

Use of Two Frequency Bands: Two frequency bands are used in this case: One band for one of the links and a different one for the rest of the links in a ring network, as shown in Figure 3.15.

The different frequency band is usually assigned to the shortest hop, using a higher frequency band in this hop. This way, frequency assignment is not a problem anymore at the expense of increasing used spectrum. The usual assignment procedure for an even number of stations is followed for the rest of the stationsthat use the other frequency band.

3.4.2 Two Frequency Assignment Plan

This plan uses only a pair of frequencies for end-to-end transmission of each bidirectional RF channel in the whole radio-relay link. Transmission and reception frequencies of one RF channel are the same on each station.

Figure 3.16 illustrates frequency assignments corresponding to a radio-relay link with a bidirectional radio-channel in linear topology, five links, two terminal stations, and four repeater stations.

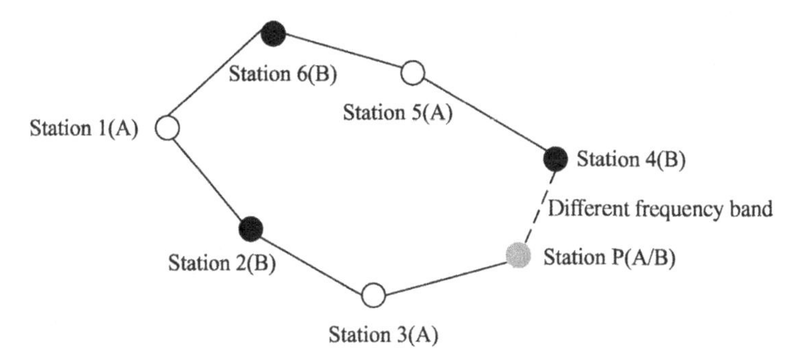

FIGURE 3.15 Diagram of an odd number of stations using different frequency bands in a ring topology.

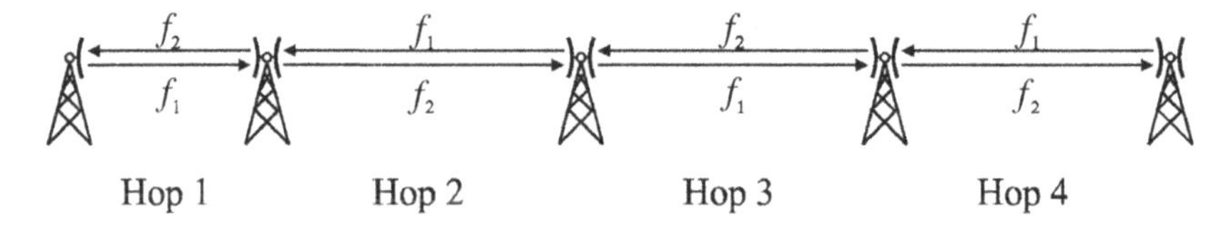

FIGURE 3.16 Two frequency assignment plan.

This type of frequency assignment has maximum frequency reuse. Nevertheless, it has to deal with a series of co-channel interferences that can make implementation unfeasible when exceeding a certain value. Co-channel interference can be produced by neighboring stations or by alternate stations due to over-shoot.

Interferences in a station that are caused by neighboring stations are due to signals radiated by the main lobe of the antenna in the contiguous station and received by the secondary lobes of its own antenna, or radiations from the main lobes of the neighboring antennas that are received by its own main lobe.

Interfering Path 1 in Figure 3.17 is caused by interferences in A from backward radiation from B and direct reception in A by its main lobe. Interfering Path 2 in the figure is caused by interferences in B of direct radiation from A and reception by side lobes of the antenna in B (Figure 3.17).

The only effective technique to keep interference values below acceptable thresholds in this scenario is using high-performance antennas, with high front/back ratios. Over-shoot interferences in one station caused from alternated link emissions that correspond with transmitters that are two hops ahead, can be relevant when hop distances are small or uncorrelated fading occurrences create adverse useful/interference ratios. Figure 3.18 illustrates an interfering path due to over-shoot, caused by direct emission from a C station that is received by the antenna in A station.

The interference caused in this situation must be calculated according to the method of calculation explained in Chapters 1 and 8 to evaluate over-shoot effects.

Wave polarization on each hop can be alternated, with the aim of reducing as much as possible the effects of these possible interferences. Figure 3.19 shows a diagram of the two frequency plan with alternated polarization every two links.

In the case of radio-relay links with several radio-channel configurations, $N + M$, frequency assignment of RF channels must be carried out taking into account long-term needs and available frequencies in the radio-relay link scenario. RF channels

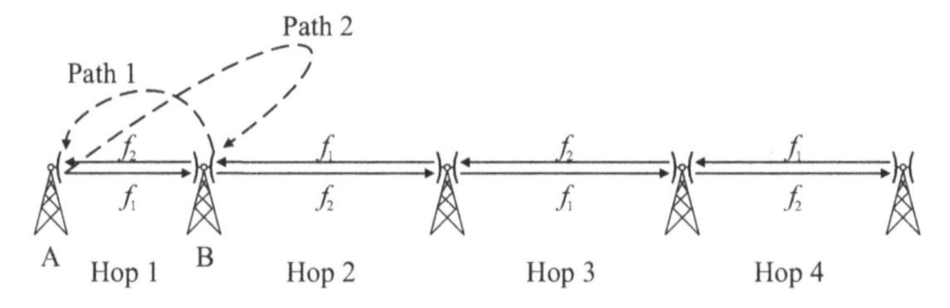

FIGURE 3.17 Co-channel interference caused by neighboring stations.

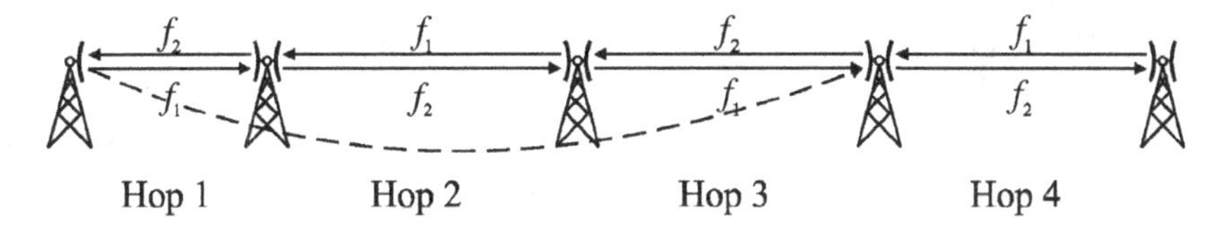

FIGURE 3.18 Over-shoot interference in two frequency plans.

must keep a certain separation and associated multiples. The separation will be defined according to the RF channel arrangement plan indicated by the corresponding ITU-R recommendation.

In the specific case of frequency assignments for frequency diversity operation, pairs should be chosen according to degree of fading correlation with frequencies in that link.

3.4.3 Assignment Plan Using More Than Two Frequencies

When XPD is not enough for mitigating the alternated link over-shoot problem, a new frequency pair should be used, alternating frequencies and polarizations every two hops. This configuration is called four frequency assignment plan and is illustrated in Figure 3.20.

If using both pairs of frequencies interference problems are not eliminated, three pairs of frequencies could be used for the radio-relay link. This configuration is called six-frequency plan and is illustrated in Figure 3.21.

This configuration uses three different frequencies in three hops, 1, 2, and 3 with one polarization, for example, H. In the next three hops the three pairs of frequencies are reused with opposite polarization.

3.5 COMMENTS ON THE FREQUENCY BAND CHOICE

The right choice of the appropriate band for a radio-relay link of the FS depends on numerous factors, not only regulatory but also technical. When examining possible frequency bands for deployment of FS radio-relay links, it must be first taken into account the national regulation on frequency band availability allocated for the FS.

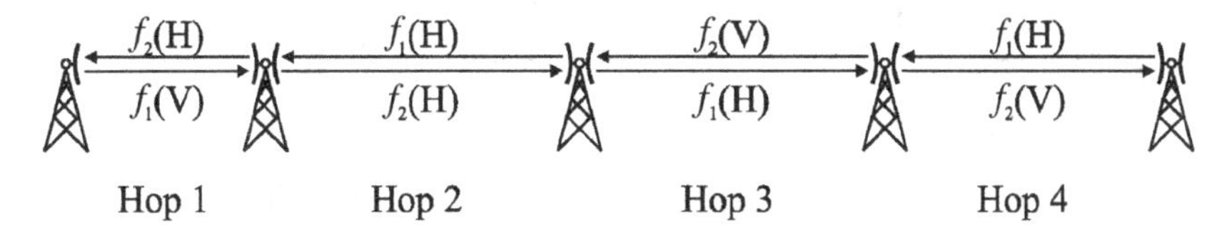

FIGURE 3.19 Two frequency assignment plan with alternated polarization every two links.

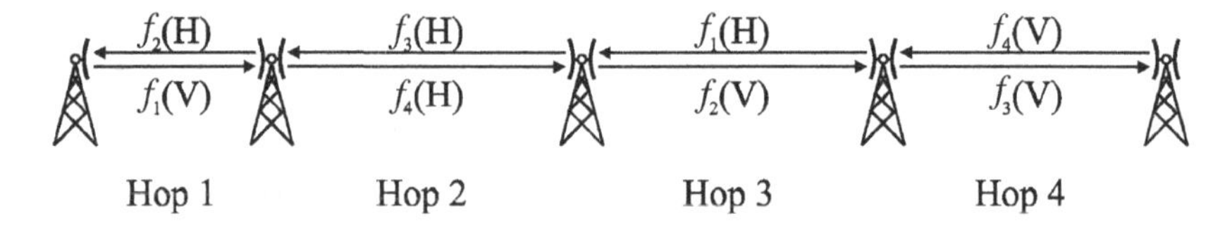

FIGURE 3.20 Four frequency assignment plan with alternated polarization every two links.

Access to specific frequency bands can be limited in some cases, due to use priorities or any other specific conditions for granting radio emission licenses.

The key factor for frequency choice is the maximum hop distance. In the early planning stage of a radio-relay link, critical lengths of the hop can be roughly estimated from the according to the calculation procedure that will be explained in Chapters 6, 7, and 8. There are available charts that make it possible to obtain the maximum hop length for a specific frequency; climate zone and downtime percentage can be drawn, by applying the previous method.

Maximum length of a hop is limited because of propagation effects, particularly due to rain at frequencies exceeding 10 GHz. In these cases, it is advisable to take climatic conditions and rain impact into account for the band choice. Obviously, that choice will largely depend on the geographical area the FS networks operate on. Hop distances and hence frequency bands are also determined by the type of service or information transmitted, by associated costs, and possible spectrum congestion in certain bands because of the current frequency assignments in each country.

Almost all actual installed radio-relay links are within the 6 and 40 GHz range, although links in very low frequencies (L Band) do already exist. There are assignments planned to be used in radio-relay links above 40 GHz and up to 86 GHz that are not currently being used, or barely used, mainly because of the problems associated with hydrometeor fading and atmospheric absorption. Allocated frequency bands are shown in Figure 3.22.

Frequency assignments can be done in different ways, depending on network complexity, future plans, etc. RF channel bandwidth, affected by the bit rate of the signal to be transmitted and the modulation scheme used, is the main feature to consider when choosing the RF channel arrangement type. Frequency planning procedure also depends on the detail level of the available information.

In any case, it will be an iterative process. Resulting interferences must be analyzed for each operation scenario. If changes would be required to enhance the protection

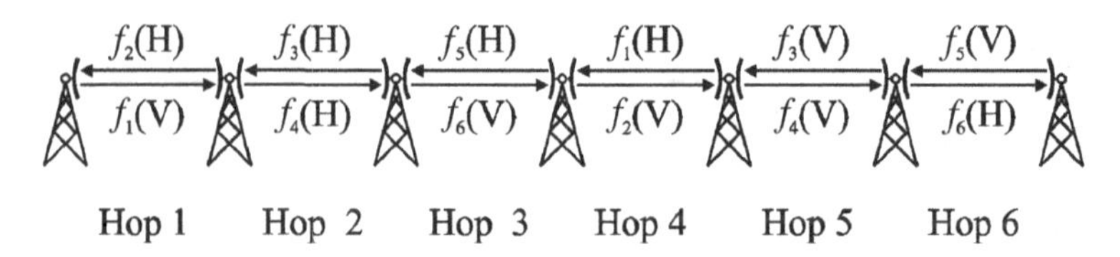

FIGURE 3.21 Six frequency assignment plan with alternated polarization every three links.

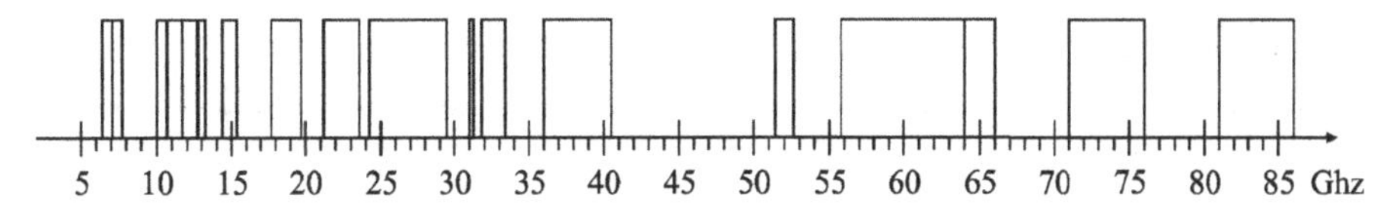

FIGURE 3.22 Frequency bands for fixed service radio-relay links.

against interferences, these modifications might involve a change in the equipment choice or a change in the channel arrangement plan.

As a general rule, the most efficient use of RF spectrum will be done, using as few frequencies as possible, beginning the planning process with those configurations involving the lowest amount of spectral resources, that will be increased, if necessary, in further design iterations.

BIBLIOGRAPHY

Radio Regulations. International Telecommunication Union. Radiocommunication Sector. ITU-R. Geneva. 2008.

ITU-R Rec. F.746: Radio-frequency arrangements for fixed service systems. International Telecommunication Union. Radiocommunication Sector. ITU-R. Geneva. 2007.

ITU-R Rec. F.758: Considerations in the development of criteria for sharing between the terrestrial fixed service and other services. International Telecommunication Union. Radiocommunication Sector. ITU-R. Geneva. 2005.

ITU R Rec. F.1245: Mathematical model of average radiation patterns for line-of-sight point-to-point radio-relay system antennas for use in certain coordination studies and interference assessment in the frequency range from 1 GHz to about 70 GHz. International Telecommunication Union. Radiocommunication Sector. ITU-R. Geneva. 2000.

ITU-R Rec. F.1399: Vocabulary of terms for wireless access. International Telecommunication Union. Radiocommunication Sector. ITU-R. Geneva. 2001.

ITU-R Rec. F.384: Radio-frequency channel arrangements for medium- and high-capacity digital fixed wireless systems operating in the upper 6 GHz (6 425–7 125 MHz) band. International Telecommunication Union. Radiocommunication Sector. ITU-R. Geneva. 2002.

CHAPTER 4

EQUIPMENT AND SUBSYSTEM TECHNOLOGY ASPECTS: A RADIO LINK DESIGNER APPROACH

4.1 INTRODUCTION

This chapter describes the different subsystems that are combined to create a radio-relay link. Technical features and configurations of microwave LOS link equipment are explained throughout this chapter, which focuses on the development of concepts and knowledge that directly affect the design of the link. The stress is put on the factors that a link planner should bear in mind when specifying and choosing equipment for a specific case.

More and more, users of radio-relay links require network implementations to be less expensive, more flexible to operate with different types of traffic, more efficient from the energy consumption perspective and with reliability and availability figures higher than 99% even under adverse propagation conditions.

On the other side of the business, equipment manufacturers offer a wide range of technical solutions, suited to protect the system against radio electric propagation problems in the different frequency bands. These technical solutions comply with requirements for different traffic types and transport capacities, in terms of spectrum efficiency and system gain.

Most of the equipment that is currently available on the market operates on frequency bands that are standardized by International Telecommunications Union, Radiocommunications Sector (ITU-R), Federal Communications Commission, and Conférence Européenne des administrations des Postes et des Télécommunications/ European Conference of Postal and Telecommunications (CEPT/ECC), below

Microwave Line of Sight Link Engineering, First Edition. Pablo Angueira and Juan Antonio Romo.

42 GHz. The majority of arrangements are based on channels with up to 56 MHz bandwidths. Since the origins of digital radio-relay link technology, the commercial choices have evolved towards high system gains by pushing the behavior thresholds close to the theoretical modulation and coding limits. Furthermore, gradual advances introduced in these digital radio-relay links have also been focused on two key areas: improvements on performance (availability and error performance) and spectrum efficiency.

Traditionally, radio-relay links have been designed to transport time division multiple/asynchronous transfer mode (TDM/ATM) signals. Due to the increasing data traffic demands and in order to integrate links easily into any transport network independent of the final application type, the tendency is to incorporate Ethernet interfaces that support Internet protocol (IP) traffic. The tendency towards IP architectures has made radio-relay link equipments evolve into the concept of microwave hybrid radio, with platforms where total bandwidth is dynamically shared with TDM traffic, including either interfaces for basic plesiochronous digital hierarchy (PDH) tributaries such as E1/T1, or Ethernet interfaces for packet switching traffic or a combination of both technologies. The need for handling a variety of IP traffic types flexibly and economically has caused these hybrid platforms to evolve further towards the so-called Ethernet radio technologies, where Ethernet frames are directly supported, whereas TDM traffic is supported by circuit emulation.

Traditional technologies used in radio-relay links have not been designed having in mind the capacities that broadband services require for access and transport network applications (above 1 Gbit/s). Consequently, one of today's technological challenges in the field of digital point-to-point radio-relay links is the increase of transmission capacity per radio channel. To achieve this objective we can either increase each radio channel bandwidth or increase spectrum efficiency, in terms of bits/second/hertz.

A response to this issue consists of adopting solutions called Gigabit Ethernet radio-relay link technologies that operate on millimeter-wave frequency bands, between 70 and 100 GHz. They are also referred to as millimeter-wave radio-relay links, with 112 MHz and 250 MHz radio channel spacing. The available commercial equipment to be used in these bands operates today on the 71–76 GHz and 81–86 GHz bands. This equipment uses high capacity modulation schemes such as m-QAM, with 256, 512, or up to 1024 levels. Nevertheless, it is necessary to consider that using high-level modulation schemes involves an increase in requirements of radio transmitters and receivers, in terms of linearity and phase noise, which makes them more expensive.

Additional methods currently used to increase capacity are radio channel aggregation, polarization multiplexing (co-channel reuse) with cross-polarization interference cancelers (XPICs), multiple input multiple output (MIMO) space multiplexing systems and adaptive modulation or dynamic modulation techniques, which maximize transmission capacity continuously, by dynamically adjusting modulation to propagation conditions.

All these functionalities have benefited from the development of processing units and software programs software defined radio (SDR). SDR enables the "radio agile" concept, where it is possible to change dynamically bit rate, modulation, and coding depending on environment conditions and operating needs.

The use of integrated circuits in the different stages of intermediate frequency (IF), baseband (BB) and auxiliary functions have continuously provided lower and lower priced implementations on the market, providing efficient solutions in bands that extend more and more toward higher frequencies. Ongoing improvements on microwave device technologies are noteworthy in this field, particularly field-effect transistors (FET) and on monolithic microwave integrated circuits (MMIC).

The following sections of this chapter revise all types of radio-relay link stations and the different blocks that are part of those stations. A summary of the widely used multiplexing and data transport technologies, starting with traditional PDH to new transport techniques based on Ethernet, is also included. Different components that can be found in any radio-relay link, not only BB equipments, modulators, transceivers, and branching systems but also antennas are discussed. Furthermore, different configurations for redundancy and diversity are explained, and finally, a section on control and monitoring systems is also included.

4.2 BASIC BLOCK DIAGRAMS

A radio-relay link is formed by terminal (nodal) radioelectric stations, and intermediate repeater stations in the most general case. A terminal radioelectric station is a hub that manages incoming and outgoing traffic, and where there may be some kind of multiplexing–demultiplexing with the aim of aggregating or disaggregating different traffic sources.

A repeater station has the function of enabling the line of sight characteristic along a specific path, usually to overcome a geographical obstacle or distance that is too large. This way, the repeater will divide the original hop into two shorter LOS hops. A repeater station receives signals from each communication direction, processes them properly and forwards them to the next hop. Processing may include BB, IF or RF stages depending on the repeater type. A specific case of repeater stations are passive repeaters, which only change signal propagation direction, based on reflecting surfaces or antennas connected back to back using a passive wave guide, usually in cases of isolated obstacles close to one of the hop ends.

Finally, a third type of station called intermediate nodal station combines the functions of the previous two. This type of station is located along the route of a specific radio-relay link operating as a repeater, but including also aggregation and disaggregation of a usually not very significant part of the total traffic of the radio-relay link. Figure 4.1 shows the overall diagram of a radio-relay link with the three types of stations.

4.2.1 Terminal Stations

Figure 4.2 shows the block diagram of a terminal station for a radio-relay link in a 1+0 configuration. This diagram intends to reflect basic features of equipment, ignoring, for the sake of simplicity, supervision functional units.

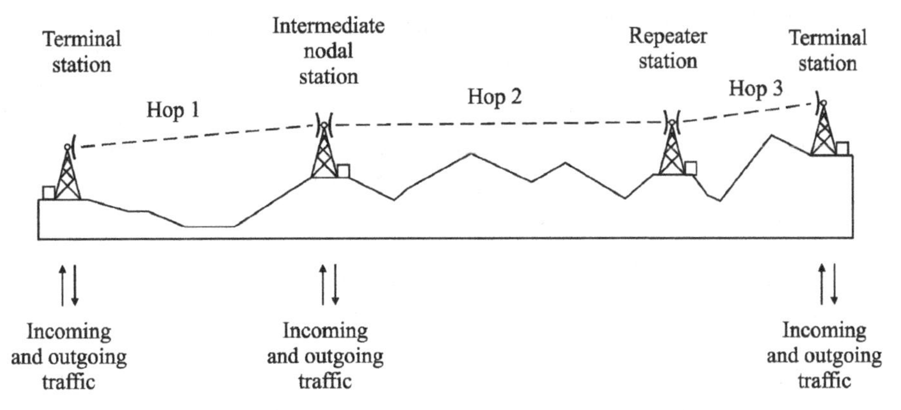

FIGURE 4.1 Overall diagram of a radio-relay link.

The following interfaces have been identified in order to simplify and provide a clearer the explanation of each block:

- BC: Basic channel digital interface.
- BB: Baseband interface.
- IF: Intermediate frequency interface.
- RF: Radio-frequency interface.

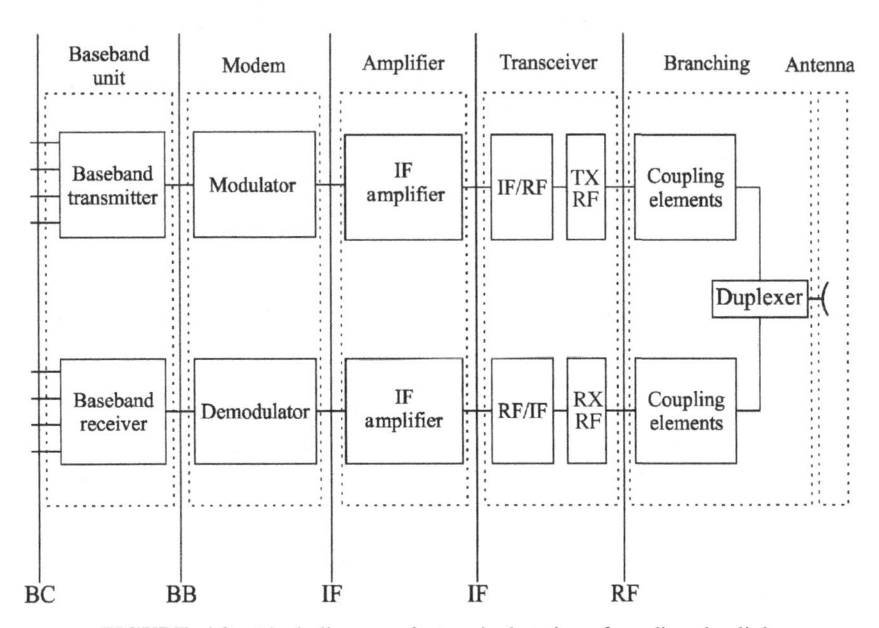

FIGURE 4.2 Block diagram of a terminal station of a radio-relay link.

All different digital interfaces for each tributary type are available in the BC. These tributaries are processed to be adapted to the transmission characteristics of the BB, forming the bit stream to feed the modulator. The features of this block will depend on the radio-relay link type and characteristics of the information to be transmitted. The modulator will take the BB bit stream to generate the modulated signal at an IF that depends on the specific implementation. 70 MHz is a typical IF value. The occupied bandwidth of the modulated signal will be the necessary bandwidth of the transmitted radio channel.

After a first IF amplification, the IF/RF converter translates the spectrum from the IF modulator output frequency to the corresponding RF carrier. The center frequency of the radio channel is called RF. The signal is then amplified and sent to the antenna distribution network input. High-power amplifiers will be usually followed by band-pass filters in order to remove unwanted signals. After amplification and possibly filtering, the RF signal goes through a microwave passive network (also called branching network) for signal distribution to the antenna. Among branching components we can find: filters, circulators, duplexers, and coupling and connection elements. It is important to note that the branching network will also include elements that isolate the transmission and reception paths with the aim of using a single antenna for transmitting and receiving. These elements are usually duplexers.

The functional sequence is very similar along the reception branch. The RF signal received at the antenna is forwarded to the receiver branch and isolated from the transmitting path. Each one of the received channels will be forwarded to a specific receiver subbranch by the microwave passive network. The blocks that will process each channel start with an amplification stage based on a low noise amplifier (LNA) specifically designed to maintain the receiving system noise figure low. It is well known that the total amount of thermal noise of the receiver will be mostly determined by the noise figure of this element.

The RF incoming signal will be mixed with the signal of the local oscillator (LO) in the frequency down-converter to obtain the frequency difference that should match the IF value where the demodulator operates. After down-conversion to IF, the signal is amplified using a variable gain IF amplifier. The gain of this amplifier is adjusted by an automatic gain control (AGC) loop that will keep the carrier level applied to the demodulator constant, within certain limits, despite the amplitude variations caused by propagation phenomena. The AGC subsystem will extract a sample signal proportional to the system input that will feed the IF amplifier back. If the level tends to decrease (e.g., because of fading), the amplifier gain will increase.

The amplified IF signal is then delivered to the demodulation unit, where the BB signal is obtained. The process at the BB stage is complementary to the one explained for transmission. Ideally, the output of the BB block will be a regenerated signal with the same characteristics as the one transmitted from the other end of the link/hop.

In reality, the frequency converters, transmitters, and receivers associated with each radio channel of the link are integrated in the same rack frame and they all share the same radiating system for transmission and reception. Groups of transmitters–receivers are called transceivers for this reason. Similarly, modulation and demodulation functions are also integrated into the so-called modulator–demodulator (MODEM). Current integration degree can even allow BB, modulation–demodulation functions, transmitter, and receiver to be packed into the same boards.

For economic reasons, when the radio-relay link frequency plan contains several radio channels, these channels usually share the same antenna and branching for transmission and reception. The block diagram would then consist of parallel branches, each one being a BB, modem and transceiver block sequence, and connected to the same branch. Except for the antenna and duplexer, the rest of the receiver and transmitter elements would be independent for each radio channel. In these cases, a high isolation level among the branches of different radio channels must be maintained, in order to avoid mutual interference.

The physical location of the functional elements of the link will change depending on the application and integration degree of each manufacturer. According to the exact installation place for each one of the system modules, there are three equipment configuration possibilities for microwave LOS systems: indoor unit (IDU), split unit (RF outdoor), and all outdoor unit (ODU).

IDU configurations are typical of high capacity radio-relay links. Units are arranged on transmission rooms and located in frames of different types and sizes: 19 in.European Telecommunications Standards Institute in Europe, 21 in. in the United States. Antennas are installed on outdoor towers and connected to the transmission shelter through coaxial cables or wave guides.

As far as split unit configurations are concerned, BB and modem blocks are located indoor, usually in shelter buildings designed specifically for this purpose, whereas the RF unit is arranged outdoor, close to the antenna. The IF connection between both units is done with coaxial cables, which also transmit the necessary power supply to the RF units. This configuration has lower transmission loss on lines than IDU configurations, particularly for higher frequencies. This arrangement is widely used in access or transport networks for mobile communications, at higher frequencies, in which room availability within the shelters is usually an important issue.

All ODU solutions, with all the radio-relay link equipment installed close to the antenna, are mainly used for transporting microcell signals in mobile communications, where base stations (BSs) are also outdoors. Bs are usually furnished with digital interfaces that are directly connected to the ODU. Connections must be made using coaxial cables: traffic, power supply, alarms, and management signals will share the transmission media.

Additionally, stations must have a shared power system, reliable enough so the service is not interrupted and within standardized limits for voltage, and frequency and electric supply signal quality.

4.2.2 Repeater Stations

There are several types of repeater stations (or simply repeaters) that might be active or passive, and intermediate nodal stations. Passive repeater stations are a type of repeater stations without any active modules. There are two types of passive repeaters.

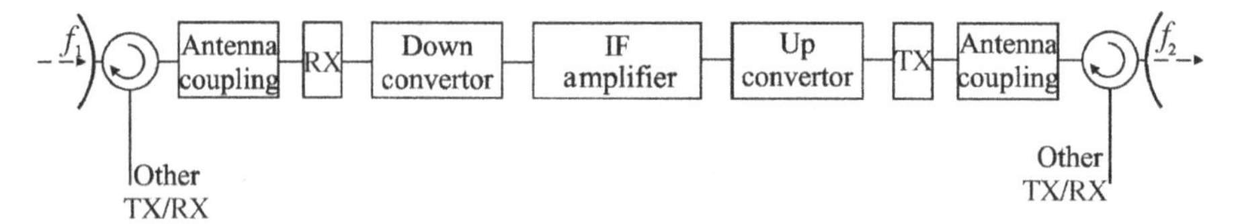

FIGURE 4.3 Block diagram of an active repeater (only the branch associated with one of the directions of a radio channel is shown.)

The simplest one is a reflecting surface located over an obstacle in order to make a change in the propagation direction of the signal. The end terminal stations of the link will point to the reflector and the system will have typically two hops: Terminal Station A to reflector and reflector to Terminal Station B. Another version of passive repeaters is composed of two antennas that are connected back-to-back by means of a transmission line, usually a wave guide. In both types of passive repeaters, the objective is to avoid isolated obstructions along the path, usually associated with obstacles located close to one of the terminal stations of the link. Passive repeaters based on back-to-back antennas can also include radio channel branching functions for different radio channels (including the necessary branching).

Active repeaters can be regenerative or non-regenerative. Regenerative repeaters demodulate, detect the BB information and remodulate it again in order to avoid as much as possible an accumulation of disturbances in links formed by various hops.

In non-regenerative active repeaters, the RF signal received by the antenna, is down-converted to an IF and then amplified, up-converted to the output frequency band and transmitted. Usually, there is no signal demodulation. These repeaters will have the necessary branching and a transceiver set for each transmission direction. Figure 4.3 shows the block diagram for one radio channel in a single transmission direction that is received in f_1 frequency and retransmitted in f_2 frequency, according to the same block and interface nomenclature that was used in the case of terminal stations.

Intermediate nodal stations are similar to regenerative repeaters. In this case, the signal is down-converted to IF, demodulated, and demultiplexed in BB. It is possible then to drop and insert channels to and from the overall traffic flow of the main link flow. Intermediate nodal stations permit to interconnect different sections of the network or different radio-relay link paths. If more than two paths are connected the radio station is called multidirectional node.

Drop and insertion of channels can be made at radio channel level or at a higher frame level of the digital interface multiplexes. Figure 4.4 shows the block diagram of a nodal station for one radio channel. The block diagrams are only shown for one of the transmission directions.

The path between a terminal station and an intermediate nodal station is called a switching section, and will be considered a control, protection and monitoring unit. The switching section term is extended to the sections between nodal stations when there are several nodal stations in the same radio-relay link.

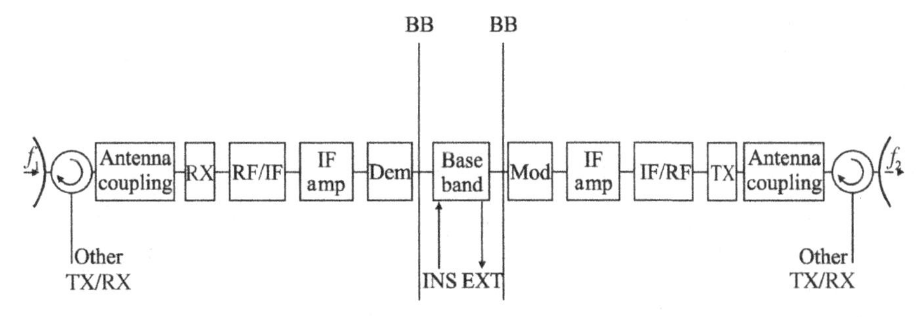

FIGURE 4.4 Block diagram of a nodal station for a single radio channel (only the branch associated with one of the directions of a radio channel is shown.)

From the point of view of traffic aggregation, repeater stations can also be classified into edge nodes, which support transmission in a single direction, and aggregation nodes, which allow to aggregating the traffic of numerous radio-relay links that converge on a single radio-relay link.

4.3 TRANSPORT TECHNOLOGIES

Regardless of the type of information exchanged among users, digital signals are grouped together (multiplexed) in order to be transmitted through the access or transport networks. Most of the frequent digital signal aggregation techniques are based on TDM, either following the PDH or synchronous optical network/synchronous digital hierarchy (SONET/SDH), the ones based on ATM and those based on the Ethernet standard.

TDM transport techniques are associated with circuit switching networks, where the circuit between both ends of the communication process is permanently kept while the communication takes place, in such a way it is possible to keep a constant information flow between those ends during the communication. This is the case of conventional telephony. Its main advantage is that once the circuit is established, it is highly available, because the path between both ends is guaranteed regardless of the information stream. Its main disadvantage is that it is associated with a significant consumption of the system resources, regardless of whether information is being sent or not while the connection is established.

The other type of networks are packet switching networks, where there is no permanent circuit between both ends, and the network just forwards the information to and from different users on a packet by packet basis. This means, in practice, that different information packets of the same flow may follow different paths. Its main advantage is that it only consumes system resources when a packet is sent or received, and the system is available to manage other packets with other information or packets associated with other users. Its drawback is the difficulty to manage "real time" information, such as voice, which requires that data packets arrive with delay restrictions and in a specific order. Nowadays, packet switching networks are able to manage "real time" information at the expense of increasing their complexity and capacity.

4.3.1 Plesiochronous Digital Hierarchy PDH

TDM techniques are based on pulse code modulation (PCM). PCM is a process to convert analog signals into digital signals through a sampling, quantification and codification subprocesses, forming basic rate signals (64 kbit/s in systems of the European hierarchy and 56 kbit/s for the American hierarchy) and primary rate signals (E1 at 2.048 kbit/s and T1 at 1.544 kbit/s respectively for the European and American hierarchies). Higher data rate signals are obtained by successive multiplexing of tributaries in secondary multiplexes. Figure 4.5 shows the PDH bit rates for Europe, Japan, and the United States.

Figure 4.6 shows the conceptual diagram multiplexing and demultiplexing 2, 8, 34, and 140 Mbits/s tributaries. The electrical interfaces associated with each level are defined by the International Telecommunication Union, Telecommunication Standardization Sector (ITU-T) G.703 recommendation.

When bitrates of all input channels in each multiplexer are not exactly the same, or there are deviations in the clocks that control the process either caused by phase noise or simply associated with circuit tolerances, it is necessary to insert padding bits (justification bits) in order to adapt the incoming bit rates of the tributaries, both between them and also with the output multiplexed bitrate. The extraction of any tributary implies, therefore, a complex physical demultiplexing of all successive levels. Clock signal can be directly extracted from E1 or T1 streams and be used as a reliable source for network synchronization, according to ITU-T G.811 recommendation. Bit

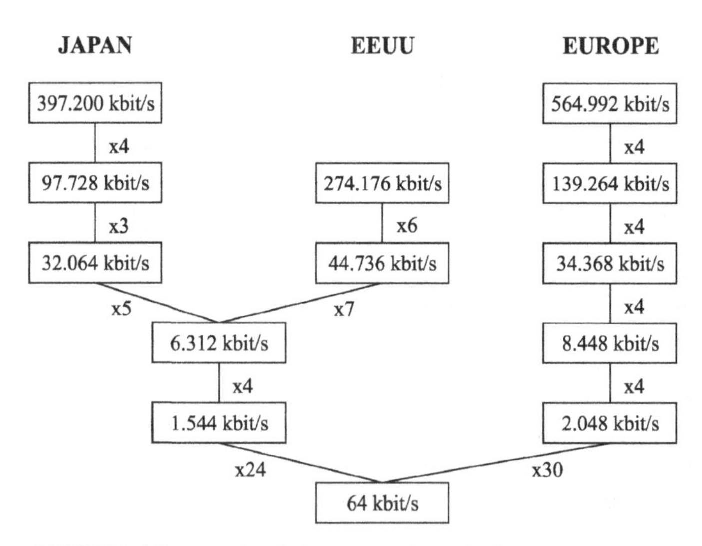

FIGURE 4.5 PDH levels in Europe, the United States, and Japan.

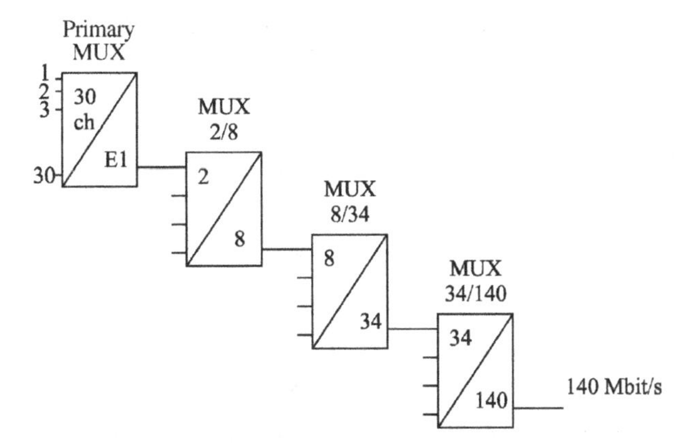

FIGURE 4.6 Multiplexing tributaries in PDH.

justification has a limit, of course. In this regard, short-term and long-term maximum phase errors of the clocks (jitter and wander) are specified by ITU-T G.823 and G.824 recommendations.

4.3.2 SONET/SDH Synchronous Networks

SONET was the first system of the so called synchronous transmission hierarchy systems. It was developed in the United States by Bellcore. SONET was standardized at American National Standards Institute (ANSI) and released as standard numbers from T1.101 to T1.107. Subsequently, the ITU-T standardized SDH which is mostly based on the ANSI T1.105 standard. Table 4.1 contains a list of Recommendations where the most important features of the SDH technology can be found.

In SDH, the multiplexed channels occupy fixed positions if the transmission frame synchronization references, so the extraction or insertion of one channel can be done individually, without requiring intermediate multiplexing and demultiplexing. Two information levels are differentiated in SDH: multiplex and signal levels. Multiplex levels are designated by virtual container level *n*, (VCn) each virtual channel (VC) being a tributary complemented with a path overhead (POH). Signal levels are virtual

TABLE 4.1 Recommendations Related to Synchronous Digital Hierarchy

Recommendation	Title	Year
ITU-T G.707	Network node interface for SDH	2007
ITU-T G.780	Terms and definitions for SDH networks	2010
ITU-T G.783	Characteristics of SDH equipment functional blocks	2006
ITU-T G.784	Management aspects of SDH transport network elements	2008
ITU-T G.803	Architecture of transport networks based on the SDH	2000
ITU-T G.810	Definitions and terminology for synchronization networks	1996
ITU-T G.811	Timing characteristics of primary reference clocks	1997

TABLE 4.2 Comparison of Levels and Bitrates in SONET and SDH

SONET			
Electric Signal	Optic Carrier	Bitrate (Mbps)	SDH Signal Levels
STS-1	OC-1	51.84	STM-0
STS-3	OC-3	155.52	STM-1
STS-9	OC-9	466.56	
STS-12	OC-12	622.08	STM-4
STS-18	OC-18	933.12	
STS-24	OC-24	1244.16	
STS-36	OC-36	1866.24	
STS-48	OC-48	2488.32	STM-16

STS, synchronous transport signal; STM, synchronous transport module

containers complemented with a signal overhead (SOH). The signal level of order n is designated as synchronous transport module level n (STM-n). As an example, the first level STM-1 allows to transport a PDH signal of 139.264 Mbps. There are interfaces that allow direct integration (mapping) of the PDH levels on SDH frames. Table 4.2 shows a comparison of levels and bitrates of SONET and SDH.

There are three types of main node functions in a synchronous network: terminal multiplexers (TM), add/drop multiplexers (ADM) and digital cross connect (DXC). These three types of elements of the SDH network are represented in Figure 4.7.

TMs are responsible for multiplexing lower capacity levels, not only plesiochronous, but also synchronous, into an equal or higher hierarchy level. ADMs allow to add or drop either plesiochronous tributaries within SDH containers or lower order synchronous transport modules (i.e., add or drop STM-1 signals to/from a STM-4 flow) at an intermediate point of the path. ADMs do not need to demultiplex a whole SDH signal, being able to access to specific tributary signals of interest. DXC

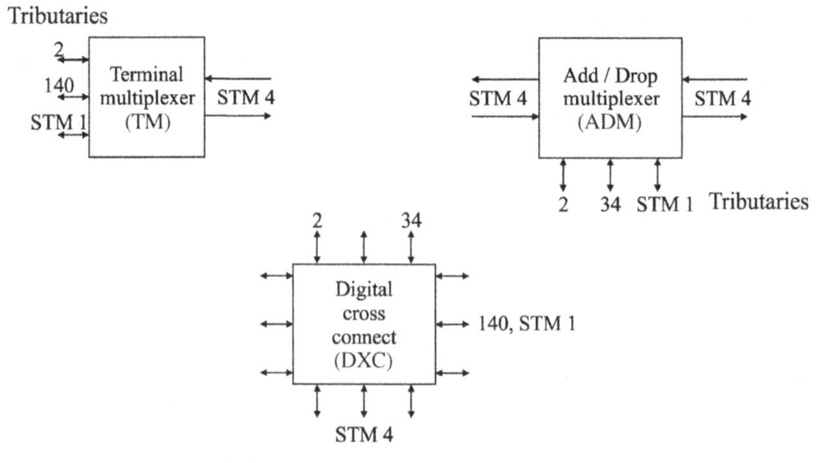

FIGURE 4.7 Elements of a SDH network.

TABLE 4.3 General ITU-T Recommendations Related to ATM

Recommendation	Title	Year
ITU-T I.150	B-ISDN ATM functional characteristics	1999
ITU-T I.361	B-ISDN ATM layer specification	1999
ITU-T I.363.1/2/3/5	B-ISDN ATM adaptation layer specification. Types 1, 2, 3, 4, and 5.	1996

multiplexers are network elements that allow interconnecting main and tributary bit streams of different levels, in order to perform rerouting or traffic protection functions. They are designated as DSX *m*/*n*, where *m* means the major hierarchy level of the SDH frame, and *n* the hierarchy level of the interconnected tributaries.

4.3.3 ATM Asynchronous Transfer Mode

ATM was endorsed by the ITU-R during the eighties and its standardization was developed in the nineties. It was initially proposed to be the supporting technology of the future broadband integrated services digital network (B-ISDN) based on packet switching. These expectations have not been consolidated in practice due to bandwidth limitations, complexity, cost, and performance, especially if those are compared with the ones of Ethernet/IP technologies. Its application field in wide area network (WAN) environment covers broadband fixed access networks, asymmetric digital subscriber line, mobile telephony networks and third generation broadband cellular networks like Universal Mobile Telecommunications Systems. Table 4.3 shows some of the most relevant recommendations related to ATM.

ATM technology transports short packets, called ATM cells. These cells have constant length and will be multiplexed in time following a synchronous scheme. These packets can be individually routed by using the so-called VCs and virtual paths (VPs) offering a connection-oriented service. VPs are the paths followed by cells between two ATM routers, being able to have several VCs.

Figure 4.8 illustrates how different information streams, of different characteristics as far as bit stream and format are concerned, will be grouped into the so-called ATM

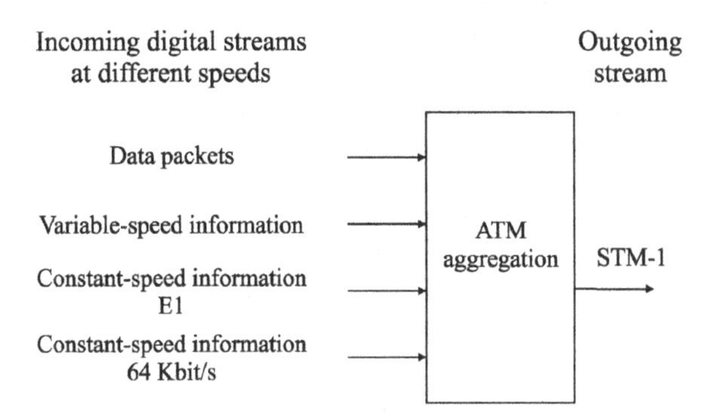

FIGURE 4.8 Simplified diagram of the ATM encapsulation process.

module in order to be transported through transmission links at bitrates of 155 or 622 Mbit/s generally relying on the SDH structure.

In the transmitter terminal equipment, information is encapsulated onto the user information field of the cell, and later an overhead section is added on the header of the cell. In the distant end, the receiver extracts the information from the incoming cells, forwarding the cell according to the data found in the header. During the initial set up stages of a communication, a virtual path (VP) will be defined. This path will be followed by cells associated to a specific service. The path will have a constant quality of service towards a fixed destination, and both features (quality of service and destination) will not be modified during the whole communication. All necessary resources to ensure the quality of service to the user during the whole connection will be then reserved during the connection. This approach combines the benefits of circuit switching (guaranteed capacity and constant transmission delay) with the benefits of packet switching (flexibility and efficiency for the intermittent traffic).

ATM devices can be ATM switches and ATM end-user equipment. The ATM switch provides cell transit functionalities through the ATM network: it accepts the incoming cells from an ATM end equipment or from an ATM switch, and switches the cell to an exit interface towards its destination. An ATM end-user, end point or end system, contains an interface adapter to the ATM network that extracts data from the cells. ATM switches support two primary types of interfaces: user to network interface (UNI) that connects ATM end systems to an ATM switch, and network to network interface (NNI) that connects two ATM switches.

4.3.4 Ethernet

Ethernet is a technology based on packet transmission. It was standardized by the Institute of Electrical and Electronic Engineers (IEEE) during the eighties. All the Ethernet equipment and protocols have been developed according to the IEEE 802.3 standard, which specifies physical and access networks. It is based on the access control protocol carrier sense, multiple access/collision detect (CSMA/CD). CSMA performance relies on the activation of a transmission that will happen only if the previous detection of an unoccupied carrier signal is successful. In the case of collision due to simultaneous transmissions, the system will wait for a random time before new retries. Ethernet frames transport application data that will be encapsulated by a higher-level transport protocol, i.e., the Transport Control Protocol/Internet Protocol (TCP/IP) family. The low cost of the equipment, achievable high bitrates, flexibility, and installation simplicity are the main advantages of Ethernet.

Originally, Ethernet was conceived as a protocol designated to meet the needs of local area networks. The achievement of higher bandwidths at low prices (Standard 802.3ae, also known as 10 Gigabit Ethernet) in the first years of the XX century, as well as the reliability and implementation flexibility, has extended the use of Ethernet to WAN levels. In this sense, Ethernet has progressively substituting ATM systems.

In summary, Ethernet is currently the most appropriate technology for the new generation broadband access and transport systems, in fixed as well as in mobile networks.

4.3.5 Synchronization in Hybrid TDM/Ethernet Networks

The undergoing evolution in telecommunications networks with new IP services and native Ethernet technologies in transport nodes requires a new paradigm for the distribution of network synchronization, not only relative to clock signals (timing), but also in frequency synchronization.

During the convergence stage towards transport networks based on packet switching, TDM and IP traffics (such as the one associated with long-term evolution, LTE) will coexist. Transport networks must be able to support the different synchronization requirements for the different traffic types being handled, regardless of the technologies in use. Consequently, a series of synchronization and interface standards have been defined. The standards apply both to the network level (network-clock) and to the services level (service-clock). Network clock can be then extracted from network interfaces UNI, NNI or from dedicated signaling interfaces.

The synchronization source where network clock is extracted from, can be either chosen from any PDH E1/T1, SDH STM1, or SONET Optical Carrier (OC)-3 signals, that will be available at any interface associated with input network traffic. Dedicated signaling interfaces can also be used to provide the clock signal, such as building integrated timing source (BITS). This technology has sync-in/sync-out ports, which can accept external synchronization sources and therefore, also provide a reference signal for the synchronization of the rest of the network nodes, connected among them through radio-relay links or cable. Time reference (timing) is transferred in the physical layer of the transmission link, ensuring immunity to the packet delay variation (PDV) or jitter.

In the frequency synchronization case, ITU-T SyncE technology has been developed in order to distribute a reference frequency over the physical layer of an IEEE 802.3 (Ethernet) network, based on synchronism techniques already used in SDH/SONET. Using this technology, insertion of any element in an existing SDH/SONET network is allowed. For that, SyncE and SDH/SONET must connect all network nodes to a time reference based on a physical layer synchronization technology.

Protocol IEEE 1588–2008, also known as "1588v2" or "PTPV2," is designed for distributing the time and/or frequency references through different networks. In opposition to SyncE, which is a physical layer technology, IEEE 158- v2 is an upper technology of upper layers, so it can operate on a network that includes nodes that do not directly support this standard. This specific functionality allows introducing IEEE 1588v2 in existing networks, including hybrid network deployments.

In Ethernet hybrid systems, TDM signals, based on E1/T1, are transported in packets emulating TDM services. TDM traffic will be transmitted as a sequence of packets on a packet-switched network. In order to recover the original frequency of the TDM bit stream in the distant end, "Service-clock" called methods are used, which provide synchronization to transport service of the packet network.

The most frequent methods for recovering the service-clock in the packet layer, over a circuit emulation service (CES) are: adaptive clock recovery (ACR) and differential clock recovery (DCR). Node resynchronization technology (node retiming)

TABLE 4.4 Signaling and Interface Technologies for Synchronization

Sync Technology	Physical Layer	Upper Layers
Network clock: UNI/NNI interfaces	ITU-T SyncE, PDH E1/T1, SDH STM1, SONET (OC)-3	IEEE 1588v2
Network Clock in synchronization dedicated interfaces	BITS	
Service-clock	Node re-timing	ACR, DCR

STM, synchronous transport module

is used in the physical layer. Node retiming technique takes as a reference an E1/T1 incoming signal and makes the new synchronization (retiming) of the rest of the E1/T1 signals in the node with that reference. All E1/T1 signals in the node are synchronous with each other in this manner.

DCR mode uses time stamps on packets and a clock frequency as shared reference, achieving recovered signals that are less vulnerable to the problems derived from delay variation in packet networks. The shared clock reference frequency can be handed either by the packet layer IEEE 1588-v2, or by a physical layer technology such as radio carrier or SyncE. Table 4.4 summarizes the most relevant synchronization technologies and interfaces mentioned in this section, as well as their fields of application.

ACM mode is used when there is not a shared common reference clock for all the network elements. Its main function consists of filtering packet delay of frames that travel through the network and recovering a stable clock source.

4.4 BASEBAND UNIT

Digital radio-relay link equipment incorporates different BB signal processing functions. The objective of these processing stages will be the conversion of input signals of digital channel interfaces, conveying different information types, into other properly formatted flows that will be handled by the modulation unit. Figure 4.9 shows the general block diagram of the BB unit on the transmitter side. The function of the radio multiplexer is to generate the aggregated radio frame. The processing carried out in the multiplexer and the arrangements made to digital interface signals will vary depending upon the type of aggregated radio frame that is going to be obtained (either based on TDM streams or Ethernet). Three types of radio technologies are possible then depending on the type of radio frame and the treatment of digital interface signals: TDM radio, hybrid radio and Ethernet radio.

In the radio multiplexer, a series of additional bits are added to the bit stream of the digital interfaces. The objective of this insertion is to correct errors, provide communication channels for internal purposes, or integrate monitoring and maintenance signals. This BB signal adaptation is proprietary of each manufacturer. The aggregated radio signal is then passed through a low-pass channel filtering, in the so-called Nyquist filter stage. The main purpose of this filter will be to limit the signal

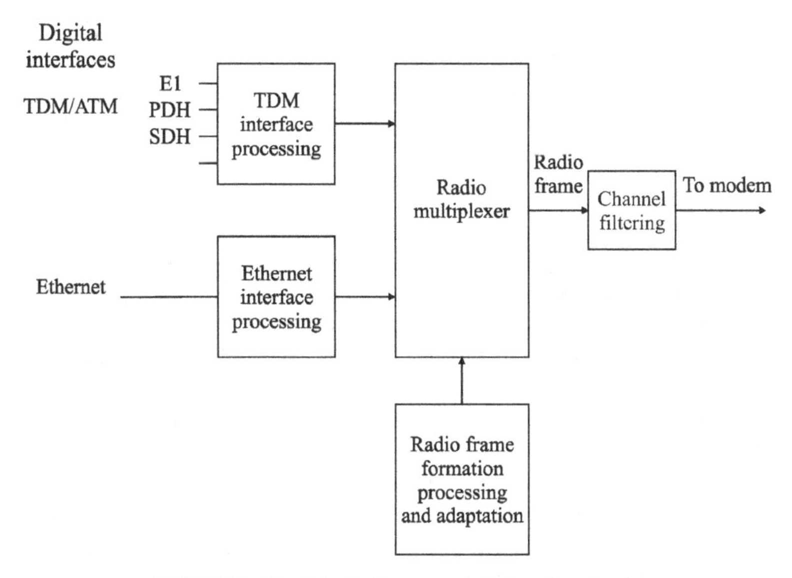

FIGURE 4.9 Block diagram of the baseband unit.

bandwidth in order to reduce inter-symbol interference (ISI). This stage is also known as the pulse-shaping block, because the signal that will be delivered to the modulator will be configured with the shape of the basic pulse used in each case.

On the receiver side, the output of the demodulator will undergo a complementary process. Consequently, the receiver stages will try to eliminate amplitude and phase distortions suffered by the signal by propagation disturbances and should also compensate the processing done on the transmission stages. The radio multiplexer is usually integrated with the demultiplexer, this group is usually referred to as multiplexer—demultiplexer (MULDEM).

4.4.1 Digital Interfaces

Commercial radio-relay link systems support a wide range of standardized digital interfaces, not only for the native Ethernet traffic, but also for TDM. Table 4.5 summarizes the most usual ones in the majority of today's equipment.

TABLE 4.5 Digital Interfaces in Radio-Relay Link Equipment

	Interface
Ethernet	IEEE 802.3 –10/100/1000 BaseT
PDH	E1/2048 kbit/s – T1/1544 kbit/s
PDH	E3/34368 kbit/s – T3/44736 kbit/s
PDH	E4/139264 kbit/s
SDH	STM-1/155.520 kbit/s
SONET	OC-3/STS-3

From the physical point of view, as far as types and characteristics of electric or optic signals are concerned, these are defined in the ITU-T G Series Recommendations. The electric interface is defined in ITU-T G.703 Recommendation and the optical interface in ITU-T G.957 Recommendation.

4.4.2 TDM Radio

Traditional TDM radio equipment is based on an aggregated radio frame with a time division multiplex structure, where a time interval dedicated to each connection is reserved, over the duration of the whole communication. This structure has its origin in circuit switching networks for voice channels. The radio multiplexer supports and processes TDM frames, so traffic that is not native TDM, must be previously adapted into a TDM structure. Figure 4.10 illustrates the BB unit of this type of equipment.

Depending on the capacity of the radio-relay link, digital TDM tributaries can be delivered directly to the radio multiplexer, or in the case of E1/T1 input tributaries, they will be previously multiplexed in a higher order TDM multiplexer before they are fed to the TDM radio multiplexer.

In TDM radios, if Ethernet traffic should be transported by the link, the incoming Ethernet traffic packets must be adapted into a TDM traffic structure for being aggregated in the radio frame. Ethernet interfaces are encapsulated and assigned to TDM frame intervals. This process involves additional overhead bits, and the efficiency of the transport is thus reduced.

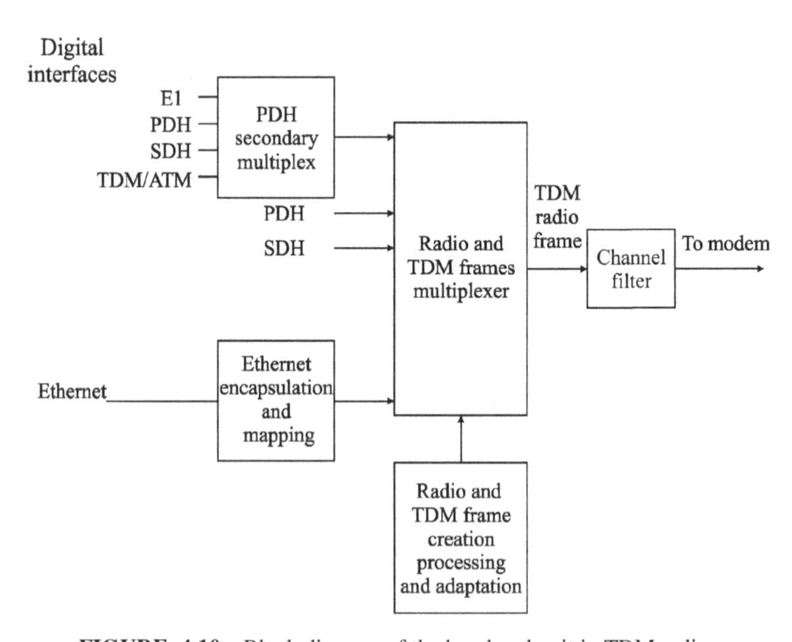

FIGURE 4.10 Block diagram of the baseband unit in TDM radio.

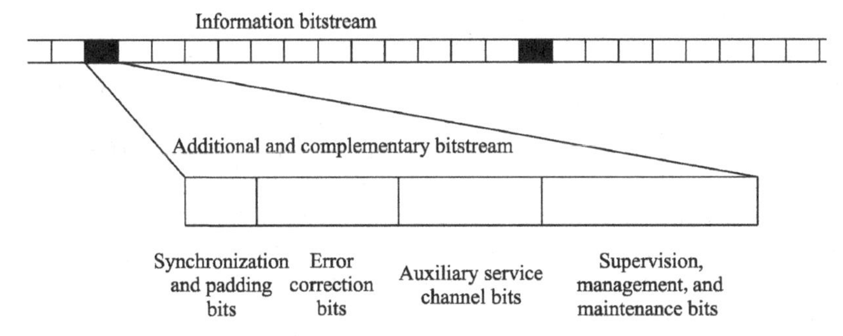

FIGURE 4.11 Example of aggregate structure of the radio frame.

4.4.2.1 *Aggregated Radio Frame Structure* The aggregated signal is obtained at the radio multiplexer unit, by adding the appropriate overhead bits to the useful information signal in each channel. The functions of the overhead bits can be error correction, auxiliary channels for internal voice and data service, and channels for monitoring, management and maintenance purposes. Additional and auxiliary channels can be drop/inserted at terminal or nodal stations.

Synchronization of the additional tributaries of multiple synchronous and asynchronous data can be done by using the technique called "stuffing or justification," which consists of inserting synchronization complementary bits, stuffing information and stuffing bits, in a similar way as the one used in the conventional digital multiplexer of the PDH. Additional and complementary bit streams are usually randomized for energy dispersal purposes. Figure 4.11 summarizes an example of radio frame structure.

Additional information that is inserted in the radio frame increases the gross bitrate. Table 4.6 shows typical values for gross bit rates in the radio frame, as well as transmitted capacity, obtained from commercial equipment.

4.4.2.2 *Coding and Error Correction* Any digital radio-relay link system uses data coding and error correction techniques. The objective of these techniques is to reduce the errors of the received bit stream in order to reduce the residual

TABLE 4.6 Gross Bit Rates of the Aggregate Radio Signal

Capacity	Useful Bit Rate (Mbit/s)	Gross Bit Rate (Mbit/s)	Overhead (%)
5*x*E1	10.24	11.50	13.2
8*x*E1	16.38	18.80	14.7
4*x* DS1	6.17	9.00	47.2
32*x* DS1	49.40	58.99	19.4
1*x*DS3	44.73	46.30	3.5
3*x*DS3	134.20	138.80	3.4
3*x*STS-1	155.52	160.20	3.0

or background bit error rate (BER), particularly in systems that use multistate modulation schemes. The error correction relaxes the receiver sensitivity requirements for a given bitrate threshold. This improvement is higher for lower bitrates.

The conventional error correction technique used in microwave LOS systems (and in many other digital ones) is the so-called forward error correction (FEC). FEC techniques are usually based on convolutional codes of arbitrary length and/or block codes that work with fixed length packets. The encoders for fixed service radio-relay link systems are based on block coding. The information data goes through an encoder that adds a certain amount of redundancy bytes (or bits, depending on the coding level). FEC also includes interleaving stages. Interleaving is a technique used in digital systems affected by error bursts that alters the order of the bytes (bits) after error coding. The objective is to avoid long consecutive errored bytes (bits) at the decoder input that exceed the capacity of the decoder to correct them. If a burst error has occurred during transmission, before decoding, the order of the bytes (bits) will be recovered, and thus, n consecutive errored bytes (bits) would be redistributed over n frames, effectively increasing the error correction capacity of the block or convolutional decoders.

After the FEC coder, redundancy and information bits are multiplexed together and will be delivered to the modulation and subsequent RF stages of the transmitter. Representative examples of such error correction codes are the widely used Reed–Solomon (RS) codes, as well as Bose–Chaudhuri–Hocquenghem codes, and Lee error correction codes. Low-density parity-check codes have been incorporated to high capacity radio-relay link systems, providing a higher coding gain and a low residual error (error floor) compared with traditional solutions based on RS codes.

4.4.2.3 BaseBand Processing The BB signal undergoes a series of signal processing stages with the main purpose of mitigating propagation impact on the signal characteristics, not only in the transmission stage, but also in the reception stage. Some of these techniques are:

- Line code conversion.
- Equalization.
- Clock recovery.
- Carrier recovery.
- Jitter reduction.
- Signal quality monitoring.
- XPIC.

Line code conversion has the aim of adapting the physical characteristics of the BB signal, when these are not appropriate for digital processing on the radio equipment. One example is the return-to-zero line formats conversion into the non-return-to-zero line codes. Moreover, line equalizers compensate the variations with frequency of the attenuation transmission cable between the radio unit and the digital multiplexer. Theoretically, the equalizer will have a characteristic that is opposite to the one on the cable, in such a way that the equivalent overall response does not depend on frequency.

The use of jitter reducers is required in parallel to the use of low cost oscillators. These oscillators produce a phase noise that is necessary to be minimized. This effect is very important in timing, and it mainly affects radio-relay links with several repeater units, where timing errors accumulate. The clock recovery circuits at each receiver will follow the associated clock input, but will also introduce more jitter due to noise and interference. The effect is increased when considering cascade clock recovery circuits in consecutive repeaters.

4.4.3 Ethernet Radio

The most important characteristic of Ethernet radio equipment is that the radio multiplex supports and processes Ethernet frames. The frame of the aggregate radio signal has a traffic structure based on Ethernet packets, where the bit stream is divided into packets and each packet goes through the network individually routed and processed towards its destination. At the destination, the information is regrouped and assembled for being presented to the user. In this case, traffic that is not native in origin must be previously adapted. Ethernet will usually allow sending more traffic capacity per each radio frame, on more flexible and less complex implementations than the ones required on TDM radio-relay links. Figure 4.12 illustrates the BB unit of Ethernet radios.

Incoming Ethernet traffic can be directly processed in the radio multiplexer. This is the usual case in interfaces with equipment that manage traffic originated in WiMax access networks, or the one that is originated in or directed to 3G or 4G BSs. In these cases, there are direct Ethernet interfaces, with low cost and easily scalable implementations, just by adding new interfaces.

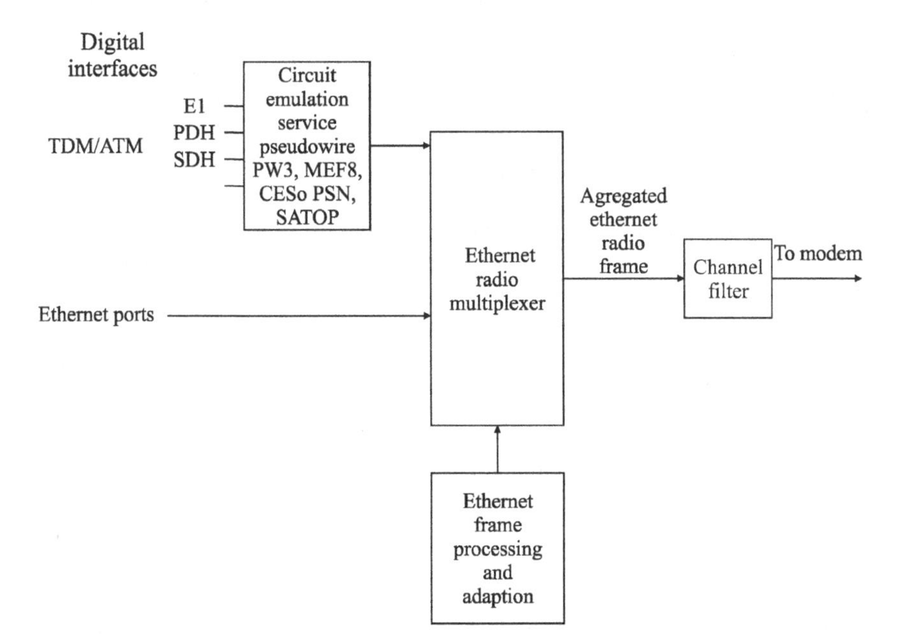

FIGURE 4.12 Block diagram of the baseband unit in ethernet radios.

TDM/ATM native traffic must be converted into Ethernet traffic in order to be processed and aggregated in the radio frame. For that, TDM interfaces include a module that provides a CES, which is a particular case of the most general case known as pseudowire. Pseudowire is intended for the emulation of the link layer of a connection oriented point-to-point service on a packet switching network (IP, multi-protocol label switching, Ethernet or others). It basically works by previously packing and encapsulating TDM/ATM native traffic into IP packets that will later be transmitted through the corresponding packet network. In order to comply with time delay requirements, pseudowire emulation of TDM traffic must be able to minimize packet delay variation, or jitter, by using jitter reduction techniques based on the use of jitter buffers. Developed by the Internet Engineering Task Force (IETF) Pseudowire Emulation Edge to Edge 3 (PWE3) is the most extended pseudowire standard and it is described on the RFC 4717. Other types of TDM pseudowires have been developed under the auspices of IETF, such as circuit emulation over PSN, and structure agnostic TDM over packet. Implementations that use Metro Ethernet Forum 8 specification have recently spread and are a consequence of an agreement for emulating TDM circuits over Metro Ethernet networks. TDM is also covered by ITU-T Y.1413 and Y.1453 Recommendations.

The Ethernet BB unit supports additional or complementary functionalities similar to the ones of TDM radios: error correction, auxiliary channels for internal voice and data service, channels for the monitoring, management and maintenance system and clock recovery and synchronization. Despite the common modules, Ethernet has its own techniques for cases where the traffic is not IP native, such as circuit emulation, or voice converters over IP and voice over IP.

The multiplex will generate the aggregate radio traffic stream. This stream will have a frame structure with the information to be transmitted that previously passed through the error correction unit. The multiplexer will add the radio frame header to the information to be transmitted. This header includes the frame alignment word, signaling for communication, synchronization and control channels, including the forward and return channel for adaptive modulation control.

The gross bit rate of the physical layer will include inter-frame gap bits, preamble, error correction as well as several overheads for monitoring purposes. The total traffic is statistically multiplexed in order to be transmitted over the Ethernet connection. The Ethernet transmission will enable the dynamic management of different types of services, according to established priorities, depending on quality levels of the radio-relay link operation, particularly in deep fading situations.

4.4.4 Hybrid Radio

Hybrid microwave radio refers to systems that enable the simultaneous transmission, using the same bandwidth, of circuits switching TDM traffic using E1/T1 ports, together with packet switching traffic using Ethernet interfaces.

Both types of traffic, Ethernet and TDM, are transmitted in their native forms, being able to share the whole bandwidth, in such a way that the total bandwidth can be

Note: RFC Stands for "Request for Comments" and identifies the documents that describe different aspects of network engineering standardized by the IETF.

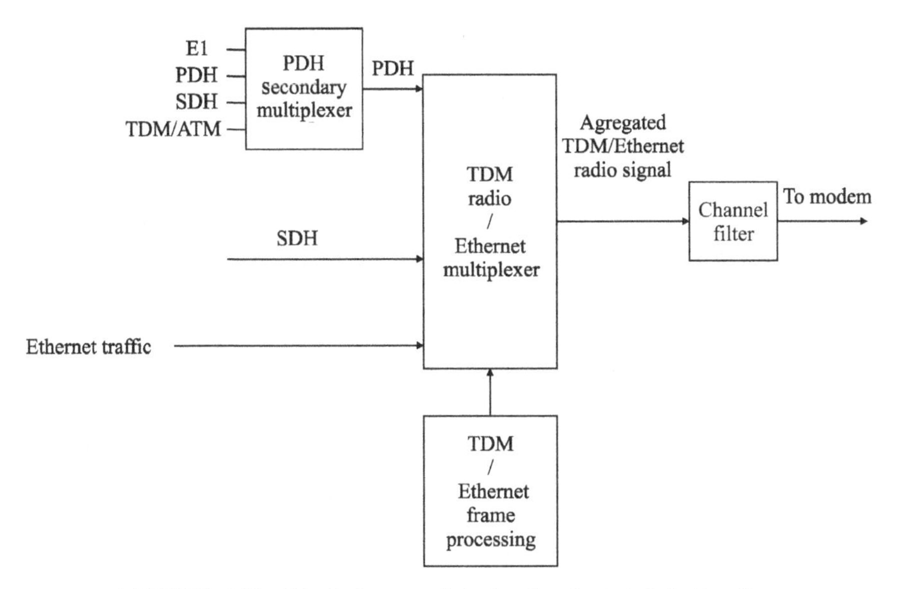

FIGURE 4.13 Block diagram of the baseband unit in hybrid radios.

entirely dedicated to TDM traffic, to Ethernet traffic, or to a mixture of both. Although there are several implementations depending on the equipment manufacturer, the most used one is to transmit TDM traffic over TDM channels of the radio frame and add Ethernet traffic to balance its capacity. Figure 4.13 shows a configuration diagram in the hybrid mode.

Radio multiplexer is formed by a switch that manages the different types of traffic and possibly bridges of the link layer.

This radio-relay link type provides an optimum performance of all existing transport technologies, improving the operation and latency required by each service, and making the evolution of existing TDM networks towards all IP networks easier. Nevertheless, when Ethernet native traffic grows, they turn out to be inefficient compared with Ethernet radios.

4.4.5 Channel Filtering

As we have seen in the different architectures for BB processing, the last block before modulation units is the channel filter or symbol shaping stage (see Figures 4.9; 4.10; 4.12, and 4.13). The aim of the channel filter is to control the overlap of adjacent spectra, and reduce ISI, produced by bandwidth limitations of the transmission channel. This is achieved ensuring that the transfer function of the channel filter is as close as possible to Nyquist frequency response.

Nyquist proved that the necessary minimum theoretical bandwidth of the system to be able to detect R_s symbols/s, without ISI, is $R_s/2$ Hz. This happens when the transfer function of the system is rectangular and constant between zero and $1/2T_s$.

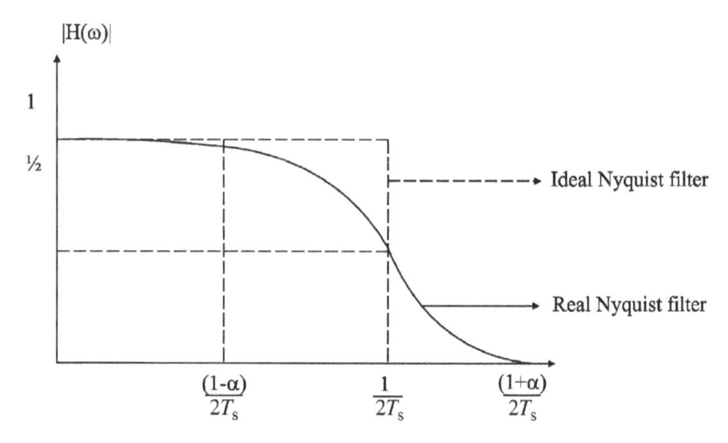

FIGURE 4.14 Transfer function of the ideal and real Nyquist filter.

Equations (4.1) and (4.2) show the frequency and time response of the theoretical Nyquist filter (T_s is the symbol period):

$$\begin{aligned} H(f) &= T_s \qquad f \leq 1/2T_s \\ H(f) &= 0 \qquad f > 1/2T_s \end{aligned} \tag{4.1}$$

$$h(t) = \text{sinc}\,(t/T_s) = \frac{\sin(\pi t/T_s)}{\pi t/T_s} \tag{4.2}$$

The impulse form obtained this way is not feasible in practice due to two reasons. First, achieving a rectangular spectrum would require a filter of infinite delay. Second, it would also require a perfect clock synchronization. In fact, a sampling deviation from the ideal value produces a slightly different ISI level. To overcome these problems, Nyquist purposes in a corollary of the criteria, to use a controlled level of bandwidth excess. A soft frequency response is obtained if we add a function with odd symmetry in relation to $1/2T_s$, Nyquist frequency, to the ideal spectrum of the equation.

A widely used real filter that meets these requirements is the so-called raised cosine filter. Transfer function and impulse response of this filter are described by equations (4.3) and (4.4):

$$\begin{aligned} H(f) &= T_s \cos\left(\frac{\pi T_s}{2\alpha}\left[|f| - \frac{(1-\alpha)}{2T_s}\right]\right) \frac{(1-\alpha)}{2T_s} < |f| < \frac{(1+\alpha)}{2T_s} \\ H(f) &= 0 |f| \geq \frac{(1+\alpha)}{2T_s} \end{aligned} \tag{4.3}$$

$$h(t) = \frac{\sin\left(\frac{\pi t}{T_s}\right)}{\frac{\pi t}{T_s}} \frac{\cos\left(\frac{\pi \alpha t}{T_s}\right)}{1 - \left(\frac{\pi \alpha t}{T_s}\right)} \tag{4.4}$$

where

$$T_s = \text{symbol period}$$
$$\alpha\ (0 < \alpha < 1) = \text{filter roll-off factor}$$

The roll-off factor represents the band excess required over ideal Nyquist bandwidth as shown in Figure 4.14. With $\alpha = 0$ the occupied bandwidth is minimum and corresponds to the Nyquist ideal one. With $\alpha = 1$, it is required twice the band, and equal to $2/T_s$. The α factor can be chosen taking into account that, in the case of a theoretical condition without interference, the following relationship is applied:

$$\alpha \leq 1 - x \tag{4.5}$$

where x is the RF channel width normalized to the symbol frequency. In practice, α values range from 15%–40%.

4.5 MODULATION AND DEMODULATION

In accordance with the block diagram reference in Figure 4.2, the next element in the chain of blocks forming a radio-relay link units is the modulator (demodulation on the receiver side). Modulation is the process of transferring the information of the BB modulating signal to any of the characteristics (amplitude, frequency, phase) of a carrier signal generated by a LO. The modulation process associates the information sequence to a set of discrete amplitudes, or carrier frequencies or phases. The modulation can also be based on a combination of those parameters (i.e., adapting both amplitude and phase of the carrier to the information sequence. Equation (4.6) gives the general representation of a digitally modulated signal:

$$s(t) = v(t) \cos\left[w_0 t + \varphi(t)\right] \tag{4.6}$$

where $v(t)$ and $\varphi(t)$ are the amplitude and phase of the modulated signal and f_0 is the frequency of the LO, called IF. Alternatively, this equation can be written as:

$$s(t) = i(t) \cos w_0 t - q(t) \sin w_0 t \tag{4.7}$$

where $i(t)$ and $q(t)$ are the in-phase and quadrature components of the information signal which modulates the amplitude of both in-phase and quadrature components of the LO carrier: $\cos \omega_0 t$ and $\sin \omega_0 t$:

$$i(t) = v(t) \cos \varphi(t) \tag{4.8}$$

$$q(t) = v(t) \sin \varphi(t) \tag{4.9}$$

Modulations can be classified according to different criteria. Most common classifications distinguish between linear or non-linear modulations and between binary or multilevel modulations.

Linear modulation applies the superposition principle to the signals transmitted in successive time intervals. An important consequence of linearity is that modulation process can be considered as a multiplication in time between the carrier and the modulating signal. The spectrum of a linearly modulated signal is exactly the spectrum of the BB signal displaced with respect to the carrier frequency.

Amplitude shift keying is an example of this modulation type, where carrier amplitude changes according to the line signal of the information bitstream. In a non-linear modulation scheme, superposition is not applied and modulated signal spectrum is related to the spectrum of the BB signal in a more complex way. Frequency shift keying is an example of a non-linear modulation, where carrier frequency changes according to the amplitudes of the line signal of the information bit stream.

In a binary modulation, each bit corresponds with a modulation level or symbol. Modulation rate, V_s (symbols/second) is expressed in bauds and coincides with bitrate. Theoretical spectrum efficiency of binary modulations is 1 bit/s/Hz.:

$$V_s\,(\text{baud}) = V_b\left({}^{\text{bit}}\!/_{\text{s}}\right) \tag{4.10}$$

In multilevel modulations, each symbol of m bits corresponds with a modulation level. The number of states or modulation levels is $M = 2^m$. Therefore, relationships between rates and symbol periods (V_s and T_s) and bits (V_b and T_b) are respectively:

$$V_s\,(\text{baud}) = \frac{V_b}{\log_2 M} \tag{4.11}$$

$$T_s = \log_2 M \cdot T_b \tag{4.12}$$

The theoretical spectrum efficiency is $\log_2 M$ bits/s/Hz. Table 4.7 summarizes the most commonly used modulations in line-of-sight radio link systems. They can be grouped in two big groups, in this case: coded and uncoded modulations.

TABLE 4.7 Modulations Used in Radio-Relay Links

Coding	Type	
Uncoded	PSK	QPSK - 4-PSK DQPSK, O-QPSK
	QAM	16, 32, 64, 128, 256, 512 and 1024
Coded	BCM TCM MLCM	

DQPSK, Differential QPSK; O-QPSK, Offset QPSK

Equipment with higher modulation systems use spectrum most efficiently than systems with lower modulation, being able to transmit information at a higher bitrate in the same bandwidth. For this reason, in order to reduce the necessary spectrum, modulation systems with a high number of levels are used in higher capacity systems. Typical applications of these methods are high capacity backbones, transport, and access networks.

On the other hand, these methods are more expensive (because they require a better high-power amplifier and, in most cases, the use of linearity preserving techniques, such as predistortion) and more sensitive to degradations (e.g., to a multipath propagation), requiring higher received signal power to ensure a similar BER.

The most simple modulation methods are more robust against propagation effects, with characteristics of a high system gain and a higher tolerance against all interference types. For medium and low capacity systems, when the efficient use of spectrum of each radio channel is not particularly important, but high tolerance against interference does matter, simple modulation methods are used, with less modulation levels.

4.5.1 Modulator and Demodulator Performance

Figure 4.15 shows the block diagram of a typical quadrature modulator/demodulator. The incoming bit stream that corresponds with the aggregate radio signal, at V_b (bits/s), is divided into two bit streams of half bitrate, $V_b/2$, which feed each

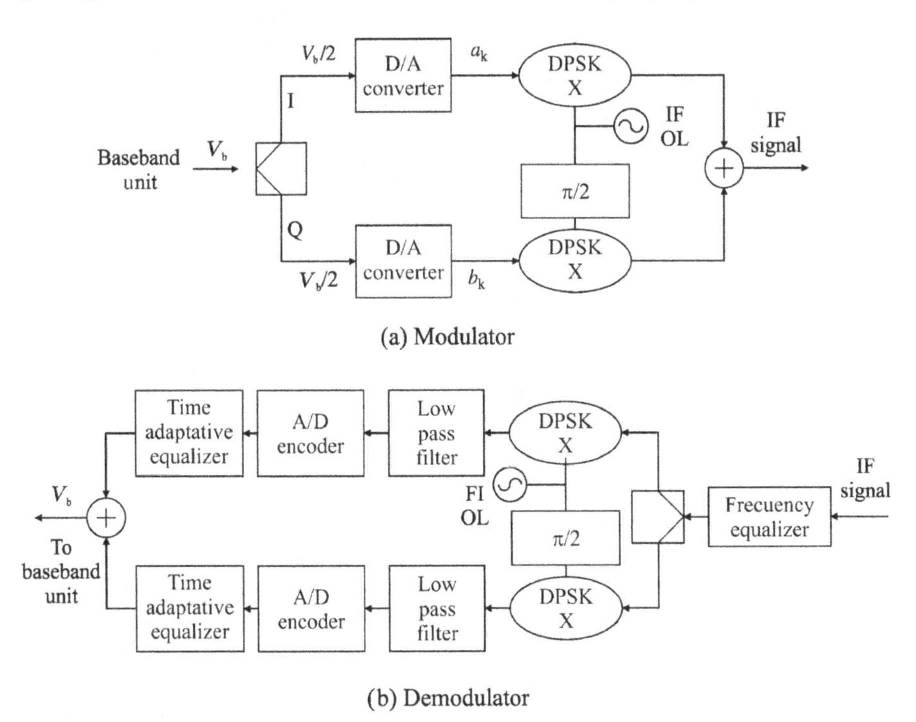

FIGURE 4.15 Block diagram of a typical quadrature modulator/demodulator.

modulation branch, I, and Q. The binary signal is converted into a multilevel sequence on each branch, generating the modulating symbols a_k and b_k respectively.

After a premodulating filtering stage, the carrier from the LO is modulated using the input symbol sequence. The process will imply the in phase component of the carrier cos $(\omega_c t)$ in the upper branch, I axis, and sin $(\omega_c t)$ in quadrature in the lower branch, Q axis. The symbol rate of the sequence is related to the modulating signal rate, depending on the modulation scheme.

The carrier frequency in the modulation and demodulation process will be the IF, and is usually 70 or 140 MHz. The LO will be usually a voltage controlled oscillator (VCO) controlled by a phase locked loop through a highly stable crystal oscillator. The device is equivalent to a VCO with automatic phase control. The outputs of phase and quadrature modulators are added together, and later the signal goes through a band-pass filter to the transmitter unit. The equation of the modulated signal is:

$$y(t) = \sum_{k} [a_k h_t (t - kT_s) \cos \omega_c t - b_k h_t (t - kT_s) \sin \omega_c t] \tag{4.13}$$

where $h_t(t)$ is the impulse response of the premodulating filter and T_s is the symbol period. Each symbol or modulation level is defined by the pair of modulating symbols (a_k, b_k), grouping $(\log_2 M)/2$ bits. As shown in Figure 4.15, the signal coming from the IF amplifier, after going through the frequency equalizer, is put into the I and Q branches, and phase and quadrature coherent demodulation takes place. x_k, y_k values are obtained during the sampling process, and based on them, a decider provides the a_k and b_k estimations of the transmitted symbols. In an interference free situation, each channel output for the general kth symbol, after sampling is:

$$\begin{aligned} \text{I Channel } x_k &= a_k h(0) + n_{Ik} \\ \text{Q Channel } y_k &= b_k h(0) + n_{Qk} \end{aligned} \tag{4.14}$$

where $h(0)$ is the response of the filter to the impulse, for $t = 0$, and n_{Ik}, n_{Qk} are noise samples at kT_s time instants, associated with I and Q channels.

Analog to digital converters transform the estimated symbols into groups of bits that are arranged serially by the parallel/serial converter. The output of this block is the binary signal at V_b bits/s. When BB digital equalization is used, equalizers are incorporated here. The purpose of the equalizers will be examined in the following sections.

The demodulation process requires that the timing when the received symbols are recovered is kept synchronized to f_s. In those instants, the receiver must decide upon the symbol value and recover the correct sequence of symbols and bits. Additional circuits for carrier and timing recovery of the transmitted symbols are necessary for the correct operation of the coherent demodulator. The choice of the sampling instant, within the symbol period, must be optimal to minimize the probability of a wrong symbol decision. The BER is related to the symbol error rate by a constant that represents the average number of information bits with errors per symbol. This constant depends on the way the information bits correspond with the symbols and it is usually a fraction of the total number of bits.

Values of a_k, b_k pairs on Cartesian axis ($x = \mathrm{I}, y = \mathrm{Q}$) are called constellation points and represent modulation states. Values of the abscissa will represent the amplitude of cos $\omega_c t$ modulating symbols and ordinates will correspond to sin $\omega_c t$ modulating symbols.

The decision distance in a constellation will correspond to half the length of the segment that joins two consecutive points of the constellation. The assignment of bits to the constellation symbols is usually done in accordance to the Gray code, which distributes bits in a way that adjacent constellation states only differ in one bit. This way, BER is minimized when a demodulation error occurs. Under error conditions, the receiver will most probably decide (erroneously) for a state contiguous to the transmitted one, which will imply only a single bit error. The module and phase of the vector that links the origin with each constellation point are, respectively, the amplitude and phase of the carrier at the modulation level corresponding to the symbol represented by this point. In real reception conditions not only noise but also additional perturbation voltages are added to the x_k and y_k recovered samples in the demodulation process. These impairments can be caused by:

- ISI within the I or Q channel, caused by an imperfect filtering or due to distortion generated by multipath propagation mechanisms.
- Mutual interference between I and Q channels, because of an asymmetric transfer function of the RF channel and selective fading.
- RF interference, co-channel, or adjacent channel.
- System nonlinearities.

Supposing that synchronization is perfect, the previously mentioned perturbations, noise included, will broaden the received constellation point. The constellation point becomes a circle with the center on the (transmitted) theoretical point, as it is shown in Figure 4.16.

The parameter called eye aperture percentage quantifies this effect and can be evaluated according to the equation:

$$e_0 = \left(1 - \frac{r}{d}\right) 100(\%) \tag{4.15}$$

where d is the decision distance and r, the circle radius that indicates the degree of eye closure.

Propagation channel introduces, furthermore, a time delay to the signal that produces a time-shift of the ideal sampling instant and a phase-shift of the carrier. The time-shift of the sampling instant must be estimated with a reasonable precision by an appropriate circuit for timing recovery (clock). The phase-shift of the carrier can be estimated by an appropriate carrier recovery circuit. In this case, it is said that demodulation is phase coherent.

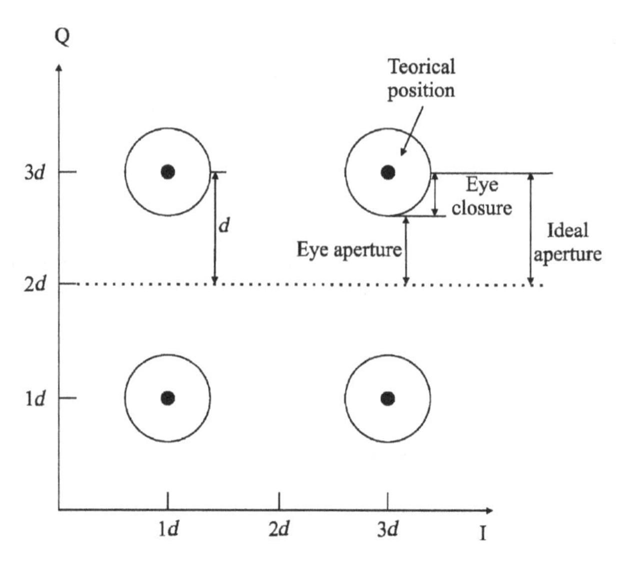

FIGURE 4.16 Decision eye aperture of a constellation.

4.5.2 Examples of Modulation Schemes

The following sections describe the features of the most representative modulation schemes used in microwave LOS links.

4.5.2.1 Phase Shift Keying 4-PSK

4-PSK modulation has four modulation states that correspond to four phases φ_k of the carrier, defined by the equation (4.16):

$$\phi_k = (2k+1) \cdot \frac{\pi}{4}\ (k = 0, 1, 2, \ldots) \tag{4.16}$$

Values of the a_k and b_k symbols are described from φ_k in the equation (4.17), where A is the carrier amplitude:

$$\begin{aligned} a_k &= A\cos\phi_k \\ b_k &= A\sin\phi_k \end{aligned} \tag{4.17}$$

As PSK is a modulation of constant amplitude, the carrier amplitude is always the same for any symbol, only changing the phase when switching from one symbol to another. Figure 4.17 shows this constellation, which consists of 4 points.

The average power of the modulated signal is the same as the average power of the carrier. Its standardized value is $A^2/2$, with:

$$A^2 = a_k^2 + b_k^2 \tag{4.18}$$

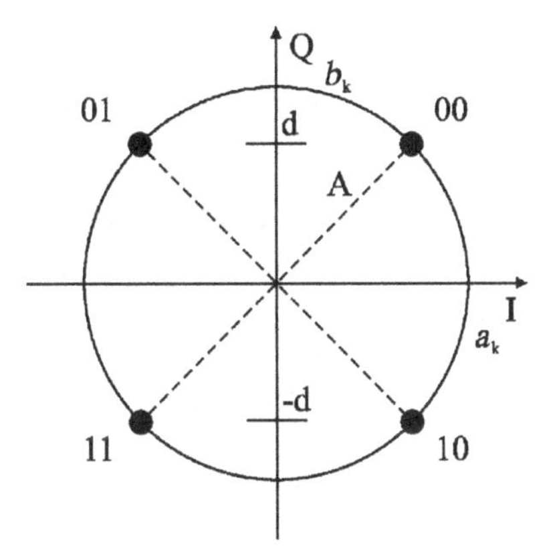

FIGURE 4.17 Constellation of the 4-PSK modulation scheme.

The average energy per bit is:

$$e_b = \frac{A^2}{2} \cdot T_b = \frac{A^2}{2 \cdot V_b} \tag{4.19}$$

4.5.2.2 16 Level Quadrature Amplitude Modulation (16-QAM) Figure 4.18 illustrates the 16 points constellation that corresponds to the 16-QAM, which consists of 4 points per each axis, located at $\pm\, d$ and $\pm\, 3d$ distances from the origin, being d the decision distance.

In 16-QAM, the amplitude and phase change according to the different symbols. Observing the constellation, the lowest and highest values of the modulated carrier

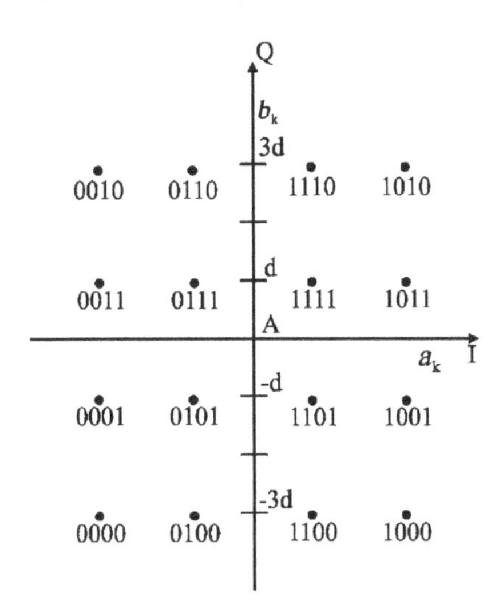

FIGURE 4.18 Constellation of the 16-QAM scheme.

amplitude are, respectively:

$$A = A_{\max} = \sqrt{18d} \qquad A_{\min} = \sqrt{2d} \tag{4.20}$$

The average power of the modulated signal is obtained from the following equation:

$$P_{\mathrm{m}} = \frac{4}{16} \cdot P_{\mathrm{mc}} \tag{4.21}$$

where P_{mc} is the average power in one quadrant. The carrier amplitudes that correspond to the point in the first quadrant are:

$$A_{11} = \sqrt{10d} \qquad A_{12} = \sqrt{18d} \qquad A_{21} = \sqrt{2d} \qquad A_{22} = \sqrt{10d} \tag{4.22}$$

Resulting:

$$P_{\mathrm{mc}} = \frac{1}{2}A^2 + 2\frac{1}{2}\left(\frac{A\sqrt{10}}{3\sqrt{2}}\right)^2 + \frac{1}{2}\left(\frac{A}{3}\right)^2 = \frac{10A^2}{9} \tag{4.23}$$

Consequently:

$$P_{\mathrm{m}} = \frac{P_{\mathrm{mc}}}{4} = \frac{5A^2}{18} \tag{4.24}$$

4.5.2.3 M level Quadrature Amplitude Modulation (M-QAM) In general, for a M-QAMtype with A maximum carrier amplitude, there are $M/4$ points per quadrant. The number of levels per axis, for each one of I and Q axis is:

$$k = \sqrt{M/4} \tag{4.25}$$

Levels on each axis are spaced $2d$, and their coordinates are:

$$\pm(2i - 1) \quad \text{with} \quad i = 1, 2, \ldots, k \tag{4.26}$$

and

$$d = \frac{A}{\sqrt{2}\,(2k - 1)} \tag{4.27}$$

Average power in a quadrant is equal to the corresponding power of the points located along the bisector of the quadrant, plus twice the power for the points of the area between the bisector and one of the axis:

$$P_{\mathrm{mc}} = \frac{1}{2} \cdot \frac{A^2}{2\,(2k-1)^2} \cdot \sum_{\mathrm{i=1}}^{k} (2i-1)^2 + 2 \cdot \frac{1}{2} \cdot \frac{A^2}{2\,(2k-1)^2} \cdot \left(\sum_{\mathrm{i=1}}^{k} \sum_{\mathrm{j>1}} \left[(2i-1)^2 + (2j-1)^2 \right] \right) \tag{4.28}$$

And the total power is:

$$P_{\mathrm{m}} = \frac{4 P_{\mathrm{mc}}}{M} \tag{4.29}$$

Generally, calculations on M-QAM are referred to the maximum power, whose equation is:

$$P_{\mathrm{mc}} = \frac{1}{2} A^2 = d^2 \cdot (2k-1)^2 \tag{4.30}$$

So, the maximum energy per bit will be:

$$e_{\mathrm{b}} = \frac{d^2\,(2k-1)^2}{V_{\mathrm{b}}} \tag{4.31}$$

4.5.2.4 Coded Modulation Coded modulation techniques combine coding and modulation in such a way that redundant bits are inserted in multilevel codes of the constellations of the transmitted signal. The difference with uncoded modulations, which also transport coded data, is that here both modulation and coding are independent processes. Coded modulation improves coding performance compared with uncoded modulation. Block coded modulation (BCM), trellis coded modulation (TCM) and multilevel coded modulation (MLCM) are representative examples of coded modulation used in microwave LOS links.

The BCM technique lies in generating multidimensional constellations that have long decision distances and, at the same time, a regular structure that allows using efficient demodulation architectures called step-by-step decoding.

TCM uses a convolutional coding optimized to achieve high values for the "Euclidean distance" between sequences of transmitted symbols. As a result, requirements of signal/noise or bandwidth decrease in order to transmit data at a defined bitrate and BER. The optimum receiver for the Trellis coded symbol sequence requires maximum likelihood sequence estimation, which is usually carried out using the Viterbi Algorithm. The BCM technique generally needs a less complex demodulator than a TCM scheme with the same performance. BCM schemes, in contrast, provide poorer improvements than TCM.

MLCM divides the constellation into subsets of symbols. Different coding levels are then applied to each subset. Each modulation level is understood as a different and independent transmission path with different minimum quadratic distances, and coding is independent on each of these levels. Convolutional codes are generally used in the lower levels, while block codes can be used in the rest of the levels. The decoding technique in this case is the so-called "multistage decoding."

4.5.3 Adaptive Modulation

Adaptive modulation or dynamic modulation techniques consist of using different modulation schemes in a radio-relay link depending on the instantaneous propagation losses derived from either troposphere refractivity changes or weather conditions along the path. These modulation changes are dynamic and automatic with the aim of maximizing radio-relay link capacity at any time, regardless of the propagation conditions.

Even though adaptive modulation is mainly used in order to offer efficient measures to avoid adverse propagation conditions, they are also appropriate to reduce interference under certain circumstances. These techniques are particularly appropriate for large bandwidth systems, with high availability requirements and using complex modulation methods. Figure 4.19 illustrates the operation of the modulation adaptation techniques.

As shown in Figure 4.19, under clear sky conditions, modulation schemes with a high number of modulation levels could be used, 256-QAM for example. This period would have a lower system gain and high spectrum efficiency to deliver a bitrate through a fixed specified bandwidth. As propagation conditions get worse, system losses would increase, so schemes with lower number of modulation states

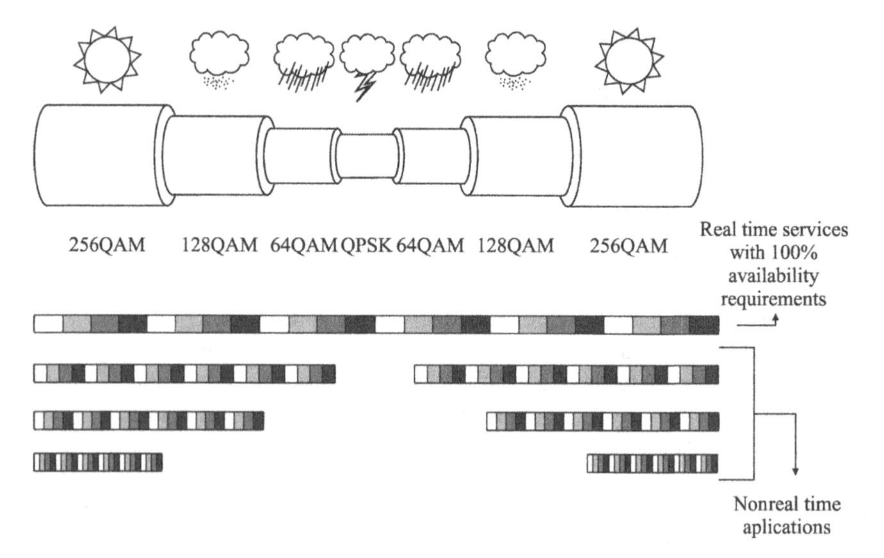

FIGURE 4.19 Adaptive modulation on varying propagation conditions.

TABLE 4.8 System Gain and Bandwidth/Capacity for Different QAM Schemes

Order	Bandwidth (MHz)/ Bitrate (Mbit/s)	System Gain for Different Frequency Bands (Bands in GHz and Gains in dB)							
		L6	U6	7	11	15	18	27	38
4-QAM	10 / 14.04	118	116	115	112.5	116	112	110.5	105.5
	30 / 43.86	114	112	113	110.5	112	108	106.5	101.5
	40 / 59.49				108.5	110	106	104.5	99.5
	50 / 4.59						105	103.5	98.5
8-QAM	10 / 21.13	113.5	111.5	110.5	108	111.5	107.5	106	101
	30 / 65.51	109	107	108	105.5	107	103	101.5	96.5
	40 / 88.81				104	105.5	101.5	100	95
	50 / 111.29						100.5	99	94
16-QAM	10 / 28.29	110.5	108.5	107.5	105.5	107.5	103.5	103	97
	30 / 87.45	106	104	105	103.5	103	99	98.5	92.5
	40 / 118.51				102.5	102	98	97.5	91.5
	50 / 148.49						97	96.5	90.5
32-QAM	10 / 35.06	106	104	103	100.5	103	99	98.5	92.5
	30 / 108.76	103	101	102	100.5	100	96	95.5	89.5
	40 / 147.37				99	98.5	94.5	94	88
	50 / 184.61						94	93.5	87.5
64-QAM	10 / 44.81	101.5	99.5	99.5	97	99.5	95.5	95	88
	30 / 138.02	98	96	98	95.5	96	92	91.5	84.5
	40 / 186.98				94	94.5	90.5	90	83
	50 / 234.23						90	89.5	82.5
128-QAM	10 / 52.87	99	97	97	94.5	97	93	92.5	85.5
	30 / 162.88	95	93	95	92.5	93	89	88.5	81.5
	40 / 220.63				91.5	91.5	87.5	87	80
	50 / 276.37						86.5	86	79
516-QAM	10 / 60.27	95.5	93.5	93.5	91	91.5	88.5	89	81
	30 / 185.37	92	90	92	89.5	88	85	85.5	77.5
	40 / 251.1				88	86.5	83.5	84	76
	50 / 314.46						83	83.5	75.5

would be used, M-QAM for example, with lower power thresholds and worse spectrum efficiencies. These conditions imply a lower capacity in the same bandwidth. Table 4.8 tries to illustrate adaptive modulation mechanism. The figure shows system gain values for a BER of 10^{-6} depending on the 10, 30, 40, and 50 MHz bandwidth, and on the estimated capacity for different types of modulation schemes, obtained from the data of commercial equipment of different manufactures.

In the case of critical weather conditions the priority services would still be delivered without propagation impairments (usually multimedia services and real time or specifically prioritized prerecorded services). Services that do not require real time transmission could be provided with different bandwidths depending on the different weather conditions. Those services, not prioritary, would suffer traffic losses in adverse propagation conditions. Adaptive modulation techniques are used in combination with priority management tecniques that will be handled at the radio multiplexer stage, choosing the traffic with the appropriate quality of service requirements for each modulation scheme in the aggregate frame.

In order to maximize dynamic capacity of traffic, radio-relay link systems usually monitor propagation channel conditions continuously in reception, combining mean square error measurements with FEC decoder information, to detect changes in channel. When channel quality is lower than a defined threshold, the distant transmitter is requested for switching to a lower modulation scheme. In a complementary manner, when channel quality increases over the established thresholds, switching to a scheme with more modulation levels is requested. Commercial equipment with adaptive modulation perform the switching using the *Hitless* algorithm, which provides near to zero response times, so changes in modulation schemes would not affect service features.

4.5.4 Error Probability

The error probability is the main parameter to evaluate the correct performance of the demodulation process. In the ideal case of a Gaussian propagation channel, an error happens in the decision if, the amplitude of the noise sample is bigger than the decision distance between the transmitted symbol and the nearest symbol in the decision instants. In this case, BER depends on:

- The decision distance, d.
- The normalized additive white Gaussian noise power in the receiver, σ, which is equal to the product of the unilateral noise spectrum density and the BB equivalent bandwidth B_n (Nyquist band).

Assuming Gray coding for modulation, BER is defined as it follows:

$$P_{eb} = k \cdot G\left({}^{d}/_{\sigma}\right) \tag{4.32}$$

where k is a constant that depends on the type of modulation and $G(t)$ is the complementary Gaussian distribution function:

$$G(t) = \frac{1}{\sqrt{2\pi}} \int_{t}^{\infty} \exp\left({}^{-u^2}/_{2}\right) du \tag{4.33}$$

BER can also be expressed in terms of the Gaussian complementary error function Erfc(t). The relationship between Erfc(t) and $G(t)$ is shown in equation (4.34):

$$G(t) = \frac{1}{2} \text{Erfc}\left({}^{t}/_{\sqrt{2}}\right) \tag{4.34}$$

The following approximation can be used for BER calculations, when $t \geq 7$, a very usual case in practice:

$$G(t) \approx \frac{1}{\sqrt{2\pi t}} \exp\left(-t^2/2\right) \tag{4.35}$$

The d/σ ratio can be normalized in terms of the bit energy to noise density ratio, e_b/n_o. In modulation schemes with variable envelope amplitudes, the energy per bit will be referred to the maximum amplitude of the modulated carrier, A. So:

$$e_b = \frac{A^2}{2 \cdot V_b} \tag{4.36}$$

$$n_0 = k \cdot T_0 \cdot f_r \tag{4.37}$$

where V_b is the bitrate of the aggregate radio frame, T_o the reference temperature (290°K), k the Boltzman constant and f_r receiver noise factor. Error probability can also be written in terms of the ratio between the average power of the received signal and equivalent input noise at the demodulator, c/n. It is immediate to establish the connection between c/n and e_b/n_o, because:

$$c = e_b/T_b \tag{4.38}$$

$$n = n_0/T_s \tag{4.39}$$

where T_b and T_s are bit and symbol periods respectively. Consequently, if we consider M as the number of modulation levels:

$$\frac{c}{n} = \frac{e_b}{n_0} \cdot \frac{T_s}{T_b} = \frac{e_b}{n_0} \cdot \log_2 M \tag{4.40}$$

For M-QAM systems, if r is the maximum to average signal power ratio of the received signal, as c/n is the average value of c, the following can be written:

$$\frac{c}{n} = \frac{1}{r} \cdot \frac{e_{b\,max}}{n_0} \cdot \log_2 M \tag{4.41}$$

In most practical systems in, the receiver equivalent noise bandwidth, B_{eq}, is wider than Nyquist bandwidth, and the c/n ratio is usually specified for that equivalent bandwidth. Set as $(c/n)_{eq}$, we have:

$$\left(\frac{c}{n}\right)_{eq} = \frac{e_b}{n_0} \cdot \frac{V_b}{B_{eq}} \tag{4.42}$$

TABLE 4.9 Bit Error Rates Under Ideal Conditions

Modulation Scheme	Bit Error Rate
Binary PSK	$P_{eb} = G\left(\sqrt{2w}\right)$
Multilevel PSK, MPSK	$P_{eb} = G\left(\sqrt{2w \cdot \log_2 M} \cdot \sin\left(\frac{\pi}{M}\right)\right)$
M-QAM	$P_{eb} \approx \frac{4}{\log_2 M} \cdot \left(1 - \frac{1}{\sqrt{M}}\right) \cdot G\left[\sqrt{\frac{3 \log_2 M}{M-1} \cdot w}\right]$

being V_b the bitrate of the aggregate radio signal. Then, the value of the bit energy to noise density ratio, set as w, is:

$$w = \frac{e_b}{n_0} = \left(\frac{c}{n}\right)_{eq} \cdot \frac{B_{eq}}{V_b} \tag{4.43}$$

If equipment manufacturer provides the $(c/n)_{eq}$ value, it must be used for BER calculations.

Table 4.9 shows BER s for the most used modulation systems in digital radio-relay links, under ideal reception conditions with an optimum receiver and supposing there is no more disturbance than thermal noise, in function of the normalized parameter $w = e_b/n_o$.

Table 4.10 provides with theoretical values of W(dB) under the previous condition, obtained for 10^{-3} and 10^{-6}BER s and different modulation systems.

Figure 4.20 shows the resulting diagrams for BER, P_{eb}, under the mentioned ideal conditions for quadrature phase shift keying (QPSK) and 16-QAM digital modulation schemes.

In real environments, transmitted samples must be recovered in the presence of thermal noise and interference. BER calculation is more complex under these conditions, particularly when the decision eye is partially closed, because there is no completely valid calculation model. Under these conditions, BER is determined by

TABLE 4.10 Bit Error Rates for Different Modulations and Thresholds

	$W = 10 \log e_b/n_o$	
Modulation Scheme	$P_{eb} = 10^{-3}$	$P_{eb} = 10^{-6}$
2-PSK	6.8	10.5
4-PSK	6.8	10.5
8-PSK	10.0	13.8
4-PSK	9.1	12.8
16-QAM	10.5	14.4
64-QAM	14.7	18.8
256-QAM	19.3	23.5

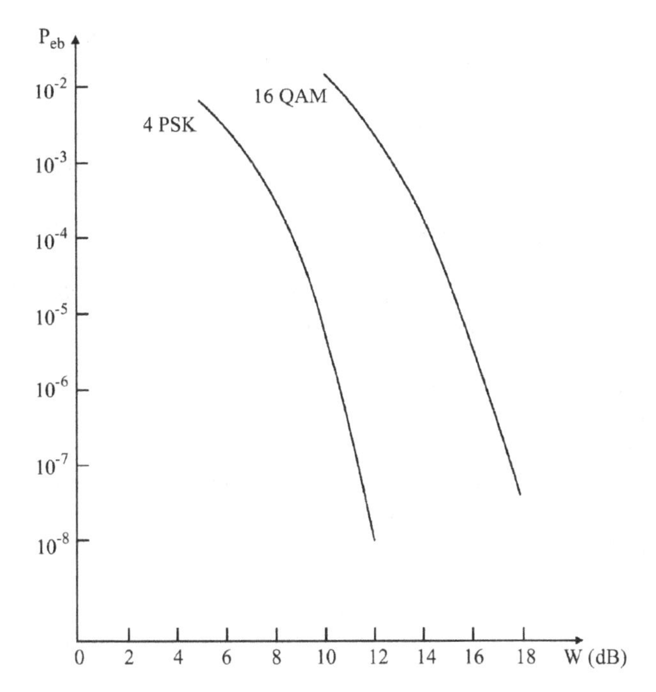

FIGURE 4.20 Bit error rate depending on W = 10 log e_b/n_o.

computer simulation or using curves obtained experimentally. In practice, the thresholds are provided by manufactures and will be based on laboratory measurements as well as on the experience of those manufacturers with respect to the equipment performance on the field.

4.5.5 Adaptive Equalization

The quality of the received signal in a radio-relay link hop is basically affected by a combination of three degradation sources: interference, thermal noise, and wave distortion due to multipath propagation.

Radio channel equalization is done at the receiver, in order to mitigate the effects of these degradations produced in the transmission RF channel, particularly the signal distortion. An adaptive equalization is required for increasing the system robustness, in such a way that transmission channel variations can be followed by the equalizer as propagation conditions change. The equalization techniques used in microwave LOS links can be classified into two groups, depending if they operate in the time or frequency domain.

4.5.5.1 Frequency Domain Equalization This type of equalizer is composed of one or various linear networks designed to produce amplitude and group delay responses that partially mitigate transmission degradation caused by multipath degradation. Frequency domain equalization is also called IF equalization because equalizers are located in the IF receiver module before the demodulator, in order to provide a reconstructed frequency response. This type of equalizer is mainly used in high capacity equipment that use radio channels with significant bandwidths.

The slop equalizer is worth mentioning the among frequency equalizers, because of its simplicity. This type compensates slope asymmetries of the frequency response amplitude within the radio channel bandwidth. The coefficients required for these equalizers are usually calculated by analyzing the frequency response in several points of the radio channel bandwidth. Frequency equalizers do not correct distortions on group delay and they are frequently combined with time domain equalizers.

4.5.5.2 Time Domain Equalization Signal equalization in the time domain has the aim to fight ISI. ISI is a typical disturbance of digital systems and is produced when the signal is affected by post-cursors ("tails") of previous symbols and preludes or precursors of the next symbols at the sampling instant. Tails and preludes are caused by bandwidth limitations of systems and transmission channels. This limitation in bandwidth spreads in the time domain the energy contained in more or less perfect basic square pulses.

Specifically, in digital radio-relay links ISI is due to two causes: radio channel bandwidth limitations associated with the channeling plan and multipath distortion. As previously mentioned, a channel filtering with Nyquist characteristic is performed in the modulation effect in order to fight the first effect, and a BB time equalization combined with space diversity reception to fight the second effect. In digital radio-relay links, time equalization is done by a BB signal processing after demodulation, on I and Q channels. Precursors and tails of each symbol are estimated in the equalizer, using this information to cancel ISI. Control information is obtained by correlating the interference appearing at the decision instant with various adjacent symbols that are producing that interference.

The architecture can be based on a combination of digital filter structures located before the decision module linear forward equalizer, and after the decision feedback equalizer. Digital coefficients can be adjusted by using several correction algorithms. Here, it is worth mentioning the Zero-Forcing method, the least mean square (LMS) algorithm, which minimizes the average square error between the real and desired output or the recursive least square, which improves the convergence of the LMS. Finally, the fractionally spaced linear transversal equalizer samples the signal several times during the same symbol interval and offers an effective solution. When a system uses quadrature modulation, the important fading effects are associated with the crosstalk generated by channel asymmetries.

Due to this crossed interferences between I and Q channels, cross equalizers between both channels should be also included.

4.5.6 Cross-Polar Interference Canceler

As it has been mentioned in Chapter 3, co-channel dual polarization (CCDP) is a solution being used more and more for increasing carrier capacity and spectrum

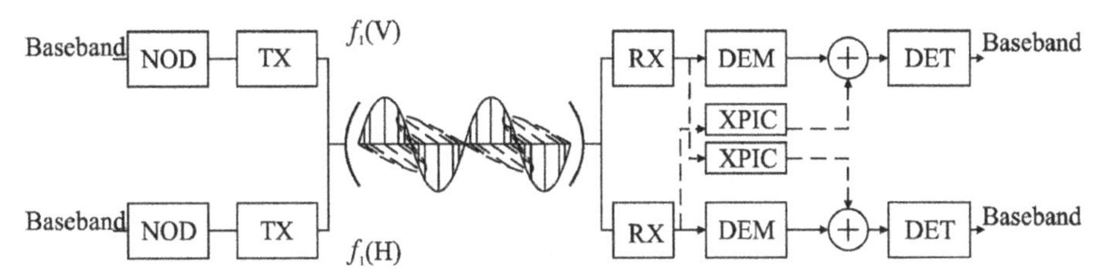

FIGURE 4.21 Operating principle of XPIC (cross-polarization interference cancellation).

efficiency. CCDP consists of transmitting two channels with different traffic in the same radio frequency, with orthogonal polarizations. This technique requires antennas with high cross-polar discrimination (XPD) characteristics for directions within the main lobe, as well as dual polarized horn feeders.

In any case, it will be difficult to achieve high isolation between crossed polarizations, because of rotational misalignments of the antennas in practical installations. Moreover, discrimination between polarizations is subject to additional deterioration associated with fading (rain or multipath). These facts require the development of digital signal processing techniques for reducing interference caused by XPD degradation.

These techniques are called XPIC algorithms (cross-polarization interference cancellation). XPIC is based on cancelling an interfering cross-polarized signal that is incorporated to a signal with orthogonal polarization. For that, the interference canceler samples interfering signals in the channel with crossed polarization and processes these samples to achieve equalized responses. The XPIC module will combine its output with the co-polar signal in the desired channel in order to compensate the interference. XPIC processing uses transverse filter structures and equalization algorithms. This is usually carried out in BB modules, although it can also be a RF or IF process. Figure 4.21 illustrates the operation of the BB XPIC canceler.

4.6 TRANSCEIVER UNIT

The transceiver unit is formed by the components that provide transmission and reception functions. Figure 4.22 illustrates the basic structure used in digital radio-relay link equipment.

Figure 4.22 shows the following blocks: Frequency Converter, Power Amplifier, RF low noise amplifier, IF Amplifier, and several band-pass filters. The most important elements are described in detail in the following subsections.

4.6.1 Frequency Converter

The IF outgoing signal from de modulator, previously amplified, is converted into a RF signal in the mixer, on the transmitter side. The received RF signal is converted

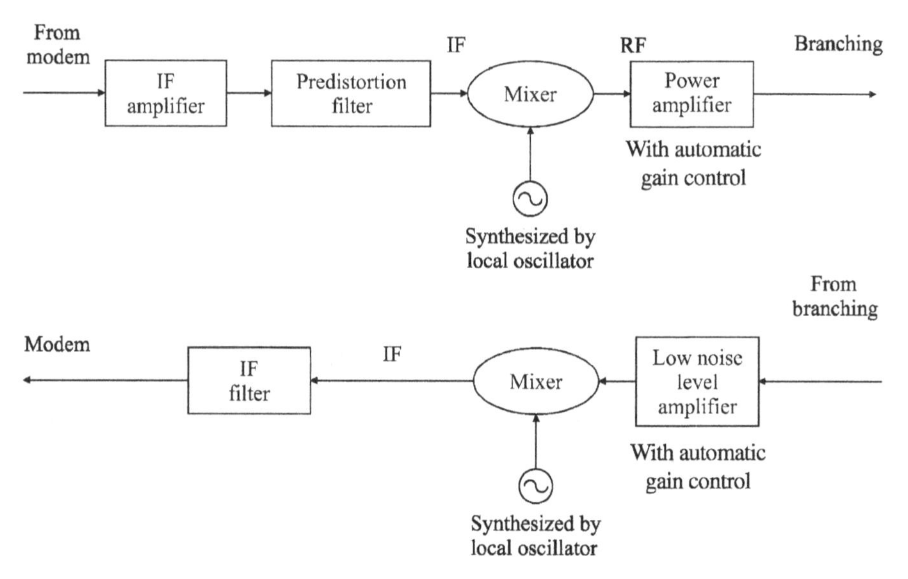

FIGURE 4.22 Block diagram of a transmitter–receiver (transceiver).

into an IF signal, on the receiver side. Ideally, this operation would neither affect the signal amplitude nor introduce noise.

Frequency conversion is achieved with non-linear devices, traditionally, diodes. The signal to be converted, as well as the reference coming from a LO, are fed into a diode, which provides the sum and difference frequency components of the input signals at its output. By band-pass filtering, the appropriate signal is chosen in each case.

In most radio-relay links, the reference signal is obtained from a synthesized or programmable VCO, which is able to supply different frequencies. To improve system performance, it is necessary to eliminate the noise and generated harmonics, by complex circuits that cancel undesired frequency components and help eliminating noise by varying the amplitude of the LO signal.

Currently, other semiconductor devices are also used as mixers, such as metal semiconductor field effect transistor, high electron mobility transistor (HEMT) or pseudomorphic HEMT. Modern equipment normally have a monitoring and test point to measure the oscillator frequency and voltage values.

4.6.2 Transmitter Power Amplifier

The RF power amplifier will receive the output of the frequency converter and will amplify the signal in order to provide an adequate power level to the antenna distribution network. Under standard performance requirements, the amplifier must operate on the linear region in order to limit intermodulation distortion. Working on the linear region implies a back-off margin that leaves room between the nominal power operation and saturation. The Back-off value increases as the number of modulation levels increase. Pre-distortion or linearity units are used in high capacity equipment, which can be located before or after the frequency converter, and their function is the correction of the intermodulation produced by the power amplifier of the transmitter.

Another important technique that has spread widely in today's equipment is automatic transmit power control (ATPC). ATPC techniques are normally used with the aim of reducing interference among radio channels in the same path or in converging paths to/from the same location. Thanks to these techniques, the output power of a RF transmitter can also be varied according to propagation conditions in the path. Under normal conditions, ATPC keeps output power in a low level and under fading conditions output power is increased. Fading is detected by the receiver in the remote end by the monitoring system indicated in the adaptive modulation section, and sent to the distant transmitter for the amplifier gain control. This type of control is usually known as remote transmit power control.

When using these power regulation techniques, one should pay attention to the effects of the power increase on transmitter linearity and an increase of overflowing over adjacent channels. Fading characteristics should also be analyzed, in order to identify the events of multipath fading in which there is a partial correlation between AGC signal level and distortion of the digital signal.

It is usual to have a monitoring and a measuring point at the amplifier output, where the average power of the RF modulated signal can be measured.

4.6.3 Receiver RF Low Noise Amplifier

The function of the receiver input amplifier is to amplify the weak signals received by the antenna, attenuated along the propagation path, and at the same time, keeping additional noise as low as possible.

As noise figure of the receiver system is largely influenced by the first element (amplifier) in the reception chain, this must be designed as a LNA, with special low noise semiconductor devices and low circuit losses.

It is worth mentioning that the overall system performance will be usually defined by specifying minimum and maximum power levels at the amplifier input. Minimum operation powers correspond to the receiver sensitivity for a defined BER, or, in other words, the required power threshold to achieve that BER. Equipment manufacturers usually facilitate reception threshold data in dBm forBER s values of 10^{-3} and 10^{-6}.

4.6.4 Receiver IF Amplifier

The aim of the IF reception amplifier is to deliver a signal to the demodulator with a constant power level within a specified optimum performance range. The IF amplifier will try to provide a constant level regardless of amplitude changes of the input signal.

The amplifier has an AGC circuit for keeping the IF output signal level stable. The AGC circuit generates a voltage value proportional to the output power of the amplifier and the amplifier gain is controlled in a feedback path by the AGC voltage. This voltage value will be related to the received power. In some equipment, this sample voltage is also used to make estimation measurements of the received power level for monitoring purposes. Usually, there will be access probe point to measure and monitor the signal at the IF amplifier output.

4.6.5 Microwave Circuit Integration

In the first generations of radio-relay links, solid-state microwave circuits were manufactured with discrete components that included active semiconductor devices such as transistors and diodes. Circuit modules were connected by cables. These design approaches produced very heavy and large equipment.

Later designs started to evolve microwave integrated circuits technology, based on planar transmission lines (i.e., stripline) in multifunction boards that reduced the need for connectors. The success of Microstrip technology and FET transistors enabled the hybrid microwave integrated circuits technology.

Significant progress has also occurred in the MMIC field, which has facilitated the evolution of radio-relay links towards compact systems formed by integrated circuits with all the system components in a single board. This technology is widely used nowadays in radio-relay link equipment manufacturing to produce a high integration level and for large-scale production. MMIC technology continues its evolution towards applications in higher frequency bands and exploring the integration of low frequency functionalities in the same RF, IF, or BB boards.

Higher and higher circuit integration level are currently under development, not only in BB, but also in IF and RF. This way, programmable architectures have been developed using technologies such as field programmable gate array and application-specific integrated circuit.

4.7 ANTENNA COUPLING ELEMENTS

The output of the power amplifier is connected to the antenna feeder through a series of transmission lines and passive devices. These elements are generally called branching circuits or branching networks. These elements enable the transmission and reception of several radio channels in one or both transmission directions using a single antenna. This function is achieved by proper combination or isolation of the output and input paths of several transmitters and receivers that operate on different channel frequencies.

The racks and frames where branching units are installed usually occupy the largest room volume of a nodal radio-relay link station. This is especially true in IDU indoor radio-relay links that usually have configurations composed of several radio channels. Cables and wave guides that transport RF signals in this type of stations are accommodated on trays along the transmission room. These trays continue horizontally from the building to the antenna tower base. The transmission lines will continue on vertical trays attached to the tower. The total distances are usually large, generally in the order of dozens of meters.

As far as ODU and split unit arrangements are concerned, antennas are very close to the transceivers, so there is little need for RF branching connections and the connection to the feeder can be made with very short length coaxial cable. Manufacturers of ODUs, in many cases, directly incorporate the antenna to the RF unit by proprietary connections, without requiring wave guides or coaxial cables. Connection between indoor and ODUs is made with coaxial cables that also convey power supply, alarms, and management signals.

For the purposes of radio-relay link design, these elements are characterized by their insertion loss. The branching losses will be relevant and should be taken into account for radio-relay link balance calculations when the interface for the power delivered by transmitter is located at the measurement point close to the amplifier output.

4.7.1 RF Transmission Lines

There are two general types of RF transmission lines used in microwave links, coaxial cable, and wave guides. As far as their influence in the radio-relay link energy balance is concerned, it will be required to know their insertion losses that will be a function of unitary attenuation per distance and the path length to the feeder of the antenna. Unitary attenuation is usually given in dB per each 100 m.

4.7.1.1 Coaxial Cables Coaxial cables are transmission lines formed by two concentric conductors spaced between each other by a without losses uniform insulating dielectric and out coated by a plastic cover that provides rigidity and external protection features. Figure 4.23 illustrates a typical coaxial cable.

Electromagnetic fields propagate along the dielectric and current flux is produced over the surfaces of the inner and outer conductors. Coaxial cables have low radiation losses as well as high protection against external interferences.

Losses in coaxial cables increase with frequency, so its use is commonly limited to 3 GHz for large connection distances between the antennae of IDUs. For higher frequencies, coaxial cable is only used for short distance connections, such as feeder connections to the outdoor transceiver units. Special cables suited for each operation frequency are used in these cases. 50-ohm impedance cables are used in most RF applications.

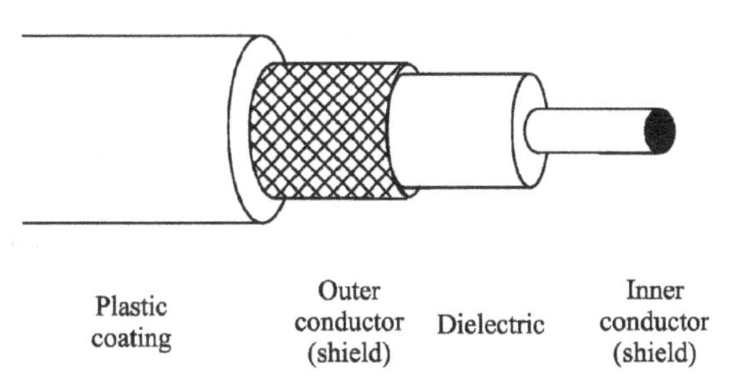

FIGURE 4.23 Diagram of a coaxial cable.

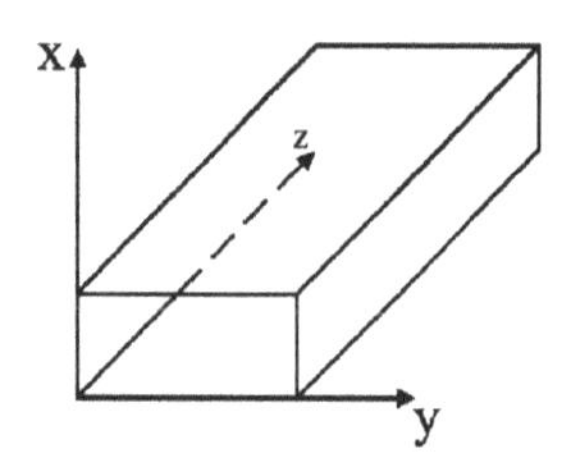

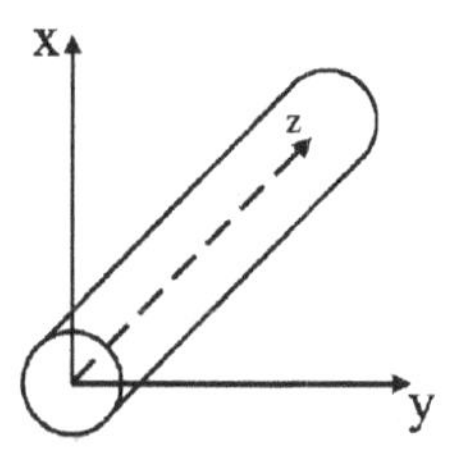

FIGURE 4.24 Rectangular and circular wave guides.

Coaxial cable can be rigid or flexible. Rigid constructions are composed of a rigid insulator and air as a dielectric. Flexible forms have dielectric formed by a flexible material such as polythene.

4.7.1.2 Wave Guides Wave guides are transmission lines formed by an outer metal structure without inner conductor. Energy propagates through variable electromagnetic fields across the dielectric, which is usually air. Wave guides confine the energy avoiding external radiation.

The section shape of a wave-guide conductors can adopt different forms, resulting in the different types of wave guides; rectangular, circular or elliptical. Elliptical and rectangular wave guides manufactured in copper or aluminium are the most frequently used ones, with silver surface layers that reduce losses at high frequencies. Their dimensions depend on frequency, in such a way that as frequency increases necessary size decreases. Figure 4.24 illustrates the shape of circular and rectangular wave guides.

Forms and dimensions of the wave guide must be chosen according to the frequency bandwidth requirements. The propagation along the wave guide is composed of different propagation modes (solutions to Maxwell equations within the wave guide). The modes have associated cut-off frequencies that will restrict the frequency ranges where the wave guide can be efficiently used.

If the system is based on dual polarized or dual band antennas, two wave guides will be required to connect to the antenna, or even better, a wave guide that supports both polarizations (circular wave guide).

Pressurizing pumps that introduce dry air or an inert gas into different wave-guide sections, called pressurization sections are used in order to keep humidity outside the wave guide.

In practice, wave guides are used for frequencies above 3 GHz. They can deliver high power with less attenuation at higher frequencies than coaxial transmission lines. The disadvantages of wave guides are the complexity of the installations. Despite the lower attenuation values, the overall loss might be too large in IDU installations. For frequencies above 10 GHz and if the branching network is long, equipment manufacturers offer alternative solutions based on ODU or split unit arrangements. The IDU section will be connected to a transceiver close to the antenna by a coaxial cable. The signals will travel on IF through the coaxial to/from the transceiver from/to the IDU modules. In summary the path travelled by the RF high-frequency signals will be much shorter (antenna–transceiver path)

4.7.2 Branching Circuits

All passive circuits and devices between the transceiver and the antenna are integrated in the branching unit. Filters, circulators, isolators, and duplexers are included in this category. As far as radio-relay link design is concerned, insertion loss is the most important parameter to consider for all of them.

The particular composition of branching circuits in each radio-relay link depends on the type of protection and diversity configurations used, which will be described later.

4.7.2.1 Filters Branching RF filters are band-pass filters that are arranged on each radio channel branches for transmission and reception. The aim is to ensure that signal spectrum is kept within the limits of the allowed frequency band, reducing the possible interference to adjacent channels in transmission. At the receiver side, their function is to reduce the influence of undesired signals and keep the influence of thermal noise within the minimum required bandwidth.

RF filters are conventionally designed with resonant cavities with reduced sizes and appropriate quality factors. When dimensioning these filters, it should be taken into account the additional filtering stage that will be usually included in the BB stages of the receiver units. When applications require a higher quality factor than those provided by conventional cavities and a reduction in equipment size, dielectric stabilized filters will be used. These devices are based on materials with high dielectric constant and low loss materials.

4.7.2.2 Circulators Circulators are devices with three ports. Each port is associated with one of the other two and the resulting pair is called coupled ports. The signal entering one port of the circulator will be delivered entirely to the coupled port, although the third port will be completely isolated. The third port is called the isolated gate. Circulators are the most widely used devices to enable sharing a single antenna by multiple radio channels.

The directivity of the circulator is the ratio between the power coupled to the corresponding port versus the power delivered to the isolated port (theoretically this factor should be infinite). In a similar way to any passive device, the insertion loss will be the ratio of the incident to the output power ratio in the coupled port.

4.7.2.3 RF Isolators Isolators are two port devices that have low insertion losses when the signal travels from the first to the second port, and high isolation, or high losses, when power enters the second port towards the first one. In this case, the power will be totally dissipated on the device. They are used in radio-relay links in order to protect the RF equipment from possible reflections that may damage transmission units.

Although there are several kinds of isolators, the ones formed by a circulator with an adapted load connected to the uncoupled gate are the most used ones.

4.7.2.4 Duplexers

Duplexers are passive devices that allow sharing a single antenna for bidirectional radio channels, by conveniently isolating the transmission and reception radio channels.

They are usually formed by a combination of circulators and band-pass filters with cut-off frequencies equivalent to the frequency band limits of the transmission and reception channels. They have reduced insertion losses.

From the transmission point of view, the duplexer must be able to manage the power ranges delivered by the RF amplifier. For reception purposes, the duplexer must provide isolation of the receiver in order to avoid coupling of transmitted signals that would cause desensitization. This effect is the reduction of the receiver sensitivity due to power coupled from transmission branches.

4.7.3 Transmission Line Selection Criteria

First, it is necessary to mention that the choice of the transmission line in a link usually depends on the type and model of RF equipment (modem), the branching network and the antenna. The model is fixed by the equipment manufacturer in most of the cases.

Anyway, it is advisable to differentiate between the two installation types: split unit–ODU, and IDU, described in previous sections. As far as ODU installations are concerned, coaxial cables are always used to connect the outdoor RF parts with the modem functional block. Radio channels in this path will travel in a first IF towards the demodulation block. In some cases, the RF unit is not physically attached to the antenna, but installed close to it. In this case, an elliptical wave guide is generally used to link the split unit with the antenna (normally a distance of a few meters).

As far as ODU installations are concerned, equipment is located in a building (shelter) next to the tower where the antenna is installed. In this arrangement, there are different cases to be taken into account in order to choose the line that links equipment and the antenna. The most important features for the choice of a line are:

- Price.
- Attenuation.
- Voltage standing wave ratio (VSWR).
- Maximum power.
- (Mechanical) flexibility.

Depending on the frequency band, some general guidelines can be provided for an optimal choice of the transmission line. We initially assume that 3 GHz is the threshold for distinguishing generic ranges of coaxial and wave guide use.

For frequencies below 3 GHz, the most widespread models are foam (polythene type) dielectric coaxials. Air dielectric coaxial cables have a better response at the expense of a remarkable increase in price, not only associated with the cable itself, but also for installation and maintenance, because, among other things, they need pressurization to keep water vapor outside the cable. When the branching network to the antenna is very long, elliptical guides can be used. These will have considerably less attenuation at the expense of a slight increase in price.

For frequencies above 3 GHz, it is advisable to use elliptical wave guides. The use of coaxial beyond this frequency is mostly restricted to very short sections, where attenuation can be assumed by the system power budget. On the other hand, other types of guides, such as rectangular guide, can be used when bend radius lower than the minimum specified by manufacturer are required. In these cases, a system with straight sections and rectangular guide elbows is usual.

4.8 ANTENNAS

Antennas used in radio-relay links must meet a series of requirements derived from the performance objectives, planning methods and associated propagation disturbances in these point-to-point fixed service systems.

First of all, a transmitter station of a microwave LOS link should be able to transfer as much power as possible to the desired receiver in the direction of the antenna axis. The transmission should avoid energy dispersion in other directions that could create interferences to other systems. Consequently, antennas must be highly directive, which is equivalent to say that their radiation patterns must have a narrow main lobe and a high directivity in the maximum radiation direction. When this characteristic is driven to the limit, and the beams are really narrow, these types of patterns are called "pencil beam" patterns. This term is applied to a model of directive antenna with a single main lobe that is contained within a small solid angle, almost circular and symmetrical with respect to the maximum radiation direction.

Nodal stations usually are convergence sites where different radio-relay links with different antennas share the same tower, and thus probable to produce interfering situations. In order to keep interference from other paths low, the antenna must receive as less quantity of signal al possible in directions different from the one of maximum radiation. This statement implies that antennas must have small secondary, side and back lobes.

Furthermore, antennas must provide a high effective aperture over incident radiation in order to be able to receive as much power as possible. Additionally, according to the wide range of possible radio channel arrangement plans and bandwidths, the performance characteristics must be stable over a considerable bandwidth. All these features should be achieved effectively and maintaining costs as low as possible.

When orthogonal dual polarizations are used in the same frequency, in order to increase communication capacity, antennas must have a high XPD.

The most widely used antennas in microwave radio-relay links, which comply with the previous features, are parabolic reflector antennas, formed by a feed element in the focus of a revolution paraboloid. Yagi and flat plane are additional types that are also used in certain microwave LOS systems.

Yagi antennas are traditionally used in some applications in the lowest frequency bands, below 2.5 GHz, due to their low cost. In the bands that range from 1.4–2.5 GHz Yagi antennas are usually protected from atmospheric inclemencies inside waterproof tubes.

Recent designs of flat-plate antennas have very reduced dimensions (less than 46 mm–2 in.), and similar radiation characteristics to parabolic antennas. For these reasons they are very attractive for applications in congested areas inside cities for frequencies above 23 GHz and short hops.

This section describes parabolic reflector antennas and analyzes the most important parameters from the point of view of the propagation and system design.

4.8.1 Parameters of Parabolic Reflectors

A parabolic antenna consists of a reflector formed by a portion of a revolution paraboloid surface and a primary radiating feeder or source with a phase center located on the geometric focus of the parabolic reflector. Figure 4.25 shows the parameters that define the geometry of the radiant structure.

The revolution paraboloid, as a spatial geometric figure has useful properties for designing and constructing reflector antennas:

- The paths starting at the F focus and reflected on the paraboloid, arrive on a surface, S, perpendicular to the axis of the parabola, with equal distance paths. One reflected wave according to path 1, takes the same time or does it with the same phase, as the waves that go through the paths 2, 3, 4, or 5 in Figure 4.25.
- The angle of incidence of any ray in one point of the parabola is equal to its own angle of reflection.

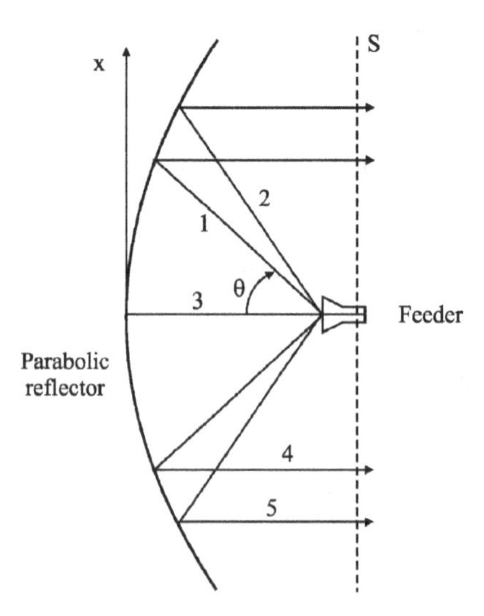

FIGURE 4.25 Geometry of a parabolic antenna.

These geometric characteristics allow transforming the spherical wave radiation (almost omnidirectional) that emerges from the feeder and is reflected on the surface of the paraboloid into a planar (flat) wave front, with parallel-reflected components that will be in-phase when arriving to the focal plane (perpendicular to the antenna revolution axis). The resulting wave front will be a collimated beam of high directivity. In reception, by reciprocity, the plane wave that falls into the reflective surface is reflected in order to focus the received power on the feeder.

From the propagation perspective, the most important parameters of the antenna are: directivity, gain, VSWR, radiation pattern, beamwidth, side and back-lobe levels, XPD, frequency band and mechanical stability.

Parabolic antennas are designed with standard values for these parameters. These parameters might vary within certain limits for each series, depending on each manufacturer. When special values for specific parameters are required, design and manufacturing are more complex and the antennas are called high performance (HP) antennas. The product portfolio of most manufacturers will be divided into these two categories, standard and HP.

4.8.2 Antenna Gain

4.8.2.1 Directivity Directivity of an antenna in a defined direction D, also called directive gain, is the relationship between radiation intensity of the antenna in that direction, and the average radiation intensity in any direction. Directivity is dimensionless and can be written as:

$$D = \frac{U(\theta, \varphi)}{U_o} \tag{4.44}$$

U is the radiation intensity, in Watt/solid angle unit, for a specific direction defined by azimuth θ and elevation φ angles. U_o is the radiation intensity of an equivalent isotropic source in Watt/solid angle unit. The average radiation intensity of an isotropic radiator is:

$$U_0 = \frac{P_{rad}}{4\pi} \tag{4.45}$$

so, directivity can be written as:

$$D = \frac{4\pi U(\theta, \varphi)}{P_{rad}} \tag{4.46}$$

If the direction is not specified for D, the direction of maximum radiation intensity will be assumed, which, for parabolic reflector antennas, coincides with the revolution axis of the antenna. The maximum value of the directivity is usually known as antenna directivity. Directivity is usually handled in logarithmic units, as the difference in dB with respect to the directivity of an isotropic radiator that will be, of course, 0 dB:

$$D_{\text{max}} = 10 \log \left[\max \left| \frac{4\pi U(\theta, \varphi)}{P_{\text{rad}}} \right| \right] \tag{4.47}$$

4.8.2.2 *Gain* The gain of an antenna, traditionally called absolute gain, is a concept that includes the losses of the antenna. It is defined as the relationship between the radiation intensity in a specified direction, and the intensity produced by an isotropic radiator fed with the same power.

As the radiation intensity associated with an isotropic radiator equals to the input power, P_{in}, divided by 4π, gain (dimensionless) can be written as:

$$G = \frac{4\pi U(\theta, \varphi)}{P_{\text{in}}} \tag{4.48}$$

If the antenna has no losses, gain would be equal to directivity for any direction. Moreover, if a direction is not specified, it is understood that gain refers to the direction of maximum radiation intensity, which coincides with the axis in the case of parabolic antennas.

Gain does not take losses due to impedance adjustment problems into account, which cause reflection losses and polarization adjustment issues, so these effects should be studied apart from gain. Gain of a parabolic antenna can be approximated by the following equation, in dBi:

$$G = 10 \log \left(\eta A \frac{4\pi}{\lambda^2} \right) \tag{4.49}$$

where

A = physical aperture area of the antenna (m^2)
η = efficiency of the antenna aperture
λ = wavelength (m)

4.8.2.3 *Aperture Efficiency* The term aperture efficiency aims at quantifying the radiation efficiency of an antenna. The radiation efficiency is associated with different loss mechanisms. Ohmic losses produced in dielectrics are one of the most significant contributions, as well as defects or non-ideal results in reflector manufacturing: misalignments and geometrical deformities on the reflective surface. Another group of causes is associated with the feeder illumination. These include losses due to spillover, produced by feeder radiation outside the physical bounds of the reflector, feeder blockage effects (the feeder element will partially block part of the reflected energy), and losses created by physical elements that hold mechanically the feeder on the focal point. The efficiency is generally in the range from 50% to 60%.

The effective aperture, A_e, of an antenna is defined as the product between the aperture efficiency and the physical aperture of the antenna:

$$A_e = \eta A \tag{4.50}$$

and replacing:

$$g = \frac{4\pi A_e}{\lambda^2} \tag{4.51}$$

The effective aperture of the antenna is the maximum radiation-receiving surface of the real antenna, in a plane perpendicular to the direction of maximum radiation. For a parabolic reflector antenna with circular aperture of diameter D, aperture surface can be written as:

$$A = \frac{\pi D^2}{4} \tag{4.52}$$

and substituting into the gain equation:

$$g = \eta \left(\frac{\pi D}{\lambda} \right)^2 \tag{4.53}$$

Assuming an efficiency of 55%, and writing the diameter of the antenna, D, in meters, and the frequency, f, in GHZ, this equation can be written in dBi as:

$$G = 17.8 + 20 \log (D \cdot f) \tag{4.54}$$

Figure 4.26 shows the gain values (dBi) obtained from equation (4.54) for different parabolic antenna diameters.

4.8.2.4 *Equivalent Isotropically Radiated Power (EIRP)*

The e.i.r.p. (mostly spelled eirp) of a transmitter antenna in a specified direction is the product of gain and net power received by the antenna.

If p_t is power measured at the IF amplifier output, l_{tt} the losses in connection circuits to the antenna, and g the gain of the antenna, in linear units:

$$\text{e.i.r.p.} \quad \textit{or} \quad \text{eirp} = p_t.l_{tt}.g \tag{4.55}$$

and in equivalent logarithmic units:

$$\text{E.I.R.P} \quad \textit{or} \quad \text{EIRP} = P_t. + L_{tt} + G \tag{4.56}$$

Units are usually dBm for E.I.R.P. and P_t, dB for L_{tt} and dBi for G. To convert dBm into dBW we have to subtract 30 dB from the previous formula.

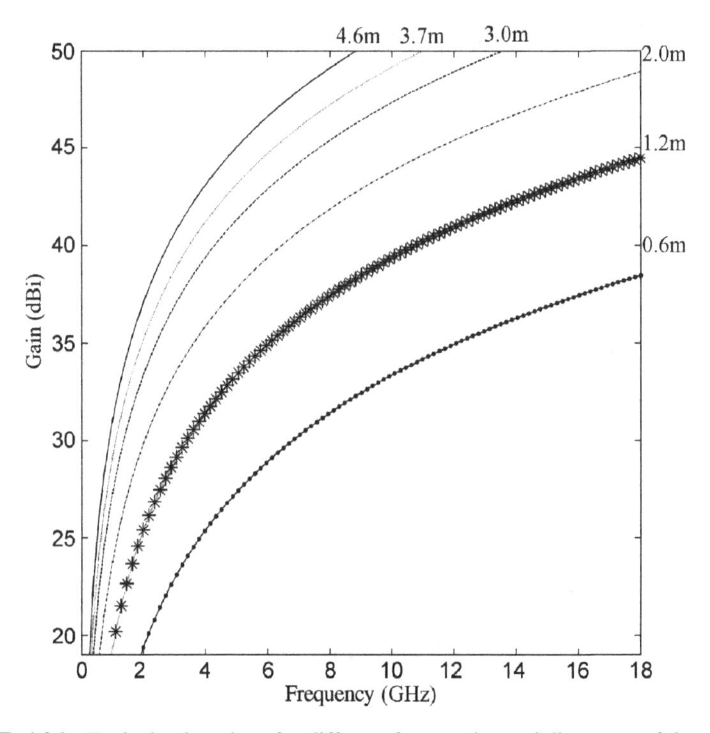

FIGURE 4.26 Typical gain values for different frequencies and diameters of the parabolic antenna.

4.8.3 Radiation Pattern

The radiation pattern of an antenna is a mathematical function or a graph representation describing the spatial variations of the radiation characteristics of an antenna. These radiation characteristics can be the amplitude of the electromagnetic field created by the antenna, power flux density, radiation intensity, directivity, phase, and polarization.

When the radiation pattern shows the amplitude, or relative amplitude, of a particular component of the electric field vector, it is called amplitude or field strength pattern. If it shows the square of the amplitude, it is called power pattern.

The graph representation of the spatial characteristics can be done in two or in three dimensions, in Cartesian or polar coordinate systems and in a linear or logarithmic scale.

Field and power diagrams are usually standardized related to the maximum value, conventionally assigning 0 dB to the maximum radiation direction. For parabolic and Yagi antennas, this will be the axis of the antenna. The rest of values associated with other directions will be negative values in dB. Theoretically, diagrams have axial symmetry, although this is not met in practice, because of irregularities and manufacturing imperfections of the reflector.

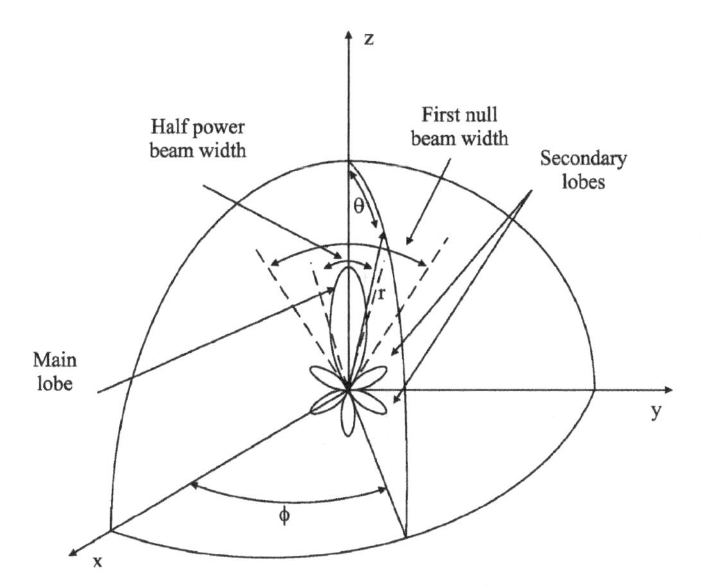

FIGURE 4.27 Coordinate system for a typical radiation pattern.

Figure 4.27 shows a coordinate system used for the three dimensional representation of a typical radiation pattern, where a main lobe and secondary lobes are identified.

The radiation of most antennas is concentrated around specific angular regions, called diagram lobes. The main lobe (also major lobe or main beam) is the one associated with the maximum radiation direction in the axis beam of the antenna. The rest of the radiation lobes of the diagram are called secondary lobes. These can be side lobes or back lobes.

Three-dimensional patterns are not very useful to obtain values and make calculations. They are usually replaced by a series of two-dimensional patterns in the vertical and horizontal planes. These diagrams are obtained by intersection of the three-dimensional radiating function with horizontal planes (usually at elevation zero) and various vertical planes (usually on several azimuths, starting from the azimuth that provides the maximum radiation). This way each one of these two dimensional diagrams will provide the gain or directivity values at any direction contained by the intersecting planes.

The most representative patterns are the ones in perpendicular planes that contain the axis of the antenna, horizontal and vertical, called azimuth and elevation or zenith plane radiation patterns respectively. Figure 4.28 shows a standardized radiation pattern in the azimuth plane typical of a parabolic antenna, with the main lobe pointing to the $\theta = 0$ azimuth. The angles of the horizontal axis represent elevation angles (0° = horizontal direction = revolution axis of the antenna).

Manufacturers of antennas usually give co-polar radiation patterns, which correspond to signals with the same polarization as the feeder, and cross-polar radiation patterns for crossed polarized signals.

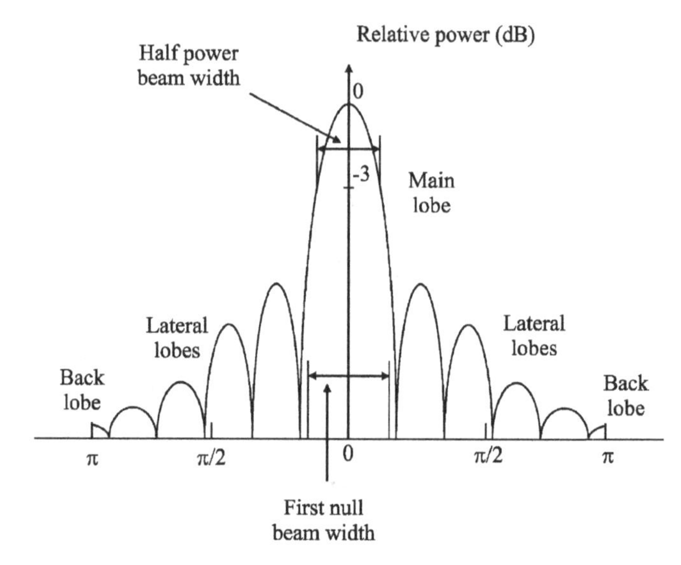

FIGURE 4.28 Typical power radiation pattern.

Antennas of the same series will show a dispersion of their radiation pattern characteristics, simply because of tolerances associated with the manufacturing process and material characteristics. Due to these tolerances and in order to be able to do general performance and interference calculations, the so-called radiation pattern envelopes (RPEs) are used. These are regular and simplified patterns obtained as envelopes of real diagrams measured at different frequencies in antennas of a specific series. A margin of some dB is added for taking into account the ageing process of the antenna. Figure 4.29 illustrates an example of an envelope pattern of a commercial antenna in the L6 band, as well as the measured radiation pattern of an individual antenna of this series. Antenna manufacturers guarantee that none of the antennas of that series model will exceed the RPE in a value that is usually established in 3 dB.

Reference radiation patterns defined in ITU-R F.699 Recommendation can be used in coordination studies and interference evaluation when there is a lack of particular information on the antenna. These patterns are given for antenna diameter to wavelength ratios (D/λ) that are higher or lower than 100. The pattern for frequencies in the range from 1 GHz to 70 GHz, and $(D/\lambda) > 100$ is shown in the following text:

$$\begin{aligned}
G(\varphi) &= G_{\max} - 2.5 \times 10^{-3}\left(\frac{D}{\lambda}\varphi\right)^2 && 0^\circ < \varphi < \varphi_{\mathrm{m}} \\
G(\varphi) &= G_1 && \varphi_{\mathrm{m}} < \varphi < \varphi_{\mathrm{r}} \\
G(\varphi) &= 32 - 25\log\varphi && \varphi_{\mathrm{r}} < \varphi < 48 \\
G(\varphi) &= -10 && 48 < \varphi < 180
\end{aligned} \tag{4.57}$$

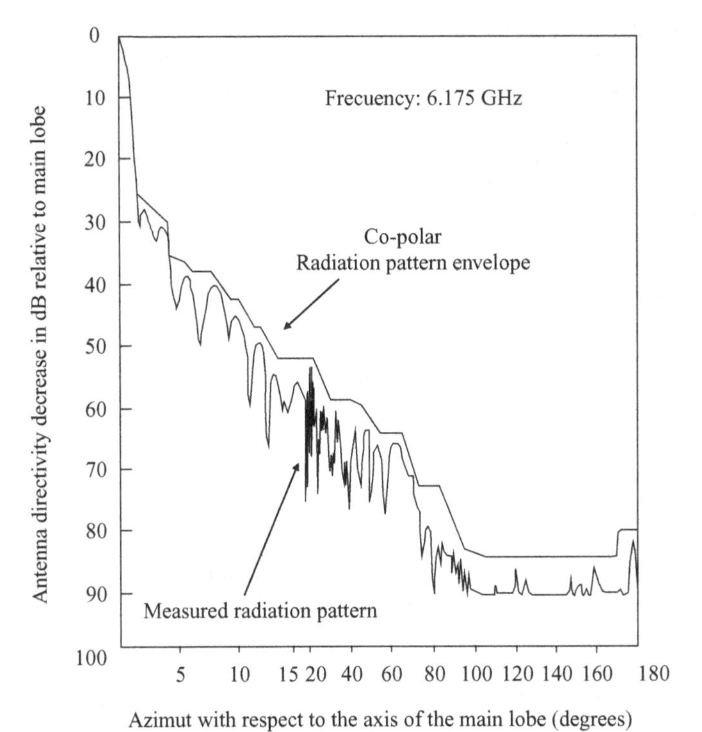

FIGURE 4.29 Radiation pattern envelope (RPE).

where

$G(\varphi)$ = relative gain (dB) of an isotropic antenna
φ = angle related to de axis (degrees)
D = diameter of the parabolic reflector
λ = wavelength
G_1 = gain of the first side lobe

$$G_1 = 2 + 15 \log \frac{D}{\lambda}$$

$$\varphi_{\mathrm{m}} = \varphi_{\mathrm{m}} = \frac{20\lambda}{D}\sqrt{G_{\max} - G_1} \text{ (degrees)}$$

$$\varphi_{\mathrm{rr}} = \varphi_{\mathrm{r}} = 15.85 \left(\frac{D}{\lambda}\right)^{-0.6} \text{ (degrees)}$$

These reference radiation patterns are based on crests envelopes of the side lobes radiation levels. ITU-R F.1245 Recommendation shows a mathematical model for patterns that represent average radiation levels of the side lobes.

4.8.3.1 Beamwidth The half-power beamwidth of the main lobe is the angle between the two directions where radiation intensity is half the maximum value. For patterns in the dB scale, it corresponds to the angle width of the main lobe at −3 dB.

This beam width of the parabolic antenna can be approximately calculated using the following equation, which gives the result in degrees from the wavelength and diameter D:

$$\alpha_{3\text{dB}} \approx 70 \cdot \left(\frac{\lambda}{D}\right) \tag{4.58}$$

Deflection angle at 10 dB at each side of the main lobe axis can be approximated by the following equation:

$$\alpha_{10\text{dB}} \approx 60 \cdot \left(\frac{\lambda}{D}\right) \tag{4.59}$$

Figure 4.30 shows the values for 3 dB and 10 dB beam widths in function of the diameter of the antenna for different frequencies.

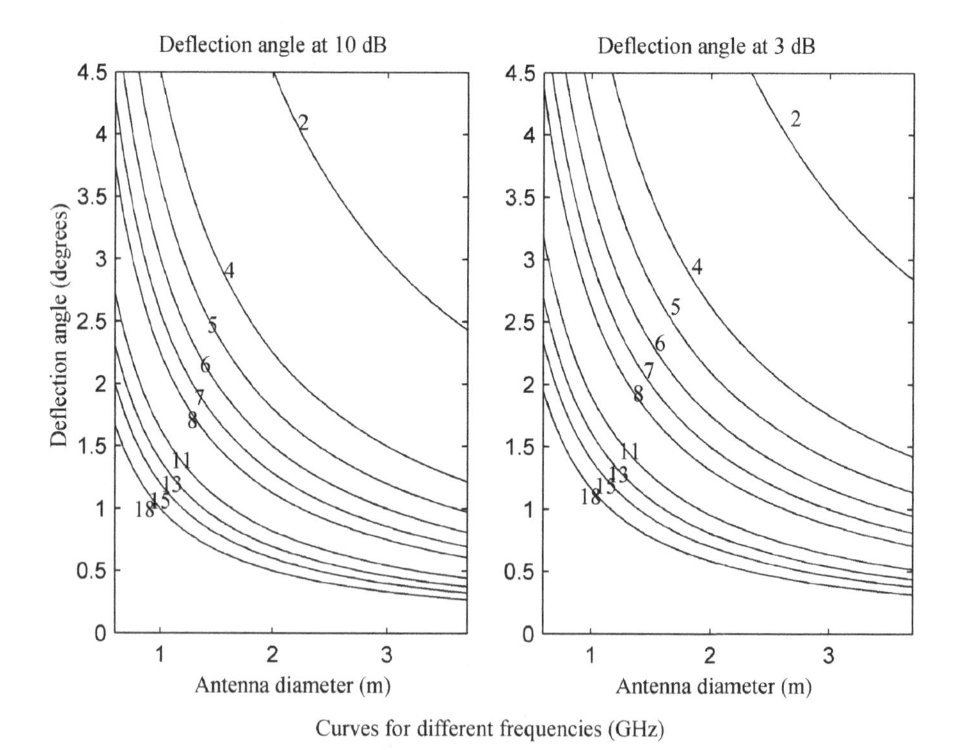

FIGURE 4.30 3 dB and 10 dB beamwidths.

4.8.3.2 Minor or Secondary Lobes

All lobes of the radiation pattern except from the main lobe are generically referred to as minor or secondary lobes. The lobes adjacent to the main one, located in the same hemisphere are called side lobes. Back lobes are radiation lobes that are at approximately 180° respect to the main lobe and, most generically, all those located in the opposite hemisphere of the main lobe.

The level of the side lobes is determined as the relationship of the level associated with the largest minor lobe with respect to the main lobe. These levels are important parameters in frequency plans and interference calculations. A low side and back lobe levels enable a more efficient use of the frequency spectrum, at the expense of more complex and expensive design and antenna manufacturing. These values are usually normalized to the maximum radiation level.

The front/back ratio is the amplitude relationship between the back lobes and the main one. The back lobes will be those associated with angles wider than 90° with respect to the axis of the main lobe. This parameter increases with the frequency and also with an increase in the diameter of the antenna. Side and back lobe level values are usually explicitly specified by the envelope radiation patterns.

4.8.3.3 Field Regions

The previous formulas are valid for distances away from the antenna, in an area called far-field region or Fraunhofer region, where the angular arrangement of electric and magnetic fields is homogeneous and independent from the distance to the antenna.

In the near field region, gain calculation is much more complex and should be provided by the manufacturer of the antenna. It generally changes with distance in an oscillatory manner and it is lower than the one corresponding to the previous formulas for the far-field region. Two near-field regions can be distinguished: reactive near field, in the immediate vicinity of the antenna where reactive fields dominate, and radiant or Fresnel near field, located between reactive and far field regions, where radiation fields dominate and angular field distribution depends on the distance to the antenna. Figure 4.31 illustrates the field regions and qualitative changes of the radiation pattern observed from each region.

The far-field distance from the antenna is:

$$d_f = \frac{2D^2}{\lambda} \tag{4.60}$$

where D is the maximum dimension of the antenna, which is the diameter in the case of the parabolic antenna. Near fields could occur in the reception antennas of short radio-relay links. For example, for a 3 m antenna at 6 GHz, far-field distance is a distance longer than 360 m.

4.8.4 Mismatch, VSWR and Returning Losses

Impedance mismatch between line transmission and antenna causes reflections that produce losses in the final radiating power at the antenna. Reflection mismatch

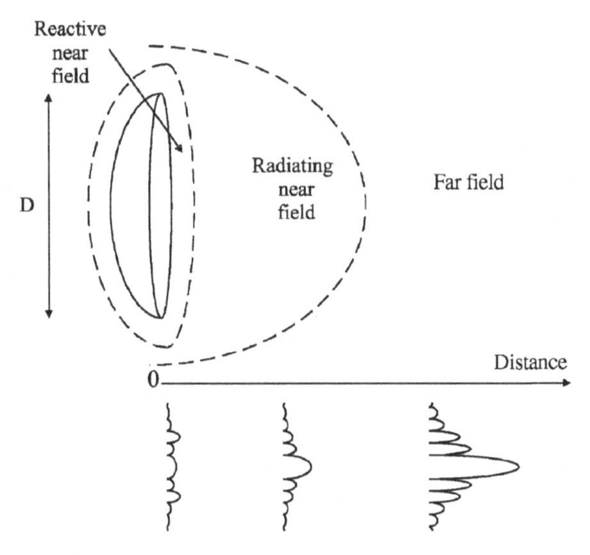

FIGURE 4.31 Field arrangement in near and far field regions.

efficiency, e_r, is usually written as:

$$e_r = (1 - |\Gamma|^2) \tag{4.61}$$

with Γ the voltage reflection coefficient at the input terminal of the antenna:

$$\Gamma = \frac{Z_{in} - Z_o}{Z_{in} + Z_o} \tag{4.62}$$

being Z_{in} the input impedance of the antenna and Z_o the characteristic impedance of the transmission line. The VSWR, is defined as:

$$\text{VSWR} = \frac{1 + |\Gamma|}{1 - |\Gamma|} \tag{4.63}$$

Mismatch is usually evaluated according to the return losses, L_R, which can be expressed in dB:

$$L_R(dB) = 20 \log \left(\frac{\text{VSWR} + 1}{\text{VSWR} - 1} \right) \tag{4.64}$$

The VSWR ratio characterizes an antenna for impedance mismatch purposes and it is important to take it into consideration for high capacity systems with strict linearity requirements. VSWR must be minimized for these systems with the aim of avoiding interferences associated with intermodulation. When VSWR equals the unity, all the power delivered to the antenna is radiated, except from, logically, the losses characterized by the aperture efficiency. When the value of VSWR value is different from the unity, part of this power will be reflected. This part will depend on mismatch level. The VSWR of the standard antennas is typically within the range from 1.06 to 1.15. HP antennas with low VSWR, have typically a VSWR within the range from 1.04 to 1.06.

4.8.5 Polarization

Polarization of an antenna in a defined direction is the polarization of the wave radiated by that antenna, thus, the time varying shape that direction and relative amplitude of the electric field vector describe. It can be represented as the curve drawn by the instant electric field vector along the propagation direction.

Polarization can be defined in terms of the transmitted wave, assuming as polarization of the antenna the polarization of a plane wave evaluated at a location in the far-field zone. In terms of the received wave, it describes the polarization of the incident plane wave at one point with a specified flux density, which provides a maximum available power on the antenna terminals.

Polarization can be classified into linear, circular or elliptical. The electric-field vector, in general, describes an elliptical figure and the field is said to be elliptically polarized. Figure 4.32 shows the diagram of elliptical polarization.

The parameters that characterize elliptical polarization are: polarization direction or rotation direction, electric-field axial ratio and inclination angle. Polarization rotation direction can be either clockwise (right-hand) or counterclockwise (left-hand).

The polarization can be separated into two orthogonal components that represent the horizontal and vertical polarization components that are orthogonal. Linear and circular polarizations can be considered as particular case of the elliptical one, where

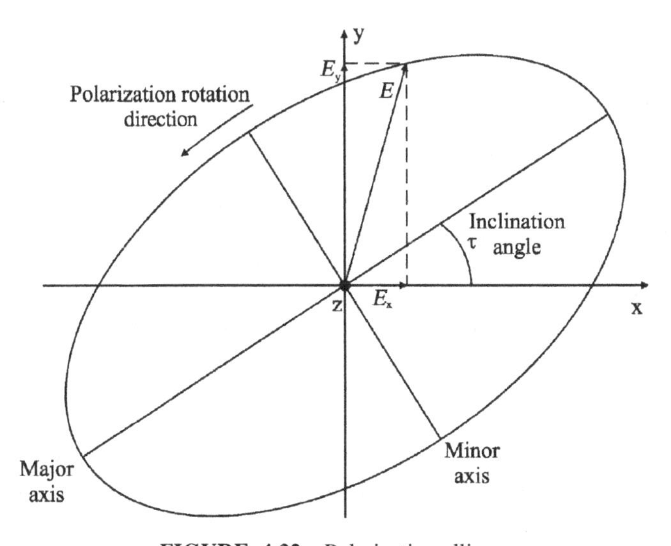

FIGURE 4.32 Polarization ellipse.

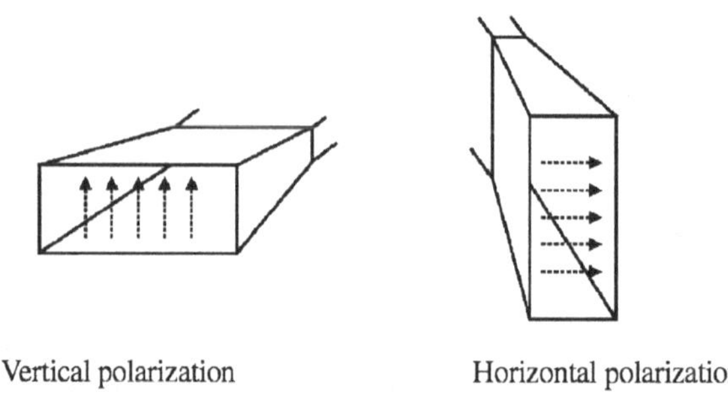

FIGURE 4.33 Polarization of a horn feeder.

the ellipse becomes a line or a circumference. The electric field vector is continuously varying on the same line in linear polarization, either vertically or horizontally. For a parabolic antenna, polarization is defined by the polarization of the feeder. Figure 4.33 shows the vertical or horizontal type of wave polarization in a rectangular horn according to its position with respect to the parabolic reflector axis.

Polarization characteristics of an antenna are represented by the polarization pattern or spatial arrangement of the polarizations, defined in terms of radiated wave as well as received wave.

At any location between transmitter and receiver, the polarization can be regarded as a combination of a pair of orthogonal polarizations: co-polar component, i.e., the polarization that we try to radiate and desire to receive, and cross-polar component, which is orthogonal to the previous one and is usually caused either by equipment non ideal characteristics or due to perturbations of the propagation channel.

If the polarization state of a receiver antenna is not adjusted to receive the maximum received power according to the transmitted polarization, the so-called polarization losses appear due to the polarization mismatch. This mismatch is measured by the polarization efficiency or polarization mismatch factor.

The received signal on the opposite polarization is attenuated by the antenna. Discrimination between the orthogonal components of the polarization (usually one desired and the other one interfering) is called XPD.

Manufacturers of antennas usually give radiation patterns for the co-polar and cross-polar polarization.

In the special case of dual polarized antennas, the patterns provided are more complex than the general single polarized case. The patterns required will be co-polar HH (response of a horizontally polarized antenna to a horizontally polarized signal) and VV (response of a vertically polarized antenna to a vertically polarized signal) as well as cross-polar HV (response of a horizontally polarized antenna to a vertically polarized signal) and VH (response of a vertically polarized antenna to a horizontally polarized signal).

4.8.6 Parabolic Antenna Types

Manufacturers offer a great variety of parabolic antennas for all radio-relay link frequency bands, with a wide range of features. Commercially available parabolic antennas can be classified according to different criteria, each of them with different characteristics and mechanical and electric specifications. Each category may also include antennas of different sizes and for different frequency bands.

Solid antennas (with solid reflectors) and grid antennas (with grid reflectors) could represent a first classification. Solid antennas that exist in the market can basically be classified into the following types: standard, focal plane, shielded, HP, ultra high performance (UHP), and multiband.

All of them can use rectangular, pyramid or conic horn antennas as primary feeders that will be chosen depending on the frequency, polarization, and desired radiation pattern. Most feeders above 3 GHz are rectangular wave-guide flanges. Coaxial feeders with air or foam dielectric are used for frequencies below 3 GHz.

4.8.6.1 Grid Antennas The reflector of these antennas is a metallic grid. These models use mostly vertical foam-filled coaxial feeders, which do not need pressurization. The polarization used is linear, horizontal or vertical depending on the orientation of feeder and grid pattern. These antennas have similar electro-magnetic features (gain, beam width, XPD, front-to-back ratio, and VSWR) to the ones of an equivalent size solid antenna.

Operation frequency bands for grid antennas are low, typically between 300 MHz and 3.5 GHz, with diameters from 1.2 m to 4 m. They are a solution when a low weight and wind load is required in the supporting structure. Typical applications are for low and medium capacity radio systems such those used for Multi-Access Rural Telephony.

4.8.6.2 Standard Antennas They are cheap antennas with solid parabolic reflector, which provide radiation characteristics that are good enough for applications in low interference environments, when a high level of side and back radiation suppression is not necessary. Reflector is usually a spun aluminum reflector, without any kind of coating. They are available for all frequencies and sizes, either single or dual polarized.

4.8.6.3 Focal Plane Antenna The effective aperture area of these antennas is increased making the focus plane of the parabolic surface spread towards the focal plane. They are usually called "deep dish" in North America because the reflector is deeper. Feeder is optimized to the reflector geometry, and they can be either single or dual polarized.

This arrangement provides efficient radiation patterns, improving side lobe suppression and front-to-back ratio if compared with Standard antennas. They are appropriate solutions for high capacity applications where a better front-to-back ratio is a requirement. They are typically used in frequency bands between 1.5 and 8 GHz.

4.8.6.4 Shielded Antenna Antennas for radio-relay links are usually furnished with different protection accessories, external to the radiant structure, which allow an appropriate performance in the outdoor environmental and meteorological conditions throughout its service life. These protection elements improve survivability and radiation characteristics in harsh conditions.

There are two types of protection, external shields and radomes. Shields are protective cylindrical surfaces attached to the reflector edge. They are formed by an RF absorbent material to reduce power reflections. This shielding achieves high values for side and back lobe suppression and optimized radiation patterns.

Radomes are external cases or covers used to protect antennas against the effects of the outdoor environment, ice, snow and dirt accumulation. They also mitigate wind load, without diminishing its electric performance. The radome usually forms an external conic or cylindrical case that covers the paraboloid aperture, in such a way its insertion losses are reduced. It is constructed in fiberglass material or sometimes rubber covered nylon.

These protections are optionally offered in Standard antennas, and are included by default for HP antennas, for all sizes and frequency bands. Currently almost all antennas include radomes, exception made for low cost installations.

4.8.6.5 High Performance Antennas Antennas designed to obtain special values for the radiation parameters are generically called HP antennas. These antennas are furnished with an outer shield and radome. HP antennas have low VSWR values, high front-to-back ratio and side lobe suppression ratios and are well suited for a better frequency coordination. They can be classified into single or dual polarized versions.

For high capacity applications with frequency reuse based on dual polarization techniques, specially designed HP antennas are used, with high XPD values in the azimuth and elevation planes, up to 40 dB below 11 GHz and up to 36 dB with frequencies up to 18 GHz. These features are obtained by using horn feeders with a ring design illumination. These feeders must be manufactured with very tight tolerances and tight quality production control.

Although they are more expensive than standard antennas, they have a low lifetime cost.

UHP antennas are mechanically similar to the HP antennas, but they include feeders with beam-forming characteristics. These antennas include specific feeders to obtain desired patterns. They are used in scenarios with special problems of frequency congestion.

Angular diversity antennas are a special type of antennas for radio-relay links, which have a dual beam either single or dual polarized. Their application field is to provide angular diversity to counteract multipath fading when there is a lack of room in the tower.

Antennas are used for an only band in most cases. There are also multiband antennas that operate simultaneously in two or more frequency bands, with independent polarizations for each band. Their application fields are frequency diversity systems

TABLE 4.11 Frequency Bands and Sizes of Parabolic Reflector Antennas

	Frequency Band							
Parabolic Antenna Diameter (m)	L6	U6	7	11	15	18	27	38
0.2						x	x	x
0.3					x	x	x	x
0.6			x	x	x	x	x	x
0.9	x	x	x	x	x	x	x	x
1.2	x	x	x	x	x	x	x	
1.8	x	x	x	x	x	x		
2.4	x	x	x	x	x			
3.0	x	x	x	x				
3.7	x	x	x					

for a simple channel or simultaneous transmission of several independent channels, when there is congestion in towers.

4.8.7 Microwave Antenna Selection Criteria

The wide range of products and configurations offered by the antenna industry ensures the possibility of an optimum choice of each installation. Factors to consider in the choice of the antenna are: electric specifications (frequency, gain, beamwidth, XPD, front-to-back ratio, and VSWR) and mechanical and environmental restrictions (ODU or IDU configurations, size, weight, wind load, or available room in the tower). Shaping features, among others, must also be analyzed. Selection criteria are an agreement between necessary features and economic cost.

4.8.7.1 Operating Frequency Band and Size The first step when choosing an antenna is to identify the size related to the frequency. Table 4.11 shows frequency and size availability data, obtained from several catalogs of different manufacturers.

4.8.7.2 Gain For a defined frequency band, gain depends directly on the size of the antenna, so, for that frequency, the antenna of smaller diameter that meets gain requirements will be chosen as a general rule. For long hops or hops with relevant losses, the diameter of the antenna will be as high as possible. Manufacturer usually gives values of gain for the lower, half and upper frequencies of the band. In the case of dual polarized antennas, gain is referred to the average value of the two ports.

4.8.7.3 Radiation Patterns The radiation pattern choice will be determined by the interference characteristics in the radio-relay link scenario, which will define the beamwidth and levels of side and back lobes. Manufacturers provide information of the envelope radiation patterns for each type of antenna, which vary with the

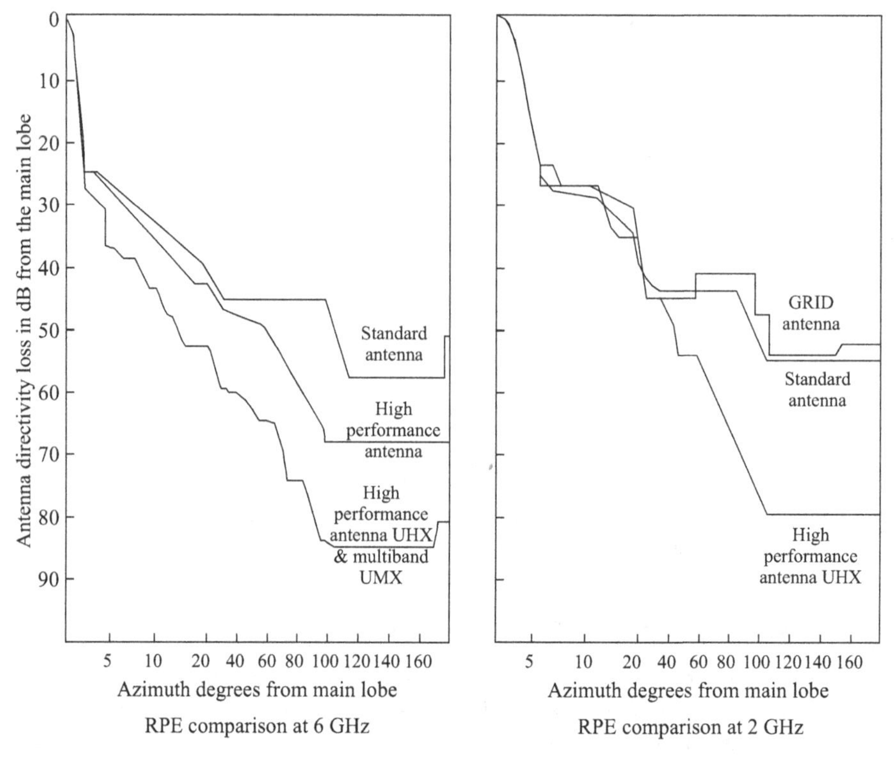

FIGURE 4.34 Comparison among RPEs of several types of antennas: 6 GHz and 2 GHz.

frequency, diameter. Figure 4.34 is a comparison among the envelopes of the radiation patterns of several series of antennas, for 6 and 2 GHz.

4.8.7.4 Polarization The polarization choice will be directly associated with the antenna model selection. Most antenna models for radio-relay links have single and dual polarization versions. It is necessary to choose the type of antenna with an appropriate value of XPD for high capacity applications with dual polarization and frequency reuse. Furthermore, it will be necessary to analyze both the cross-polar RPE and co-polar one in congested environments.

4.8.7.5 VSWR Branching design and the choice of adequate antennas are carried out aiming at minimum return losses in any RF system. On one hand, this criterion allows maximizing power that is delivered to the antenna, and, on the other hand, protecting RF transmission chain against reflections. Furthermore, the lack of remarkable reflections in the antenna is important in order to ensure a maximum isolation between transmission and reception modules.

For these reasons, a limitation in the maximum VSWR value that the antenna will introduce in the system will should be a specification of the installation. Catalogs usually provide the range of VSWR values for the different types of antennas.

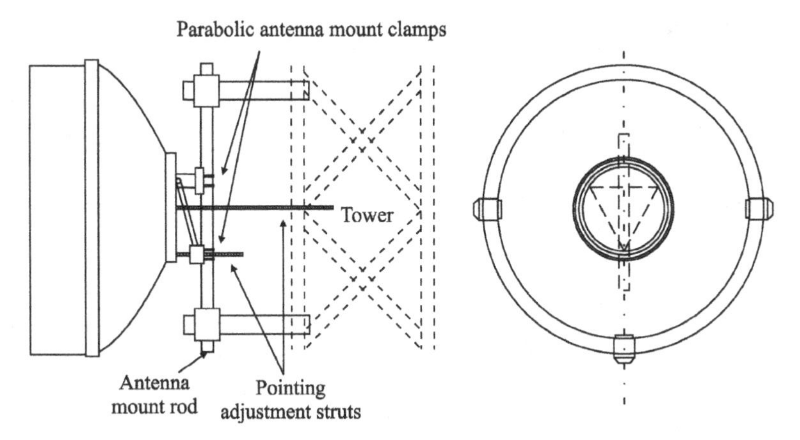

FIGURE 4.35 Installation diagram and attachments of a parabolic antenna.

4.8.7.6 Tower Loading and Environmental Considerations For a specific location and supporting structure, wind load and weight, ice load, and corrosion effects must be taken into account for the choice of the antenna. Placement conditions differ significantly depending on the dimensions of the antenna and the indoor or outdoor arrangement type.

Figure 4.35 illustrates an example of outdoor installation and attachment of a parabolic antenna.

Parabolic antennas are usually furnished with a supporting structure (mounting ring) where fixed and adjustable struts are provided in order to attach the antenna to a vertical rigid cylinder and to rigid parts of the tower where it will be installed. The function of the rigid cylinder is to hold the antenna weight, whereas the adjustable struts provide the mechanism to adjust the orientation of the parabolic antenna adequately. The rigid cylinder will be attached to the tower by adequate brackets and struts.

Wind load is caused by the forces resulting from wind pressure on the antenna and its supporting structure. The swinging and rotation of the antenna and tower structure (tower/mast and antenna) will produce modifications in the pointing direction of the axis (main lobe) of the antenna. Considering that most installations will have high directivity, this effect can cause significant variations of the received level. A maximum value for the signal attenuation due to misalignment of the antenna should be specified prior to the design. According to this value and based on a specific radiation pattern or RPE, the maximum acceptable deviation angle will be determined.

Ice and wind load are usually considered together for the calculation of loads in towers or supporting structures. The amount of ice on the antenna is usually measured by the "equivalent radial ice thickness", which consists of the equivalent thickness that the accumulated ice would have around a cylinder, supposing that the ice would be evenly spread around the cylinder.

TABLE 4.12 Deflection and Survival Wind Characteristics

Parabolic Antenna	Survival Wind (km/h) (mph)	Radial Ice (mm) (In)	Maximum Deflection (degrees) (Wind Speed 110 km/h)
With standard radome	200 (125)	25 (1)	0.1
Without radome	200 (125)	25 (1)	0.1

Table 4.12 shows tolerated values for survival wind and maximum deflection, guaranteed by antenna manufacturers. Values in the table can be considered a standard in use. Accordingly standard solid parabolic antennas must be designed to exceed survival winds up to 200 km/h (125 mph), with an ice radial load of 25 mm (1 inch). This values cause a temporary deflection of the main lobe equivalent to one third of the beam width of the antenna, which involves a decrease of approximately 1 dB on the received signal level.

Some manufacturers offer antenna models specially designed for supporting high wind speeds.

Finally, another aspect to take into account when choosing an antenna and associated options is the corrosive environment the installation can be exposed to, that can long-term affect the good condition of the structure of the antenna. For these cases, special anticorrosive covers must be taken into consideration. Among the used techniques, it is worth mentioning the use of steel parts, covers based on synthetic resins or even galvanized.

4.9 REDUNDANCY ARRANGEMENTS

Microwave LOS links must have high availability and performance characteristics in order to be a competitive option against other alternative transmission systems. In many cases, in order to achieve these availability and performance objectives, it will be required to use redundancy and backup systems for specific elements of the radio-relay link.

Redundancy techniques can be classified depending on the main objective being sought. If the system is being protected against propagation perturbations and anomalies, particularly multipath fading, we will refer to the redundancy as "diversity." On the other hand, whenever the target is counteracting performance and availability anomalies caused by equipment failures we refer to redundancy as "equipment protection." This classification is merely conceptual. Most of the current redundancy arrangements in practice can be used not only to provide diversity, but also to protect equipment against failures and breakdowns, being difficult to establish clear frontiers between both functionalities.

This way, some equipment arrangements created with the conceptual aim of providing diversity, can also provide equipment protection and vice versa.

Diversity techniques are based on transmitting the same information through two different radio paths, which must have poor mutual correlation. If this condition is fulfilled, diversity signals are independently affected by propagation mechanisms.

The radio paths refer to any of the parameters of a radio-relay link (path, frequency, angle or polarization). The effectiveness of a diversity system depends on the achieved correlation degree of the signals that propagate through the different paths. In reception, there will be two or more signals with the same information that will be simultaneously received and must be processed to generate a single output signal.

Diversity techniques can be classified according to the type of diversified radio electric path and the type of choice of the redundant signals. The most common options are space, frequency, angle, time, route, or polarization diversity.

Depending on the type of processing applied to the different signals arriving from the diversity paths, we can talk about switching diversity and combination diversity. Chapter 8 describes these techniques from the point of view of the diversity gain that is obtained in the system.

On the other hand, equipment protection mechanisms have the aim of increasing system reliability. There are several levels of protected elements. Specific sensitive units can be doubled to be replaced in case of degradation or complete protected arrangements can be designed between both ends of the radio-relay link. Complete protection of a radio channel, power supply included, can be directly made over another radio channel of the radio-relay link or over other different physical routes, generally arranged in rings, by network layer protection mechanisms. These two protection mechanisms are illustrated in Figure 4.36.

Generally, a radio-relay link that has M active and N redundant radio channels is called "M + N." In the case of a single active radio channel without redundancy, it is called radio-relay link in "1 + 0" configuration.

To arrange and execute diversity and protection solutions, a series of algorithms and monitoring, telemetry and remote control mechanisms are needed, as well as an action logic with a process unit that controls the performance of switches and combiners. In diversity combination systems, received are processed together in order to obtain a single combined output. The signal processing for combining signals arriving by the diversity paths can be made either in predetection or in postdetection BB. In predetection, combination can be made in RF, or in IF. Combiners may use gain control, maximum power or minimum dispersion algorithms.

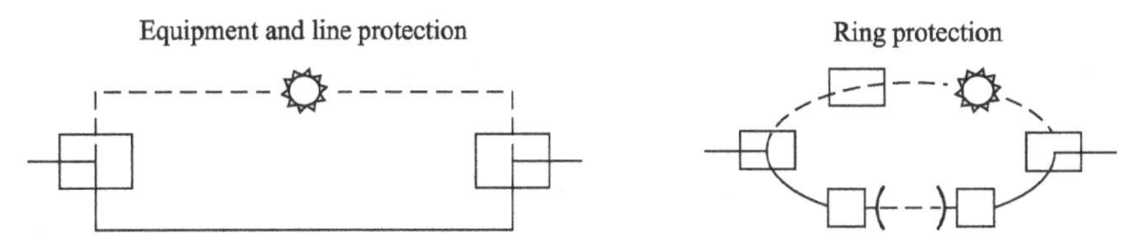

FIGURE 4.36 Different types of complete protection (a) Equipment protection (b) Route protection (ring).

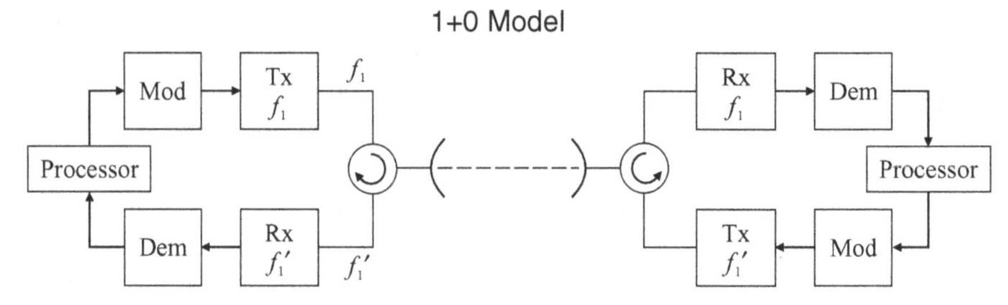

FIGURE 4.37 Diagram of a 1 + 0 configuration.

Active to reserve radio channel switching can be automatic, by the action of a switching logic activated by the signal failure, signal degradation supervision signals or simply by the alarms originated due to local failures in equipment. It can also be started manually, for example in cases where the equipment does not fail, but there is a need to perform maintenance operations without service interruption.

Switching is usually carried out separately for each communication direction and can affect transmission or reception. Switching in reception can be in IF, or in BB. The following subsections develop these concepts and describe the most frequently used configurations and arrangements in systems with equipment redundancy.

4.9.1 Unprotected Configurations

It is the simplest equipment configuration. It is usually referred to as N + 0 and does not provide any kind of equipment protection or diversity. One transmitter, TX, is used for each radio channel, and one receiver, RX, in each direction. Figure 4.37 illustrates a 1 + 0 configuration.

4.9.2 Hot Standby Configuration

In this arrangement a back-up (protection) radio channel equipment operates in the same frequency as the active radio channel. Both transmitters operate permanently, although only one of them will be connected to the antenna. The other transmitter is ready to transmit in case of failure on the active equipment. In failure or breakdown conditions, a switch will change the antenna feed from the active to the reserve radio channel in case of failure or breakdown. As both equipment work on the same frequency, this configuration is also called isofrequency redundancy.

At the receiver side, the system will operate in the so-called active reserve mode. This mode implies two permanently operating redundant receivers on the same frequency. The desired signal will be extracted either by switching or combining the output of the two receivers. Combination methods will usually provide enhanced S/N ratios at the processing unit if compared with a switching technique that just selects the best S/N at each moment. This arrangement can be extrapolated for one reserve and N active channels, N + 1 configurations, and in its most general N + M

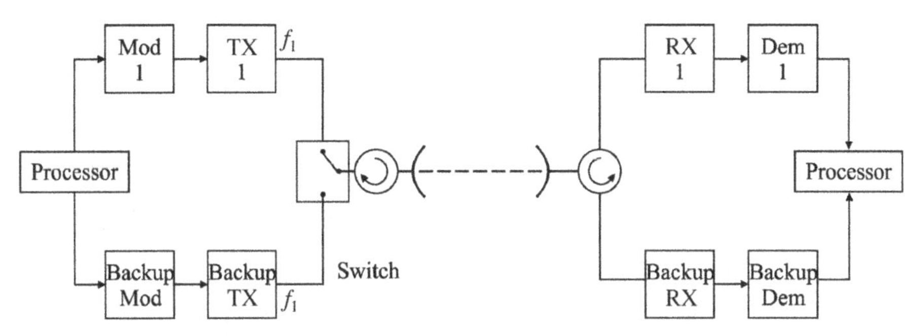

FIGURE 4.38 Diagram of a hot standby 1+1 configuration.

case where N radio channels are active and M will be reserve. Figure 4.38 shows the 1 + 1 configuration.

Each radio channel, in each one of the directions will require two transmitters, two receivers, a switching system in transmission, and a switching or combination system at the receiver side. The system will require only one antenna in the transmitter station, and one antenna in the receiver station.

The switches used are multiline, in such a way that when one of the N operating channels is interrupted, one of the protection channels will immediately recover the signal of the interrupted channel.

4.9.3 Frequency Diversity Configuration

In frequency diversity configurations, the active BB is fed to two transmitters, each one operating in a different frequency. The station will send both signals using the same antenna. This type of arrangement is also called heterofrequency or working standby. On the receiver side, there are two receivers, each one tuned at each frequency of the diversity system. The output signal will be chosen either by switching or by processing, in a very similar way to hot-standby systems. In case of failure of fading in any of the paths, there will be an effective active reserve (the other channel). In the specific case of fading it is expected that this phenomenon does not occur simultaneously on both frequencies. This will be true if there is enough spacing between the pair of frequencies used for diversity. Configurations can also be 1+1 or in general N + 1.

Figure 4.39 illustrates an example of 1+1 system. Chapter 8 will discuss this scheme from the point of view of the protection against fading, referring to the gain that this arrangement introduces. In fact, this arrangement requires doubling RF and IF equipment, exception made for the antenna.

4.9.4 2 + 0 Protection with Radio-Link Bonding

Radio-link bonding systems are high capacity systems that transport a specific amount of traffic by permanently using two or more carriers. This configuration, in practice,

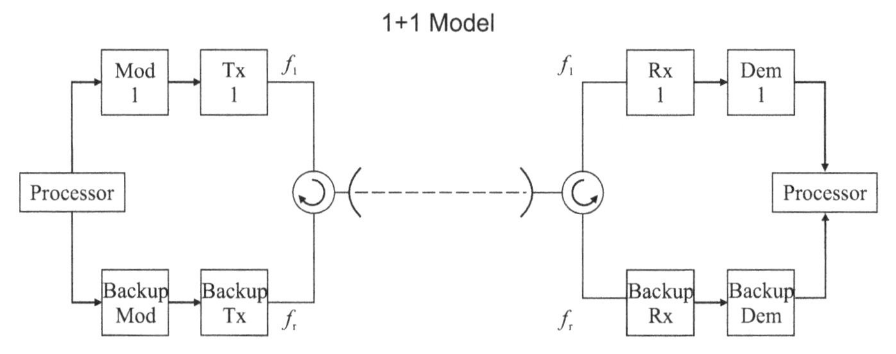

FIGURE 4.39 1+1 frequency diversity.

consists of delivering efficiently the total amount of traffic using two totally independent radio channels. This arrangement can be configured as an efficient line protection. If a radio channel is degraded, the second radio channel will support high priority services. For the two-carrier case, the protection arrangement is referred to as $2 + 0$.

This configuration will be based on carrier frequencies well separated between them in order to avoid any correlation of both radio channels. If the system is being configured for protection against multipath fading, the frequency spacing criteria will be the same as in frequency diversity. For protection against rain, in intense rain regions, different frequency bands must be chosen, for the lower frequency band to ensure high priority services and for the upper frequency band to support less priority services, with a more reduced availability.

These arrangements of priority control of traffic streams usually appear in combination with adaptive modulation schemes, transmission with agile frequency and power control and radiant structures that support it. The group of intelligent functionalities is supported by anomalous situation detection elements and by an acting logic usually based on SDR techniques.

4.9.5 MIMO Configurations

MIMO (multiple input multiple output) systems are techniques widely used in wireless and mobile communications services (3G, 4G, WiMAX, WiFi). MIMO techniques applied to fixed service point-to-point microwave links is known as LOS (line of sight) MIMO.

A MIMO NxN system is formed by N transmitters and N receivers usually transmitting N different signals. In the simplest case of a 2×2 configuration, the system is formed by two transmission antennas, spaced between each other d_1, and two in reception with a d_2 spacing between each other. The MIMO system will be composed by two transmitters and two receivers that will transmit and receive two different signals through different propagation paths at the same frequency. The basic operation principle of the LOS MIMO technique is to achieve a difference between the

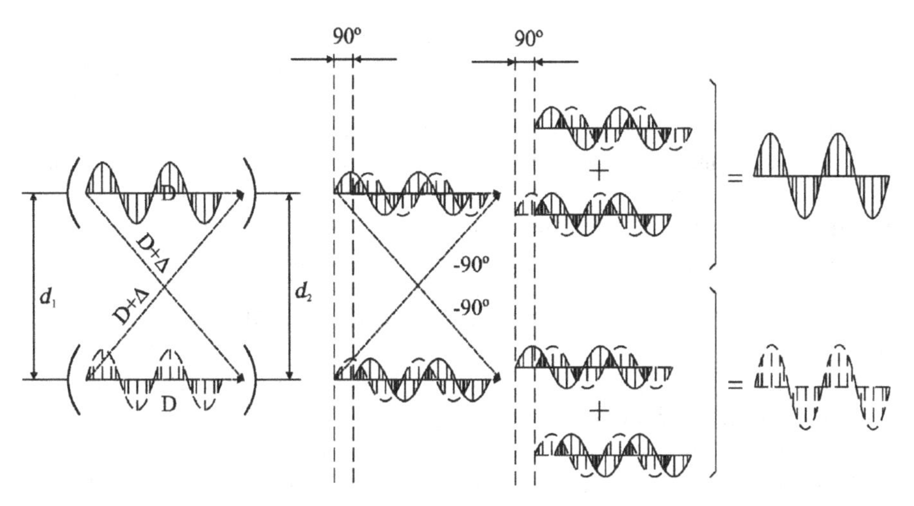

FIGURE 4.40 Basic operation principle of 2 × 2 LOS MIMO.

travelled routes that make the two signals be orthogonal in the receiver. For a 2 × 2 configuration, the aim is that the relative phase between each other is 90°. This way interference would be theoretically cancelled and the transmission capacity would be doubled, if compared to the one of a simple link. In LOS MIMO systems based on directive antennas, a phase difference of 90° is achieved by an appropriate spacing between the antennas. Figure 4.40 illustrates the basic operation principle of the LOS MIMO technique for a 2 × 2 configuration.

The following equation shows the necessary relationship between the distances of transmitter and receiver antennas:

$$d_1 d_2 = \frac{Dc}{2f} \tag{4.65}$$

where f is the operating frequency, c is the speed of light and D is the hop length.

With the aim of increasing the capacity of radio-relay link systems while maintaining the channel bandwidth, combined techniques for increasing spectrum efficiency can be used. An example would be that of dual carrier configurations, where MIMO LOS techniques are combined with dual polarization and XPIC processing techniques.

4.10 SYSTEM MONITORING AND MANAGEMENT

Radio-relay link systems have a group of monitoring and management tools and mechanisms in order to operate properly. The aim is to obtain as much information as possible about the state of the radio-relay link at a specific moment as well as make maintaining operations easier.

Traditionally, these systems have been centralized, particularly in networks with a great number of radio-relay link hops, monitoring the network from a central point by different proprietary protocols of each manufacturer, with client/server architectures. These management systems automatically collect continuous alarm data, equipment state and performance from each one of the units forming the radio-relay link. This information can be either stored to be processed when maintenance and re-planning are required or it may be used to take immediate action. The management systems demand a continuous stream of telemetry information with access to the different monitored remote elements from the control system.

Integration with other superior management systems is made through interfaces with standardized protocols.

Operation and maintenance applications include different functions that are organized in modules: failure management, performance management, configuration management, and BER verification.

The failure management function must enable the identification of the specific location where an element or unit has anomalous performance. This function should also include the activation of the necessary correction measures. The status of network alarms can be shown in many different ways depending on its severity and priority, as graphics or lists.

The system performance management function collects, stores and processes performance data of every element of the radio-relay link transmission network and produces quality reports adapted to the operation, maintenance and planning functions. BER testers are used to manage the performance quality. These allow to measure performance of E1 frames according to ITU-T O.151 specifications, with the aim of verifying system performance or locating failures.

The configuration management function allows configuring network elements remotely, giving a permanent inventory of the network equipment.

All these management functions can be integrated in an embedded application to control elements on a local basis. The access to this embedded application could be done directly from a primary center with Ethernet or universal serial bus connections. Modern radio-relay links more and more use remote access to the element local management function through data communications network based on IP technology, by using a standard web browser. Authorized users and accessibility profile control system are used for guaranteeing access security.

BIBLIOGRAPHY

Error Correction Coding: Mathematical Methods and Algorithms. T. K. Moon . John Wiley & Sons, Hoboken, NJ. 2005.

IEEE Standard for Information technology. Telecommunications and information exchange between systems. Local and metropolitan area networks. Specific requirements. Part 3: Carrier Sense Multiple Access with Collision Detection (CSMA/CD) Access Method and Physical Layer Specifications. Institute of Electrical and Electronic Engineers. IEEE. Piscataway, NJ. Modified in 2008.

ITU-R Rec. F.699: Reference radiation patterns for fixed wireless system antennas for use in coordination studies and interference assessment in the frequency range from 100 MHz to about 70 GHz. International Telecommunication Union. Radiocommunication Sector. ITU-R. Geneva. 2006.

ITU-R Rec. F.1101: Characteristics of digital fixed wirelcss systems below about 17 GHz. International Telecommunication Union. Radiocommunication Sector. ITU-R. Geneva. 2004.

ITU-T Rec. G.703: Physical/electrical characteristics of hierarchical digital interfaces. International Telecommunication Union. Telecommunication Standardization Sector. ITU-T. Geneva. 2001.

ITU-T Rec. G.811: Timing characteristics of primary reference clocks. International Telecommunication Union. Telecommunication Standardization Sector. ITU-T. Geneva. 1997.

ITU-T Rec. G.823: The control of jitter and wander within digital networks that are based on the 2048 kbit/s hierarchy. International Telecommunication Union. Telecommunication Standardization Sector. ITU-T. Geneva. 2000.

ITU-T Rec. G.824: The control of jitter and wander within digital networks thatare based on the 1544 kbit/s hierarchy. International Telecommunication Union. Telecommunication Standardization Sector. ITU-T. Geneva. 2000.

ITU-T Rec I.150: B-ISDN Asynchronous transfer mode functional characteristics. International Telecommunication Union. Telecommunication Standardization Sector. ITU-T. Geneva. 1999.

ITU-T Rec I.361: B-ISDN ATM layer specification. International Telecommunication Union. Telecommunication Standardization Sector. ITU-T. Geneva. 1999.

ITU-T Rec. I.363.1/2/3/5: B-ISDN ATM adaptation layer (AAL) specification. Types 1, 2, 3/4 and 5. International Telecommunication Union. Telecommunication Standardization Sector. ITU-T. Geneva. 1996.

ITU-T Rec.Y.1413: TDM-MPLS network interworking – User plane interworking. International Telecommunication Union. Telecommunication Standardization Sector. ITU-T. Geneva. 2004.

ITU-T Rec. Y.1453: TDM-IP interworking – User plane interworking. International Telecommunication Union. Telecommunication Standardization Sector. ITU-T. Geneva. 2006.

MEF 8. Implementation Agreement for the Emulation of PDH Circuits over Metro Ethernet Networks. Metro Ethernet Forum. MEF. -. 2004.

RFC 4717. Encapsulation Methods for Transport of Asynchronous Transfer Mode (ATM) over MPLS Networks. Internet Engineering Task Force. IETF. -. 2006.

The ComSoc Guide to Next Generation Optical Transport: SDH/SONET/OTN. Huub van Helvoort. Wiley-IEEE Press. 2009.

ITU-R Rec. F.1245: Mathematical model of average radiation patterns for line-of-sight point-to-point radio-relay system antennas for use in certain coordination studies and interference assessment in the frequency range from 1 GHz to about 70 GHz. International Telecommunication Union. Radiocommunication Sector. ITU-R. Geneva. 2000.

Understanding SONET/SDH and ATM: Communications Networks for the Next Mellennium. Stamatios V. Kartalopoulos. Wiley-IEEE Press. 1999.

CHAPTER 5

PERFORMANCE OBJECTIVES AND CRITERIA FOR FIXED SERVICE MICROWAVE LINKS

5.1 INTRODUCTION

The quality of service provided by a microwave LOS link system can be evaluated by different parameters and associated calculations and design algorithms. Depending upon the source taken as reference, the parameters used to measure the quality of service of a link are the outage probability, fidelity, performance, availability, or reliability. All of them provide a measure of the system performance at different levels and under different failure conditions. This chapter describes the system performance specific parameters and objectives elaborated by the International Telecommunications Union, Telecommunication Sector (ITU-T) and International Telecommunications Union, Radiocommunications Sector (ITU-R) whereas Chapter 8 will include these parameters and objectives in the system design process.

The International Telecommunications Union (ITU) has defined a set of quality objectives for end-to-end connections that are based on the concepts of availability and error performance. The objectives are used both as requirements for system performance prediction in the design phases and quality measurement parameters when the link is operative. The ITU provides also the distribution criteria to allocate performance objectives to links that compose different sections of the whole end-to-end system (access, inter-office, international link, etc) according to predefined transmission models that will be also described in this chapter.

Availability accounts for the periods where the microwave link system is either working well below minimum performance thresholds and thus, transmitted data

Microwave Line of Sight Link Engineering, First Edition. Pablo Angueira and Juan Antonio Romo.

are completely corrupted and also periods where any of the link equipment fails. Unavailability could be described as the measure of the "out of service" condition. Obviously, ITU defines also the criteria to evaluate when a communications link and specifically is "in-service" and "out of service," conditions that will strongly depend on the transport protocols and technology being used.

The second parameter associated with ITU objectives is error performance. Error performance objectives (EPO) define the minimum system quality during the "in-service" conditions. The system performance observation time is divided into two parts: the time periods where the connection provided by the microwave LOS link is available and the time periods where the connection is unavailable. The EPO is only evaluated during availability periods, and thus, establishes a performance threshold for the "in service" condition. In both cases, availability and system performance, the definition of objectives will be based upon maximum time percentages where the parameters utilized for measuring quality exceed the degradation thresholds.

The first ITU Recommendation relative to error performance was the ITU-T G.821 published in 1980 that included parameter definitions and system thresholds based on bit error rate (BER) values. This recommendation defines the error performance of an international digital connection, part of the Integrated Service Digital Network (ISDN) and operating below the primary rate (2048 Kbit/s).

The updated version of Rec. ITU-T G.821 is still on use for links based on equipment designed before the revision made on December 14, 2002, that originated the new Recommendation ITU-T G.826. Any link working below the primary rate based on equipment designed after the date mentioned above will be dimensioned based on ITU-T G.826.

Recommendation ITU-T G.826 also defines performance objectives for bitrates equal or higher than the primary rate, exception made for synchronous digital hierarchy (SDH) paths that use equipment designed after March 2000. In that case, ITU-T Rec. G.828 applies. Both ITU-T G.826 and G.828 provide error performance limits based on errored bit blocks and not on BERs.

ITU-T Series G recommendations mentioned in previous paragraphs apply to any digital link, independent of the physical media used to deliver the information. The ITU-R develops the general conditions provided by ITU-T and applies the system performance criteria to the specifics of microwave LOS links and other fixed systems in the Series F Recommendations.

The error performance events and objectives in microwave LOS links with equipment designed before December 2002, are developed by Recs. ITU-R F.594, ITU-R F.634, ITU-R F.696 and ITU-R F.697. In any other case, the system performance objectives are described by ITU-R F.1668.

Regarding availability, initial criteria were based on the same parameters developed by ITU-R G.821. The content of this recommendation was upgraded by ITU-T G.827 that included availability parameter definitions that apply to any digital fixed link, independently of the transmission media (fiber, metallic cables or wireless links). For specific use in real microwave LOS links, the recommendations ITU-R F.1492 and ITU-R F.1493 established the availability objectives and associated distribution criteria for national and international sections of the link path. Finally,

Recommendation ITU-R F. 1703 unified the availability criteria, and remains as the only reference today for designing any fixed wireless link according to availability objectives.

In real practice, there are also cases where the quality and availability objectives will be defined by the operator of the link according to criteria different from those provided by ITU and usually based on the specifics of the application and traffic conveyed by the link. Chapter 8 will include alternative objective requirements as another option for design procedures.

The objective of this chapter is to explain and provide synthetic use guidelines of the concepts and objectives related to error performance and system availability contained by ITU recommendations that will be later used for system design purposes (Chapter 8) and system monitoring and operation (Chapter 9).

5.2 ERROR PERFORMANCE OBJECTIVES AND CRITERIA BASED ON RECOMMENDATION ITU-T G.821

5.2.1 Application Scope of Related Recommendations

The Recommendations relative to ITU-T G.821 for microwave LOS link design are ITU-R F.594, F.634, F.696 and F.697. They are all based on the parameter and error event definitions described in ITU-T G.821 and defined as a function of the BER. All these recommendations apply to systems where the bitrates are below the primary rate of the ISDN (2048 kbit/s). Table 5.1 summarizes the regulatory framework defined by these recommendations.

The recommendation ITU-T G.821 defines the parameters and EPOs for international digital connections that operate below the primary rate of the digital hierarchy, that is, N $\times$ 64 kbit/s, and $1 \leq N \leq 24$ or $1 \leq N \leq 31$. This recommendation applies currently only to systems based on hardware designed before December 14, 2002.

The Recommendation ITU-R F.594 provides EPOs of the hypothetical reference digital path (HRDP) for microwave LOS systems in connections at a bit rate below the primary rate and forming part or all of the high grade portion of an ISDN.

The recommendations for digital real links forming part of international digital connections below the primary rate are ITU-R F.634 for links on the high-grade portion, ITU-R F.696 for those on the medium-grade portion and finally F.697 for links part of the local-grade portion.

5.2.2 Error Performance and Associated Definitions

A relevant contribution of Recommendation ITU-R G.821 is the description of the error events and parameters for evaluating error performance.

5.2.2.1 Error Performance Events An event is associated with a specific error condition. An event is defined by the intensity, characteristics or degree of the

TABLE 5.1 Regulatory Framework Defined by Recommendations Derived from ITU-T G.821

Recommendation	Year	Application	Superseded
ITU-T G.821	1988 2002[a]	HRX	ITU-T G.826 (2002)
ITU-R F.594	1997	HRDP systems providing connections at a bit rate below the primary rate and forming part or all of the high-grade[a] portion of an ISDN	ITU-R F.1668 (2005)
ITU-R F.634	1997	Real digital radio-relay links forming part of the high-grade[b] portion of international digital connections at a bit rate below the primary rate within an ISDN	ITU-R F.1668 (2005)
ITU-R F.696	1997	HRDSs forming part or all of the medium-grade[c] portion of an ISDN connection at a bit rate below the primary rate utilizing digital radio-relay systems.	ITU-R F.1668 (2005)
ITU-R F.697	1997	Real digital radio-relay links forming part of the local-grade[d] portion at each end of an ISDN connection at a bit rate below the primary rate	ITU-R F.1668 (2005)

[a] The last update of this recommendation was approved in 2002.
[bcd] Local-grade, medium-grade, and high-grade definitions can be found in Rec. ITU-T. G.821 and will be described in detail in the further sections of this chapter.

error condition over a specific observation period (usually 1 s). Recommendation G.821 defines two different events:

- *Errored second (ES)*: It is a one-second period in which one or more bits are in error or during which loss of signal (LoS) or alarm indication signal (AIS) is detected.
- *Severely errored second (SES)*: It is a one-second period which has a bit error ratio $\geq 1 \cdot 10^{-3}$ or during which loss of signal (LoS) or AIS is detected.

5.2.2.2 *Error Performance Parameters* The error parameters are those used to determine the availability and/or EPOs. The error parameters are based on error events. According to the events described in previous paragraphs there are two different types of error parameters:

- Errored second ratio (ESR): The ratio of ES to total seconds in available time during a fixed measurement interval.
- Severely errored second ratio (SESR): The ratio of SES to total seconds in available time during a fixed measurement interval.

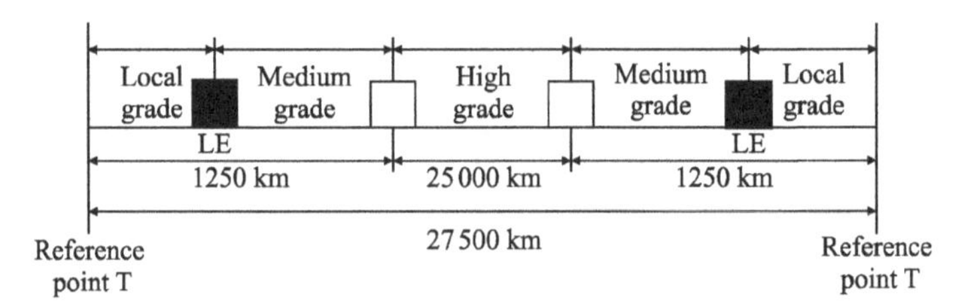

FIGURE 5.1 ITU-T G.821: Circuit model for defining error performance (longest possible connection 27 500 km).

Previous versions of the recommendation included an additional event: the degraded minute. A degraded minute is a period of 1 min where the BER exceeds 1.10^{-6}. Degraded minutes did not include the SES periods. In practice this event was hardly used in reality, and consequently it was removed from the recommendation in 2002.

5.2.3 Reference Transmission Models for Performance Objective Allocation

Recommendation ITU-T G.801 defines a digital transmission model called hypothetical reference connection (HRX). HRX is an end to end connection with a maximum length of 27 500 km intended for the study of performance, objective degradations and quality objectives that will apply to the connection between end terminals.

Figure 5.1 shows a HRX, completely digital connects two terminals according to ITU-T G.801. The model was created in the context of ISDN, for circuit switching connections operating at 64 kbit/s as the reference to allocate EPOs to the different sections of any end-to-end connection.

This model includes three different sections of the digital circuit. Those sections are independent of the transmission systems involved, and are designated as local-grade, medium-grade and high-grade. Each one of the sections will be defined as either one of the three categories depending upon its location along the circuit connection (local offices, international centers, etc).

Once the sections are defined, the allocation of performance objectives can be done as a function of the distance or as a straight allocation, whichever the distance. In the case of high-grade sections, the EPOs is done as a function of the link distance (between 2500 and 280 km), where as for medium-grade and local-grade sections, the EPOs are fixed allocations independent of the circuit length.

5.2.4 Objectives and End-to-End Error Performance Allocation in Real Links

ITU-R recommendations F.594, F.634, F.696 and F.697 were created from ITU-T G.821 with the objective of providing guidelines for system design and performance evaluation in real microwave LOS links. Table 5.2 summarizes the EPOs (using SER, SESR) of the different portions in an international ISDN connection HRX. The

TABLE 5.2 Error Performance Objectives and Allocation Over an International ISDN Connection

Circuit Classification	Error Performance Recommendation ITU-T G.821		Error Performance Digital Radio Relay System	
	ESR	SESR	ESR	SESR
Local-grade (global end-to-end margin)	0.012	0.00015	0.012 Rec. ITU-R F.697	0.00015 Rec. ITU-R F.697
Medium-grade (global end-to-end margin)	0.012	0.00015	0.012 Rec. ITU-R F.696	0.0004 Rec. ITU-R F.696
High-grade *25 000 km* *2 500 km*	0.032 0.0032	0.0004 0.00004	0.0032 Rec. ITU-R F.594, ITU-R F.634	0.00054 Rec. ITU-R F.594ITU-R F.634
International ISDN Connection 27 500 km	< 0.08	< 0.002 (0.001 + 0.001)		

table also contains the objectives that were specified for digital radio relay systems (DRRSs) (microwave LOS links) and the associated recommendation.

As an example, equation (5.1), shows one of the cases included in ITU-R F.634 that provides the following EPO values for a real microwave LOS link of length L, part of the high-grade section of the HRX model (values expressed as time percentage of any month):

$$\mathrm{SESR} = \begin{cases} \dfrac{L}{2500} 0.000054 & 280 < L(\mathrm{km}) \leq 2500 \\ \left[0.0005 + \dfrac{L}{2500} 0.00004\right] & 2500 < \mathrm{L(km)} \end{cases}$$

$$\mathrm{ESR} = \frac{L}{2500} 0.0032 \tag{5.1}$$

5.3 ERROR PERFORMANCE OBJECTIVES AND CRITERIA BASED ON RECOMMENDATIONS ITU-T G.826 AND ITU-T G.828

5.3.1 Application Scope

Recommendations ITU-T G.826 and ITU-T G.828 were elaborated to replace ITU-T G.821, widening at the same time the application scope of the latter. There are two situations where ITU-T G.826 applies:

- International digital connections at bitrates below the primary rate, that is, bitrates lower than N $\times$ 64 kbit/s ($1 \leq N \leq 24$ or 31) based on equipment designed before December 14, 2002.
- International digital paths operating at constant bitrates equal or higher than the primary rate. In this case, a second classification is made depending upon the system type: Plesiochronous digital hierarchy systems (PDH), or synchronous digital hierarchy systems (SDH). In the case of SDH systems, ITU-T G.826 only applies if the equipment was designed before March 2000.

The connections refer to the same concept as ITU-T G.821 in HRX (ISDN N $\times$ 64 kbit/s connections between end-to-end terminal equipment), whereas the digital paths refer to the model called hypothetical reference path (HRP). A HRP is defined as the whole means of digital transmission of a digital signal of specified rate including the path overhead (where it exists) between equipment at which the signal originates and terminates. An end-to-end HRP spans a distance of 27 500 km. HRDP are composed of all the transmission equipment and infrastructure required to convey a specific bitrate between signal source and destination.

ITU-T G.828 applies to transmission paths of the SDH that use equipment designed after the recommendation publication date (March 2000). Both ITU-T G.826 and G.828 can be applied to any digital reference connection, independently of the transmission technology or media.

In order to adapt the criteria provided by G.826 and G.828 to microwave LOS links, the ITU-R developed in 2005 the ITU-R F.1668. This recommendation contains updated information on EPOs for real digital fixed wireless links used in HRX and HRDP models. ITU-R F.1668 is currently the only recommendation defining EPOs for all real digital fixed wireless links designed after 2000 (after 2002 for links below the primary rate) and substitutes Recommendations ITU-R F.1397 and ITU-R F.1491.

5.3.2 Error Performance and Associated Definitions

The definitions of parameters in digital connections operating below the primary rate of the digital hierarchy are based on BER measurements and error measurements with observation times of 1 s. In the case of higher bitrates and where performance specifications are very demanding, BER values are not accurate enough to evaluate the quality, performance or availability of the connection.

Digital transmission systems are usually based on the coding, switching, and transmission of blocks of bits. In conclusion, the use of error performance criteria based on blocks and errored block rates is inherently advantageous. The definitions of the error parameters in digital paths at rates higher than the primary rate will be based on block rates. Thus, the erroneous blocks will be easily detected by the error detection and correction modules, that, additionally will be a relevant advantage for "in service" measurements. This fact constitutes a relevant advantage if compared with criteria based on BER values that require the transmission of dedicated frames for accurate measurements and cannot be done without interrupting the service.

TABLE 5.3 Anomalies and Defects in SDH Systems According to ITU-T G.826

Anomaly	
a_1	Errored frame alignment signal
a_2	EB as indicated by an EDC
Defect	
d_1	LoS
d_2	AIS
d_3	LOF

EB, errored block; LOF, loss of frame alignment, LoS, loss of signal

A block is defined as the set of consecutive bits associated with a certain path, where each bit belongs to one and only one block. The bits consecutive bits that form a block may not be contiguous in time. A common case is that of systems using time interleaving in cases where long dropouts are expected. Each block is monitored by means of an inherent error detection code (EDC), e.g., bit interleaved parity or cyclic redundancy check. The EDC bits are physically separated from the block to which they apply. It is not normally possible to determine whether a block or its controlling EDC bits are in error. If there is a discrepancy between the EDC and its controlled block, it is always assumed that the controlled block is in error.

The definition of error performance events in ITU-R G.826 includes the concepts of "anomaly" and "defect," in addition to generic "erroneous bit." Anomalies and defects are introduced in order to provide information about the relevance of different error occurrences for the system performance. An anomaly consists of a discrepancy between the expected and received value of a bit or a bit block. An anomaly does not necessarily produce a degradation of any of the functions of the system. If the frequency of anomalies is high or if any of the functions of the system is affected by anomalies, the problem is referred to as "defect." Anomalies and defects are usually defined in relation to specific transmission systems. As an illustrating example, Table 5.3, shows the anomalies and defects associated with SDH systems as described by ITU-T G.826

5.3.2.1 Error Performance Events for Connections (HRX Model)

The definitions provided by ITU-T G.821 for error events in end-to-end digital connections are also part of the content in ITU-T G.826. It should not be forgotten that N $\times$ 64 kbit/s systems below the primary rate after 2002 will not be longer designed on the basis of G.821 but G.826. Thus, similarly to ITU-T G.821 the following parameters are defined:

- *ES*: It is a one-second period in which one or more bits are in error or during which loss of signal (LoS) or AIS is detected.
- *SES*: It is a one-second period that has a bit error ratio $\geq 1 \cdot 10^{-3}$ or during which loss of signal (LoS) or AIS is detected.

5.3.2.2 Error Performance Events for Paths (HRDP Model) In the case of HRDP the events are based on block errors. The associated definitions are as follows:

- *Errored Block (EB)*: A block in which one or more bits are erroneous.
- *ES*: A one-second period with one or more errored blocks or at least one defect. Defects will be error events that make any of the functions of the system fail.
- *SES*: A one-second period that contains at least a 30% of errored blocks or at least one defect. SES is a subset of ES.
- *Background block error (BBE)*: Erroneous block not part of a SES.

5.3.2.3 Error Performance Parameters Once the events have been defined, the error performance parameters are described as a function of their appearance rate on the system. It is important to note that these parameters will only be applied in error performance studies only in the case of system availability. Unavailable periods will be eliminated from the path performance analysis and will be treated separately with a specific associated availability objective:

- *ESR*: The ratio of ES to total seconds in available time during a fixed measurement interval. This parameter is applicable to both paths and connections.
- *SESR*: The ratio of SES to total seconds in available time during a fixed measurement interval. This parameter is applicable to both paths and connections.
- *BBER*: The ratio of BBEs to total blocks in available time during a fixed measurement interval. The count of total blocks excludes all blocks during SESs. This parameter is applicable only to paths.

Although path oriented ESR and SESR in ITU-T G.826 are similar in concept to the equivalent connection oriented ESR and SESR in ITU-T G.821, BBER is a new concept. The objective of this parameter is to provide a tool that describes the system performance reference in absence of relevant perturbation sources, whichever the origin of the perturbation might be.

Microwave LOS link manufacturers, when referring to BBER, specifications usually refer to this condition as unfaded system performance.

5.3.3 Reference Transmission Models for Performance Objective Allocation

Reference transmission models are hypothetical entities (networks) with defined length and composition. These models are intended to generalize the links and paths in any network and thus enable the provision of parameters and quality performance criteria that can be easily adapted to the variety of network architectures, transmission links and paths that can be found in reality.

Additionally, transmission models provide a framework to differentiate generic sections in communication networks (access sections, international paths, interoffice

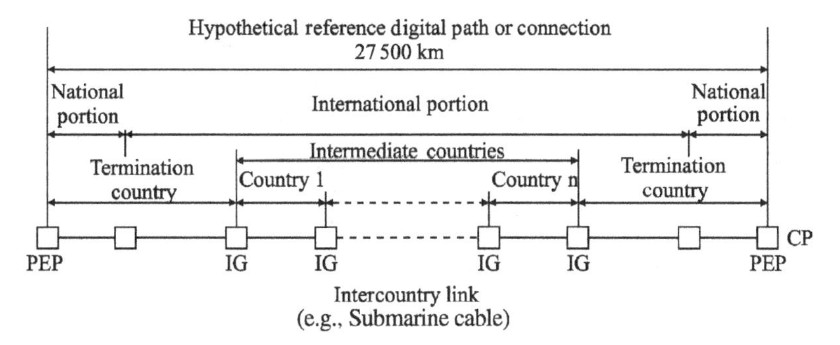

FIGURE 5.2 Hypothetical reference connection or path.

paths), in order to make an allocation of the global EPOs to the different sections of a generic network. Recommendations ITU-T G.826 and G.828 are based on the following models:

- *HRX*: Hypothetical reference connection.
- *HRDP*: Hypothetical reference digital path.
- *HRDS*: Hypothetical reference digital section.

HRXs are used to define and distribute the EPOs for any of the directions of the communication in a N $\times$ 64 kbit/s ($1 \leq N \leq 24$ or ≤ 31) circuit switched connection. Figure 5.2 shows a representation of the hypothetical reference model for EPOs in circuit switching connections (HRP or HRX depending on the system bitrate, above and below the primary bitrate respectively).

The boundary between the national and international portions in ITU-T G.826 is defined to be at an international gateway (IG) which usually corresponds to a cross-connect, a higher-order multiplexer or a switch. The model assumes a maximum of four intermediate countries and sections between countries (i.e., submarine cables). IGs are always terrestrial equipment physically resident in the terminating (or intermediate) country.

The HRDP is composed of all the digital transmission media to convey a signal with a defined bitrate, including the path overhead (in case it is present) between the origin and destination equipment. An end-to-end HRX spans a distance of 27 500 km. The design objectives provided by the ITU-T for transmission equipment are usually expressed in terms of the maximum allowable degradation in the HRDP. There are some variations from the general definition:

- *Digital paths*: A digital path can be unidirectional or bidirectional and may include the infrastructure in customer premises and the path owned by the network operator.
- *Digital PDH paths*: Digital paths part of a PDH network, where ITU-T Rec. M.60 applies.

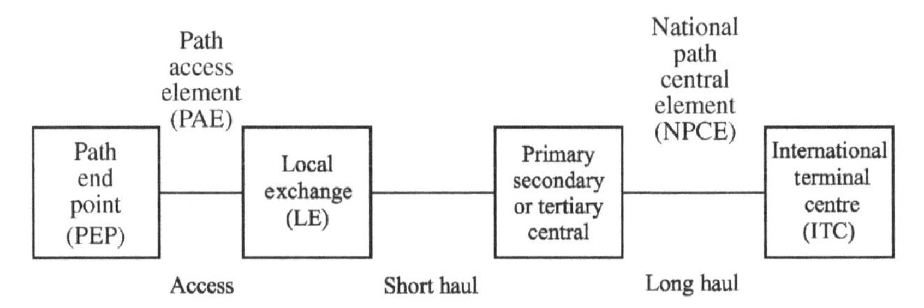

FIGURE 5.3 Basic sections in a national portion of a HRDP.

- *Digital SDH Paths*: A SDH digital path is a trail carrying the SDH payload and associated overhead accross the layered transport network between the terminating equipment.

The national portion of a 27 500 km HRPs or HRXs can be subdivided into three basic sections: access, short haul and long haul. Figure 5.3 represents these sections with usual interfaces. These interfaces that compose the network model, usually exchanges, vary from country to country.

The ITU provides a general definition of the access network from the connection end path end point (PEP) to the first node of the network local access switching center/cross connector. The next network section, short haul, refers to the connections between the local exchange and higher order switching centers—primary center (PC), secondary center (SC), or tertiary center (TC)—that will depend on each network architecture. Finally the long-haul section includes the connections between a PC, SC or TC and anIG.

The third generic network structure is the HDRS. This concept is defined in ITU-T G.801 and is associated with a limited part or section of a HRDP. HDRS usually have lengths that are similar to the real link lengths (50–280 km) and do not include multiplexation/demultiplexation stages. This structure is not used by current designs based on ITU-R F.1668.

In order to provide a summary that illustrates the concepts of HRX and HRDP associated with different real networks, Figure 5.4 shows the different parts of a generic path between two terminating countries. The architecture of each one of the networks has been simplified in order to identify each one of the sections of the path (access, short haul and long haul).

5.3.4 Objectives and End-to-End Error Performance Allocations in Real Links

Recommendation ITU-R F.1668 is the reference for EPOs in real links based on equipment designed after December 2002. This recommendation substitutes ITU-R F.1397 and ITU-R F.1491.

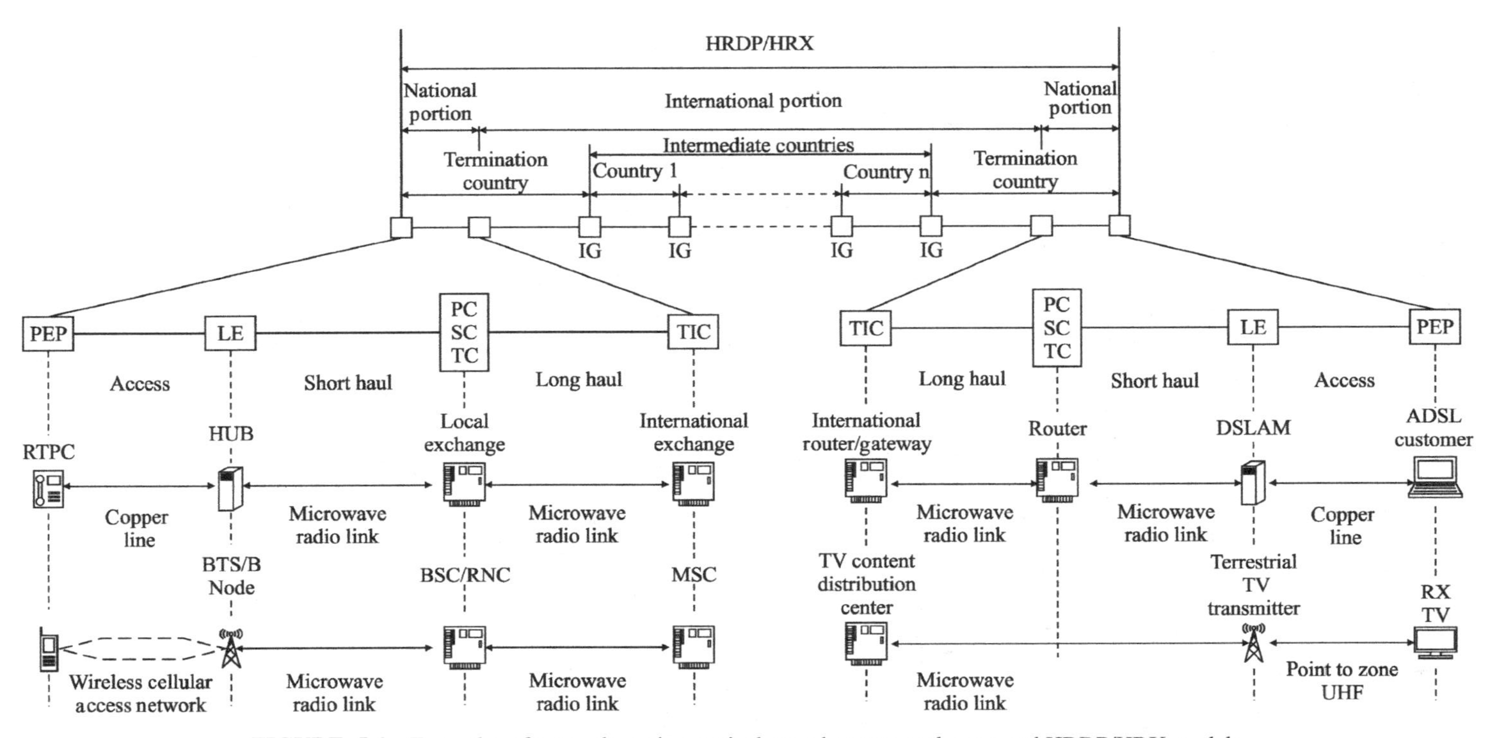

FIGURE 5.4 Examples of network section equivalences between real cases and HRDP/HRX models.

The quality objectives are defined in this Recommendation as EPO and are based on the event and parameter definitions provided by ITU-T G.826 and G.828. The EPO are calculated and allocated following different criteria according to a number of factors:

- *Link positions in the HRDP*: links part of the national or international portions of the HRDP.
- *Links position in the international portion*: international portion within the terminating country or in an intermediate country.
- *Links part of the national portion*: access, short haul and long haul.
- *Transport technology*: ITU-T G.826 and ITU-T G.828 (PDH and SDH).

In any of the possible cases, the EPO are described in terms of error performance parameters ESR, SESR, and BBER.

5.3.4.1 Links in the International Portion If the link being designed or under supervision is part of the International Portion, the EPO will depend on the link distance. The procedure to obtain EPO values is summarized in Tables 5.4 and 5.5. The values will vary depending on the ITU-T recommendation that applies to the specific design: ITU-T G 826 and G.828. Table 5.4 applies to designs based on ITU-T G.826, that will be used for all PDH links and SDH links that use equipment designed before 2002.

Table 5.4 contains an assignation that is a function of the link length (L_{link}). This length has a minimum value of $L_{\min} = 50$ km. If the real length is shorter, the value in equations (5.2) and (5.3) will be 50 km. Parameters B and C are constants defined within the recommendation (see Appendix I). L_{R} is the reference length, assumed 2500 km. An additional tolerance adjustment is applied to B and C though the parameter B_{R} that is known as the block tolerance. B_{R} can take any value between 0 and 1.

In the case of PDH, each hierarchy level has its own particular EPO. Specifically the ESR will vary depending upon the bitrate ranges 5–15, 15–55, and 55–160 Mbit/s.

TABLE 5.4 Recommendation ITU-R F.1668: Application Guidelines for Links in the International Portion in the Scope of ITU-T G.826

ITU-T G.826 – International Portion			
Intermediate Country		Terminating Country	
EPO depends on the link distance: $\text{EPO} = B_{\text{j}} \times (L_{\text{link}}/L_{\text{R}}) + C_{\text{j}}$ (5.2)		EPO depends on the link distance: $\text{EPO} = B_{\text{j}} \times (L_{\text{link}}/L_{\text{R}}) + C_{\text{j}}$ (5.3)	
$L_{\min} < L_{\text{link}} < 1000$	$L_{\text{link}} > 1000$	$L_{\min} < L_{\text{link}} < 500$	$L_{\text{link}} > 500$
$j = 1$	$j = 2$	$j = 3$	$j = 4$
Table 2a in Rec. ITU-R F.1668 (See Appendix I to this chapter)		Table 2b in Rec. ITU-R F.1668 (See Appendix I to this chapter)	

TABLE 5.5 Recommendation ITU-R F.1668: Application Guidelines for Links in the International Portion in the Scope of ITU-T G.828

ITU-T G.828– International Portion			
Intermediate Country		Terminating Country	
EPO depends on link distance: $\text{EPO} = B_j \times (L_{\text{link}}/L_R) + C_j$ (5.4)		EPO depends on link distance: $\text{EPO} = B_j \times (L_{\text{link}}/L_R) + C_j$ (5.5)	
$L_{\min} < L_{\text{link}} < 1000$	$L_{\text{link}} > 1000$	$L_{\min} < L_{\text{link}} < 1000$	$L_{\text{link}} > 1000$
$j = 1$	$j = 2$	$j = 3$	$j = 4$
Table 1a in Rec. ITU-R F.1668 (See Appendix I to this chapter)		Table 1b in Rec. ITU-R F.1668 (See Appendix I to this chapter)	

On the contrary, SESR and BBER do not depend on system throughput. The maximum bitrate for which the specifications apply in this case is 400 Mbit/s.

SDH systems after 2002 are designed according to the procedure on Table 5.5. Parameters B_j, C_j, L_{link}, L_R, $L_{\min}$ and B_R have the same definitions as in the previous case. The EPO values will also depend on the bitrate for SDH systems. The dependency in this case is as follows:

- *ESR*: Different values for each one of the bitrates: 1664, 2240, 6848, 48 960, 150 336 Mbit/s.
- *SESR*: A single EPO for the whole range 1664–150 336 Mbit/s.
- *BBER*: Different values for two ranges: 1664–48 960 and 48 960–150 336 Mbit/s.

5.3.4.2 Links in the National Portion

If the microwave LOS link system being designed or under supervision is part of the national portion of the HRDP, the first factor to take into account is the section of the networks where the link is installed. The EPO for the access and short haul use a block-based allocation that is independent of the link distance. In the case of long haul, there is a mixed allocation procedure, with a portion function of the distance and another one that depends on the link length. Following a similar scheme to the one in the previous section, Table 5.6 shows a summary of the EPO calculation procedure, for PDH and SDH equipment before 2002 (G.826) and later SDH systems (G.828).

The EPO of the access section will depend on the bitrate of the system under study as well as on the parameter C that varies between 0.075 and 0.085. The value of C in each case depends on each administration or operator, and has implications on the overall allocation of performance objectives over the whole HRDP.

The case of short-haul links is very similar. In this case the EPO will depend on the link bitrate for systems based on procedures of the G.826 and G.828. There is also an adjustment parameter for this case, B, identical to the one in access links (ranges from 0.075 to 0.085).

Links part of the long-haul section are composed of two contributions. The first part is a constant factor that depends only on the system technology and bitrate. A part

TABLE 5.6 Recommendation ITU-R F.1668: Application Guidelines for Links in the National Portion in the Scope of ITU-T G.828 and G.826

National Portion					
Access		Short Haul		Long Haul	
EPO independent of the distance		EPO independent of the distance		EPO depends on link distance, bitrate, and parameter A $A = (A_1 + 0.002)\ L_{\text{link}}/100$; for 50 km $\leq L_{\text{link}} \leq$ 100 km $A = A_1 + 2 \times 10^{-5}\ L_{\text{link}}$; for 100 km $< L_{\text{link}}$	
G.826	G.828	G.826	G.828	G.826	G.828
Table 5b ITU-R F.1668	Table 5a ITU-R F.1668	Table 4b ITU-R F.1668	Table 4a ITU-R F.1668	Table 3b ITU-R F.1668	Table 3a ITU-R F.1668

is referred to as A, and in turn is composed of a correction factor A_1 between 0.075 and 0.085. In the same way as previous cases, there will be different EPO values for each system technology (PDH, SDH). In both cases, ESR and BBER depend on the bitrate, whereas SESR objectives are the same for any rate. Additionally, there will be two scenarios in long-haul links: L_{link} longer or shorter than 100 km, with specific values for both cases.

5.4 AVAILABILITY CRITERIA

Unavailability is defined by ITU as the time percentage in which the link is out of service due to equipment failures or propagation perturbations with significant duration in time. Interruptions of the energy supply and simultaneous failure of backup subsystems are also a cause of unavailability if they are long enough in time.

In the case of degradation caused by propagation anomalies, those must be severe enough so the link connection is considered interrupted. The criteria to define availability and unavailability conditions are defined in Recommendations ITU-T G.826, G.827 and G.828.

5.4.1 Availability Criteria for Unidirectional Systems

A period of unavailable time begins at the onset of ten consecutive SES events. These 10 s are considered to be part of unavailable time. A new period of available time begins at the onset of ten consecutive non-SES events. These 10 s are considered to be part of available time. Figure 5.5 illustrates the concept.

5.4.2 Criteria for Bidirectional Paths/Connections

A bidirectional path or connection is in the unavailable state if either one or both directions are in the unavailable state. Figure 5.6 shows an example.

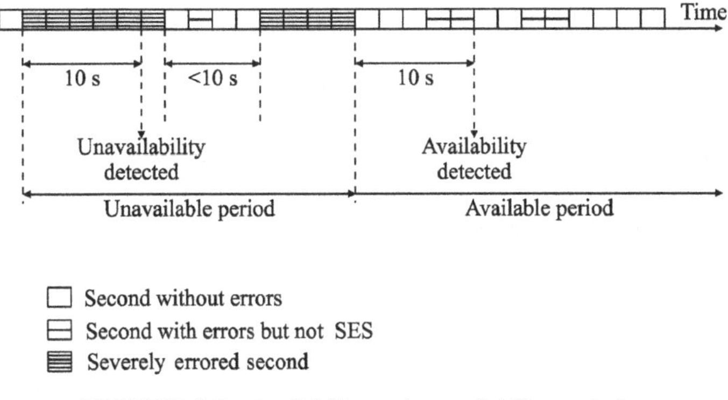

FIGURE 5.5 Availability and unavailability periods.

The available time is obtained adding all the available time periods starting from the beginning of the observation period, for any observation period choice.

5.4.3 Availability Parameters

Availability is described by two inter-related parameters. The first one is the availability or availability ratio (AR) defined as the time percentage in which the system is available with respect to the total generic observation period.

The second parameter is the unavailability ratio (UR) which is defined as the portion of the time in which the system is unavailable during an observation period. The UR is obtained dividing the total unavailable time over the duration of the observation period. Any of both parameters, UR and AR are used for defining design, measurement and maintenance objectives. In any case, the sum of UR and AR over the same observation period should equal one.

$$\mathrm{AR} + \mathrm{UR} = 1 \tag{5.6}$$

During the stages of the link design it will be necessary to differentiate between the unavailability portions due to propagation anomalies U_{P} and those degradations caused by equipment failures U_{E}, so the total unavailability can be expressed as a

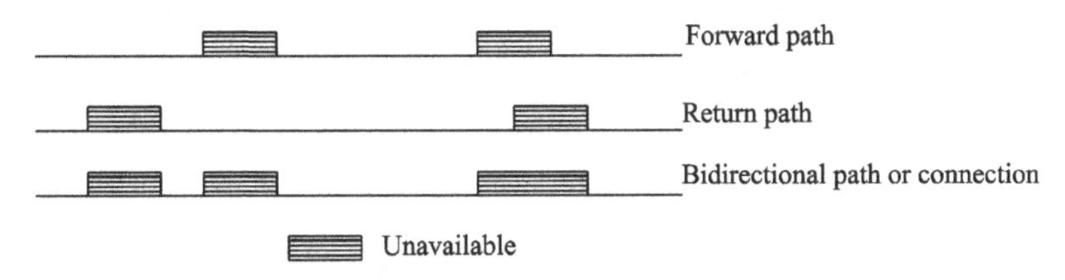

FIGURE 5.6 Availability and unavailability periods in bidirectional systems.

function of both contributions:

$$U_{\mathrm{TOTAL}} = U_{\mathrm{P}} + U_{\mathrm{E}} \tag{5.7}$$

The mean time between interruptions of the digital path, mean outage (MO), is the mean duration of any continuous available time period. The outage intensity (OI), is obtained from MO as the inverse of the MO:

$$\mathrm{OI} = \frac{1}{\mathrm{MO}} \tag{5.8}$$

The total AR of the microwave link, AR can be obtained following the expression (5.9):

$$\mathrm{AR} = 1 - \left[\frac{(T_1 + T_2 - T_{\mathrm{b}})}{T_{\mathrm{e}}}\right] \tag{5.9}$$

where

T_1 = overall unavailable time in one of the transmission directions
T_2 = overall unavailable time in the other transmission direction
T_{b} = bidirectional unavailable time
T_{e} = observation time

for unidirectional systems $T_{\mathrm{e}} = T_2 = 0$.

When estimating availability, any relevant nonintentional failures causes that can be statistically predicted have to be taken into account, including those originated in the radio equipment, power sources, propagation, interference, auxiliary infrastructure and human activity. The estimation of unavailability will include in the calculation the estimation of the mean time to bring the system back into service after failures.

5.5 AVAILABILITY OBJECTIVES FOR MICROWAVE LOS LINKS DESIGNED BEFORE 2005

5.5.1 Basic Concepts and Definitions

Recommendations ITU-R F.557, ITU-R F.695, ITU-R F.696 and ITU-R F.697 are the references for availability calculations in microwave LOS link systems designed before the approval of ITU-R F.1703 in 2005.

Specifically, availability objectives for HRXs and HRDPs are described by ITU-R F.557 according to the same definitions for available and unavailable times as shown in previous sections. The model used for availability allocation purposes is shown in Figure 5.7. It should be noted here that the model is exactly the same as the one intended for EPOs and can be found in ITU-T G.821.

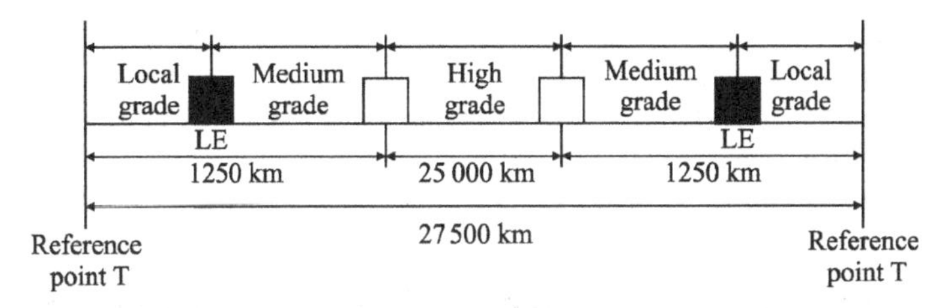

FIGURE 5.7 HRX model for availability calculations (ITU-T G.821).

Recommendations ITU-R F.695, F.696 and F.697 specify the availability objective allocations to different sections of the connection or the path (local grade, medium grade and high grade). The definitions of events and parameters for error characterization have already been described in detail in previous sections and correspond to those on Recommendations ITU-T G.821 and ITU-T G.826.

5.5.2 Availability Objectives in Real Links

This section presents a summary of the availability objectives on use for microwave LOS links designed before the publication of ITU-R F.1703 in 2005. As in sections describing EPO values for links before 2002, the application of ITU-R F.557, F.695, F.696 and F.697 is very scarce today and even less probable in the future.

Digital microwave LOS links part of the High Grade of the ISDN are defined by ITU-R F.557 and F.695. These recommendations propose a preliminary availability value for high grade sections equivalent to the 99.7% of the time, with a remark about usual practice values will be in the range of 99.5%–99.9%.

In real links, ITU-R F.695 describes the availability objective of 99.7% proportional to the link length. This percentage will be evaluated over a period long enough to provide statistically accurate values. This period will be probably longer than 1 year. The definition of the observation time will depend on the specific case, and should be selected considering the usual outage times and characteristics associated with the local conditions (propagation, geographic factors, population distribution, management of maintenance operations, etc).

In the case of links forming part of the medium grade section of the ISDN, the availability objectives are described in IUT-R F.696. This recommendation defines bidirectional objectives for HRDSs. These sections represent a generic model of a network section close to the structure of real links. HDRS are classified into four different categories from Class 1 to Class 4 sections. Each class corresponds to a specific quality grade according to ITU-T G.801. Unavailability objectives range from 0.033% for Class 1 up to 0.1% for Class 4 links.

Finally, the title of Recommendation ITU-R F.697 refers to availability and EPOs for digital microwave LOS systems forming part of the local grade of the ISDN, but in fact, no values are provided in this recommendation, except for specifying that the values should be in the range of A%–B%, being both parameters under study. Currently, neither ITU-T nor ITU-R has provided further information about availability objectives for local grade ISDN. The only known reference is an Appendix to Recommendation ITU-R F.697 that contains some examples of values used by different administrations.

5.6 AVAILABILITY OBJECTIVES FOR MICROWAVE LOS LINKS DESIGNED AFTER 2005

5.6.1 Recommendations and Application Scope

Microwave links designed after 2005 are based on definitions and general criteria of recommendations ITU-T G.827 and ITU-R F.1703. Recommendation ITU-T G.827 (approved in 2003) defines network performance parameters and objectives for the path elements (PEs) and end-to-end availability of international constant bit-rate digital paths. These parameters are independent of the type of physical network supporting the end-to-end path, e.g., optical fiber, microwave LOS links, or satellite.

Recommendation ITU-T G.827 applies to paths based on the PDH as defined in ITU-T Rec. G.705, the SDH as defined in ITU-T Rec. G.707 or the optical transport network technology as defined in ITU-T Rec. G.709. The availability attribution is done in all cases with two contributions: a fixed block allocation plus an additional part based on the distance of the link being designed, measured or maintained.

In the specific case of microwave LOS Links, ITU-R F.1703 is the only recommendation that describes the availability objectives for links alter 2005. The objectives apply to any microwave link forming part of any HRDP not exceeding 27 500 km.

5.6.2 Reference Models for Availability Objective Allocation

The reference models used for allocating availability objectives are very similar to those already described in previous sections for EPO for links based on equipment after 2000/2002 within the scope of ITU-R F.1668. The model divides the HRDP into international and national portions. The national portion is composed of three different sections: access, short haul and long haul. The international portion includes a section in terminating countries, and might include sections in intermediate countries as well as inter-country links such as fixed satellite service or submarine cable links.

Despite the similarities between the models used for EPO and availability objectives, there are slight differences, especially with regard to terminology used to define some of the PEs. The model for availability studies is composed of:

- *PE*: A portion of an end-to-end path defined for availability specification purposes.
- *Inter-country path core element (ICPCE)*: The ICPCE is the PE carried on the highest order digital path across the geographical border between two countries. An ICPCE may be transported on a satellite, a terrestrial or an undersea cable

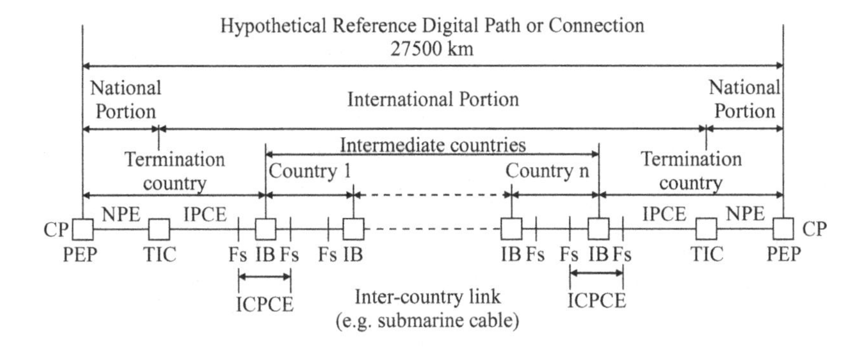

CP Customer Premises
Fs Frontier Station
IB International Border
ICPCE Inter-Country Path Core Element
IPCE International Path Core Element
NPE National Path Element
PEP Path End Point
TIC Terminal International Center

FIGURE 5.8 HDRP for availability allocation purposes as defined in Recommendation ITU-T G.827.

transmission system. In the case of a satellite transmission system, the frontier station is considered to be located at the earth station.

- *International path core element (IPCE)*: The IPCE is the PE used in the core network within one country. The boundary of this element depends on its application. For a transit country, this element is limited by the two frontier stations. For a terminating country, this element is limited by the IG and the frontier station.
- *National path element (NPE)*: The NPE is a PE used in a terminating country to connect the international portion and the PEPs.

Figure 5.8 shows the situation of each one of the above described elements within the international path.

5.6.3 Error Performance Concepts Related to Availability

Availability will be defined by error events and error performance parameters occurring during periods equal or longer than 10 s. The relevant parameter for availability analysis will be the SESR defined according to Recommendation ITU-T G.821.

5.6.4 Availability Objectives in Real Microwave LOS Links

The ITU-R, in Recommendation F.1703 specifies the availability objectives for microwave LOS links forming part of a HRDP following a procedure similar to the method explained for EPO.

First, the parameters that will be used to specify the objectives are defined: AR, MO and OI. The definitions are followed by values objective depending on the location of the link: international or national portion, access, short haul and long haul, etc.

The general availability objectives for the overall HRDP are composed of two contributions, a fixed block allocation and a variable allocation as a function of the distance. Nevertheless, the distribution of the part depending on the distance is not homogeneous along the HRDP.

Links on the access and short-haul sections are allocated on a fixed block basis independently of the link distance. On the contrary, links forming part of the international portion or belonging to the long haul of the national portion will have a fixed block allocation and an additional contribution that depends on the link distance. The equations used to obtain the AR and the mean time between outages MO are:

$$AR = 1 - \left(B_j \frac{L_{link}}{L_R} + C_j \right) \tag{5.10}$$

$$MO = 1/OI = \frac{1}{D_j \frac{L_{link}}{L_R} + E_j} \tag{5.11}$$

The reference length L_R is 2500 km and the lower limit for the link length L_{link} is $L_{mín} = 50$ km. In cases where the link is shorter than this value L_{link} is assumed 50 km. The subscripts of the parameters in equations (5.10) and (5.11) differentiate each specific case. The subscript *j* stands for:

Link within the International Portion	
1	$L_{min} < L_{link} \leq 250$ km
2	250 km $< L_{link} \leq 2500$ km
3	2500 km $< L_{link} \leq 7500$ km
4	$L_{link} > 7500$ km
Link within the National Portion	
5	Access section
6	Short-haul section
7	Long-haul section

The values for B_j, C_j, D_j and E_j can be found in different tables of the Recommendation ITU-R F.1703. The tables to facilitate the consultation of these constants have been included in Appendix II to this chapter.

5.7 ITU ERROR PERFORMANCE AND AVAILABILITY USAGE GUIDELINES

Error performance and availability requirements constitute one of the most important parts of the required specifications in order to design a microwave LOS link. Chapter 8 will explain the inclusion of EPO and availability into the link design

TABLE 5.7 ITU Recommendation Map Related to Error Performance and Availability for Microwave LOS Link Design

<table>
<tr><th>Bitrate</th><th>Application Date (Equipment Design Dates)</th><th>Rec. ITU-T</th><th>Rec. ITU-R</th><th>Error Parameter</th><th>Reference Transmission Model</th></tr>
<tr><td colspan="6">ERROR PERFORMANCE</td></tr>
<tr><td rowspan="4">Below primary rate</td><td rowspan="4">Before December 14, 2002</td><td rowspan="4">G.821</td><td>F.697</td><td>BER</td><td>HRX local grade</td></tr>
<tr><td>F.696</td><td>BER</td><td>HRX medium grade</td></tr>
<tr><td>F.594</td><td>BER</td><td>HRX high grade</td></tr>
<tr><td>F.634</td><td>BER</td><td></td></tr>
<tr><td>Below primary rate</td><td>After December 14, 2002</td><td rowspan="3">G.826</td><td rowspan="4">F.1668</td><td>BER</td><td>HRX</td></tr>
<tr><td>Constant rate at primary or higher rates–PDH</td><td>Any</td><td rowspan="3">Errored blocks</td><td rowspan="3">HRDP</td></tr>
<tr><td>Constant rate at primary or higher rates–SDH</td><td>Before March 2000</td></tr>
<tr><td>Constant rate at primary or higher rates–SDH</td><td>After March 2000</td><td>G.828</td></tr>
<tr><td colspan="6">AVAILABILITY</td></tr>
<tr><td>Any</td><td>Before March 2005</td><td></td><td></td><td>BER</td><td></td></tr>
<tr><td>Any</td><td>After March 2005</td><td>G.827</td><td>F.1703</td><td>Errored blocks</td><td>HRDP</td></tr>
</table>

process and their relation to factors associated with the services carried by the link, the equipment characteristics and the influence of the propagation phenomena. Table 5.7 provides a summary of the characteristics and applicability of each one of the recommendations described in this chapter.

5.8 DEGRADATION DUE TO INTERFERENCES

Interferences are one of the most relevant degradation sources of the quality of service in a microwave LOS links. The Recommendation ITU-R F.1565 provides the quality degradation objectives and criteria for links forming part of a certain HRDP that might suffer interferences from other systems in either the international or national portions. This recommendation follows the scheme provided by ITU-R F.1668, defining an expression for the EPO as a function of the link length and a series of constant values for each calculation scenario (national and international portions, etc) organized in tables.

The allocation of availability and EPOs depend on the interference source type. Thus, the first systems to be taken into account are those using the same frequency bands on a primary attribution basis. In those cases, the interference should not degrade the error (ESR, SESR, and BBER) or availability (UR and OI) parameters more than a 10% of the original error performance and availability objectives. In cases where the interference arrives from either systems without a primary service attribution or unwanted radiated emissions, the degradation caused should not be higher than a 1%, or any close value.

As a summary of these criteria, and if multiple interference sources are present, Recommendation ITU-R F.1094 defines the following distribution percentages for error performance and availability objectives:

1. X% will be allocated to the interference portion caused by fixed services that share the frequency band (i.e., other links in the same area). The recommended value for X is 89.
2. Y% for frequency sharing on a primary basis. This case is referred to as inter-service sharing (i.e., Fixed Service Satellite links). The recommended value for Y is 10.
3. Z% for the rest of interfering sources. The recommended value for Z is 1.

The sum of X% + Y% + Z% should not exceed either the EPOs given in Recommendation ITU-R F.1668 or the availability objectives of Recommendation ITU-R F.1703.

BIBLIOGRAPHY

ITU-R Handbook: Digital Radio-Relay Systems. International Telecommunication Union. Radiocommunication Sector. ITU-R. Geneva. 1996.

ITU-R Rec.F.594: Error performance objectives of the hypothetical reference digital path for radio-relay systems providing connections at a bit rate below the primary rate and forming part or all of the high grade portion of an integrated services digital network. International Telecommunication Union. Radiocommunication Sector. ITU-R. Geneva. 1997.

ITU-R Rec. F.634: Error performance objectives for real digital radio-relay links forming part of the high-grade portion of international digital connections at a bit rate below the primary rate within an integrated services digital network. International Telecommunication Union. Radiocommunication Sector. ITU-R. Geneva. 1997.

ITU-R Rec. F.696: Error performance and availability objectives for hypothetical reference digital sections forming part or all of the medium-grade portion of an ISDN connection at a bit rate below the primary rate utilizing digital radio-relay systems. International Telecommunication Union. Radiocommunication Sector. ITU-R. Geneva. 1997.

ITU-R Rec. F.697: Error performance and availability objectives for the local-grade portion at each end of an ISDN connection at a bit rate below the primary rate utilizing digital radio-relay systems. International Telecommunication Union. Radiocommunication Sector. ITU-R. Geneva. 1997.

ITU-R Rec. F.1397: Error performance objectives for real digital radio links used in the international portion of a 27 500 km hypothetical reference path at or above the primary rate. International Telecommunication Union. Radiocommunication Sector. ITU-R. Geneva. 2002.

ITU-R Rec. F.1491: Error performance objectives for real digital radio links used in the national portion of a 27 500 km hypothetical reference path at or above the primary rate. International Telecommunication Union. Radiocommunication Sector. ITU-R. Geneva. 2002.

ITU-R Rec. F.1492: Availability objectives for real digital radio-relay links forming part of international portion constant bit rate digital path at or above the primary rate. International Telecommunication Union. Radiocommunication Sector. ITU-R. Geneva. 2000.

ITU-R Rec. F.1493: Availability objectives for real digital radio-relay links forming part of national portion constant bit rate digital path at or above the primary rate. International Telecommunication Union. Radiocommunication Sector. ITU-R. Geneva. 2000.

ITU-R Rec. F.1565: Performance degradation due to interference from other services sharing the same frequency bands on a co-primary basis with real digital fixed wireless systems used in the international and national portions of a 27 500 km hypothetical reference path at or above the primary rate. International Telecommunication Union. Radiocommunication Sector. ITU-R. Geneva. 2002.

ITU-R Rec. F.1668: Error performance objectives for real digital fixed wireless links used in 27 500 km hypothetical reference paths and connections. International Telecommunication Union. Radiocommunication Sector. ITU-R. Geneva. 2007.

ITU-R Rec. F.1703: Availability objectives for real digital fixed wireless links used in 27 500 km hypothetical reference paths and connections. International Telecommunication Union. Radiocommunication Sector. ITU-R. Geneva. 2005.

ITU-T Rec. G.821: Error performance of an international digital connection operating at a bit rate below the primary rate and forming part of an Integrated Services Digital Network. International Telecommunication Union. Telecommunication Standardization Sector. ITU-T. Geneva. 2002.

ITU-T Rec. G.826: End-to-end error performance parameters and objectives for international, constant bit-rate digital paths and connections. International Telecommunication Union. Telecommunication Standardization Sector. ITU-T. Geneva. 2002.

ITU-T Rec. G.827: Availability performance parameters and objectives for end-to-end international constant bit-rate digital paths. International Telecommunication Union. Telecommunication Standardization Sector. ITU-T. Geneva. 2003.

ITU-T Rec. G.828: Error performance parameters and objectives for international, constant bit-rate synchronous digital paths. International Telecommunication Union. Telecommunication Standardization Sector. ITU-T. Geneva. 2000.

ITU-T Rec. G.829: Error performance events for SDH multiplex and regenerator sections. International Telecommunication Union. Radiocommunication Sector. ITU-R. Geneva. 2002.

The Impact of G.826. M. Shafi and P. Smith. IEEE Communications Magazine. *IEEE, Piscataway, NJ*. September. 1993.

APPENDIX I: TABLES FOR ERROR PERFORMANCE OBJECTIVE CALCULATIONS

RECOMMENDATION ITU-R F.1668

The tables included in this appendix contain the data required for calculating the constants B_j and C_j corresponding to the error performance objectives in Recommendation ITU-R F.1668. (*Tables courtesy of ITU-R*).

TABLE 5.8 Parameters for the EPO for Intermediate Countries According to ITU-T Rec. G.828

		$L_{min} \leq L_{link} \leq 1000$ km		1000 km $< L_{link}$	
Parameter	Bitrate (kbit/s)	B1	C1	B2	C2
ESR	1664	$5 \times 10^{-4}(1 + B_R)$	0	5×10^{-4}	$2 \times 10^{-4} \times B_R$
ESR	2240	$5 \times 10^{-4}(1 + B_R)$	0	5×10^{-4}	$2 \times 10^{-4} \times B_R$
ESR	6848	$5 \times 10^{-4}(1 + B_R)$	0	5×10^{-4}	$2 \times 10^{-4} \times B_R$
ESR	48 960	$1 \times 10^{-3}(1 + B_R)$	0	1×10^{-3}	$4 \times 10^{-4} \times B_R$
ESR	150 336	$2 \times 10^{-3}(1 + B_R)$	0	2×10^{-3}	$8 \times 10^{-4} \times B_R$
SESR	1664–150 336	$1 \times 10^{-4}(1 + B_R)$	0	1×10^{-4}	$4 \times 10^{-5} \times B_R$
BBER	1664–48 960	$2.5 \times 10^{-6}(1 + B_R)$	0	2.5×10^{-6}	$1 \times 10^{-6} \times B_R$
BBER	150 336	$5 \times 10^{-6}(1 + B_R)$	0	5×10^{-6}	$2 \times 10^{-6} \times B_R$

TABLE 5.9 Parameters for the EPO for Terminating Countries According to ITU-T Rec. G.828

		$L_{min} \leq L_{link} \leq 500$ km		500 km $< L_{link}$	
Parameter	Bitrate (kbit/s)	B3	C3	B4	C4
ESR	1664	$5 \times 10^{-4}(1 + B_R)$	0	5×10^{-4}	$1 \times 10^{-4} \times B_R$
ESR	2240	$5 \times 10^{-4}(1 + B_R)$	0	5×10^{-4}	$1 \times 10^{-4} \times B_R$
ESR	6848	$5 \times 10^{-4}(1 + B_R)$	0	5×10^{-4}	$1 \times 10^{-4} \times B_R$
ESR	48 960	$1 \times 10^{-3}(1 + B_R)$	0	1×10^{-3}	$2 \times 10^{-4} \times B_R$
ESR	150 336	$2 \times 10^{-3}(1 + B_R)$	0	2×10^{-3}	$4 \times 10^{-4} \times B_R$
SESR	1664–150 336	$1 \times 10^{-4}(1 + B_R)$	0	1×10^{-4}	$2 \times 10^{-5} \times B_R$
BBER	1664–48 960	$2.5 \times 10^{-6}(1 + B_R)$	0	2.5×10^{-6}	$5 \times 10^{-7} \times B_R$
BBER	150 336	$5 \times 10^{-6}(1 + B_R)$	0	5×10^{-6}	$1 \times 10^{-6} \times B_R$

TABLE 5.10 Parameters for the EPO for Intermediate Countries According to ITU-T Rec. G.826

		$L_{min} \leq L_{link} \leq 1000$ km		1000 km $< L_{link}$	
Parameter	Bitrate (Mbps)	B1	C1	B2	C2
ESR	<Primary rate	$2 \times 10^{-3} (1 + B_R)$	0	2×10^{-3}	$8 \times 10^{-4} \times B_R$
ESR	1.5–5	$2 \times 10^{-3} (1 + B_R)$	0	2×10^{-3}	$8 \times 10^{-4} \times B_R$
ESR	>5–15	$2.5 \times 10^{-3} (1 + B_R)$	0	2.5×10^{-3}	$1 \times 10^{-3} \times B_R$
ESR	>15–55	$3.75 \times 10^{-3} (1 + B_R)$	0	3.75×10^{-3}	$1.5 \times 10^{-3} \times B_R$
ESR	>55–160	$8 \times 10^{-3} (1 + B_R)$	0	8×10^{-3}	$3.2 \times 10^{-3} \times B_R$
ESR	>160–400	N.A.	N.A.	N.A.	N.A.
SESR	≤400	$1 \times 10^{-4} (1 + B_R)$	0	1×10^{-4}	$4 \times 10^{-5} \times B_R$
BBER	1.5–400	$1 \times 10^{-5} (1 + B_R)$	0	1×10^{-5}	$4 \times 10^{-6} \times B_R$

TABLE 5.11 Parameters for the EPO for Terminating Countries According to ITU-T Rec. G.826

		$L_{min} \leq L_{link} \leq 500$ km		500 km $< L_{link}$	
Parameter	Bitrate (Mbit/s)	B3	C3	B4	C4
ESR	<Primary rate	$2 \times 10^{-3} (1 + B_R)$	0	2×10^{-3}	$4 \times 10^{-4} \times B_R$
ESR	1.5–5	$2 \times 10^{-3} (1 + B_R)$	0	2×10^{-3}	$4 \times 10^{-4} \times B_R$
ESR	>5–15	$2.5 \times 10^{-3} (1 + B_R)$	0	2.5×10^{-3}	$5 \times 10^{-4} \times B_R$
ESR	>15–55	$3.75 \times 10^{-3} (1 + B_R)$	0	3.75×10^{-3}	$7.5 \times 10^{-4} \times B_R$
ESR	>55–160	$8 \times 10^{-3} (1 + B_R)$	0	8×10^{-3}	$1.6 \times 10^{-3} \times B_R$
ESR	>160–400	N.A.	N.A.	N.A.	N.A.
SESR	≤400	$1 \times 10^{-4} (1 + B_R)$	0	1×10^{-4}	$2 \times 10^{-5} \times B_R$
BBER	1.5–400	$1 \times 10^{-5} (1 + B_R)$	0	1×10^{-5}	$2 \times 10^{-6} \times B_R$

TABLE 5.12 EPOs for Real SDH Fixed Wireless Links Belonging to the Long-Haul Inter-Exchange Network Section of the National Portion of the HRP According to ITU-T Rec. G.828

Bitrate (Mbps)	1664 (VC-11. TC-11)	2240 (VC-12. TC-12)	6848 (VC-2. TC-2)	48 960 (VC-3. TC-3)	150 336 (VC-4. TC-4)
ESR	$0.01 \times A$	$0.01 \times A$	$0.01 \times A$	$0.02 \times A$	$0.04 \times A$
SESR			$0.002 \times A$		
BBER		$5 \times 10^{-5} \times A$			$1 \times 10^{-4} \times A$

TABLE 5.13 EPOs for Real Fixed Wireless Links Belonging to the Long-Haul Inter-Exchange Network Section of the National Portion of the HRP and HRC According to ITU-T Rec. G.826

Bitrate (Mbit/s)	<Primary Rate	1.5–5	>5–15	>15–55	>55–160	>160–400
ESR	0.04 A	0.04 A	0.05 A	0.075 A	0.16 A	N.A.
SESR	0.002 A	0.002 A	0.002 A	0.002 A	0.002 A	0.002 A
BBER[a]	N.A.	$2A \times 10^{-4}$	$2A \times 10^{-4}$	$2A \times 10^{-4}$	$2A \times 10^{-4}$	$1A \times 10^{-4}$

[a]The parameter BBER is only applicable to HRDPs.

TABLE 5.14 EPOs for SDH Fixed Wireless Links Forming All of the Short-Haul Inter-Exchange Network Section of the National Portion of the HRP and HRC According to ITU-T Rec. G.828

Bitrate (Mbit/s)	1664 (VC-11. TC-11)	2240 (VC-12. TC-12)	6848 (VC-2. TC-2)	48 960 (VC-3. TC-3)	150 336 (VC-4. TC-4)
ESR	$0.01 \times B$	$0.01 \times B$	$0.01 \times B$	$0.02 \times B$	$0.04 \times B$
SESR	$0.002 \times B$				
BBER	$5 \times 10^{-5} \times B$				$1 \times 10^{-4} \times B$

TABLE 5.15 EPOs for Fixed Wireless Links Forming All of the Short-Haul Inter-Exchange Network Section of the National Portion of the HRP and HRC According to ITU-T Rec. G.826

Bitrate (Mbit/s)	<Primary Rate	1.5–5	>5–15	>15–55	>55–160	>160–400
ESR	0.04 B	0.04 B	0.05 B	0.075 B	0.16 B	N.A.
SESR	0.002 B	0.002 B	0.002 B	0.002 B	0.002 B	0.002 B
BBER	N.A.	$2B \times 10^{-4}$	$2B \times 10^{-4}$	$2B \times 10^{-4}$	$2B \times 10^{-4}$	$1B \times 10^{-4}$

TABLE 5.16 EPOs for SDH Fixed Wireless Links Forming All of the Access Network Section of the National Portion of the HRP According to ITU-T Rec. G.828

Bitrate (Mbit/s)	1664 (VC-11. TC-11)	2240 (VC-12. TC-12)	6848 (VC-2. TC-2)	48 960 (VC-3. TC-3)	150 336 (VC-4. TC-4)
ESR	$0.01 \times C$	$0.01 \times C$	$0.01 \times C$	$0.02 \times C$	$0.04 \times C$
SESR	$0.002 \times C$				
BBER	$5 \times 10^{-5} \times C$				$1 \times 10^{-4} \times C$

TABLE 5.17 EPOs for Fixed Wireless Links Forming All of the Access Network Section of the National Portion of the HRP and HRC According to ITU-T Rec. G.826

Bitrate (Mbit/s)	<Primary Rate	1.5–5	>5–15	>15–55	>55–160	>160–400
ESR	0.04 C	0.04 C	0.05 C	0.075 C	0.16 C	N. A.
SESR	0.002 C	0.002 C	0.002 C	0.002 C	0.002 C	0.002 C
BBER	N.A.	$2C \times 10^{-4}$	$2C \times 10^{-4}$	$2C \times 10^{-4}$	$2C \times 10^{-4}$	$1C \times 10^{-4}$

APPENDIX II: TABLES FOR AVAILABILITY OBJECTIVE CALCULATIONS

RECOMMENDATION ITU-R F.1703

The tables included in this appendix contain the data required for calculating the constants B_j and C_j corresponding to the error performance objectives in recommendation ITU-R F.1703. (*Tables courtesy of ITU-R*).

TABLE 5.18 Parameters for AR Objectives for Links Forming Part of an International Portion of Constant Bit-Rate Digital Path

	$L_{min} \leq L_{link} \leq 250$		$250 < L_{link} \leq 2500$		$2500 < L_{link} \leq 7500$		$L_{link} > 7500$	
Length (km)	B_1	C_1	B_2	C_2	B_3	C_3	B_4	C_4
International portion	1.9×10^{-3}	1.1×10^{-4}	3×10^{-3}	0	3×10^{-3}	0	3×10^{-3}	0

TABLE 5.19 Parameters for AR Objectives for Links Forming Part of a National Portion of Constant Bit-Rate Digital Path Element

Access Portion		Short-Haul Portion		Long-Haul Portion	
B_5	C_5	B_6	C_6	B_7	C_7
0	5×10^{-4}	0	4×10^{-4}	3×10^{-3} if 250 km $\leq L_{link} < 2500$ km 1.9×10^{-3} if $L_{min} \leq L_{link} < 250$ km	0 if 250 km $\leq L_{link} < 2500$ km 1.1×10^{-4} if $L_{min} \leq L_{link} < 250$ km

TABLE 5.20 Parameters for OI Objectives for Links Forming Part of an International Portion of Constant Bit-Rate Digital Path

	$L_{min} \leq L_{link} \leq 250$		$2250 < L_{link} \leq 2500$		$2500 < L_{link} \leq 7500$		$L_{link} \geq 7500$	
Length (km)	D_1	E_1	D_2	E_2	D_3	E_3	D_4	E_4
International portion	150	50	100	55	100	55	100	55

TABLE 5.21 Parameters for OI Objectives for Links Forming Part of a National Portion of Constant Bit-Rate Digital Path Element

Access Portion		Short-Haul Portion		Long-Haul Portion	
D_5	E_5	D_6	E_6	D_7	E_7
0	100	0	120	100 if 250 km $\leq L_{\text{link}} < 2500$ km 150 if $L_{\text{min}} \leq L_{\text{link}} < 250$ km	55 if 250 km $\leq L_{\text{link}} < 2500$ km 50 if $L_{\text{min}} \leq L_{\text{link}} < 250$ km

APPENDIX III: CASE STUDIES FOR ERROR PERFORMANCE OBJECTIVE CALCULATIONS

RECOMMENDATION ITU-R F.1668

Case 1 International Portion

Table 5.22 contains the calculations and results for different bitrates in a link located on an intermediate country of an international HRDP.

TABLE 5.22 EPO Calculations in an International Path, for Links in Intermediate Countries

Case 1	International Portion. One Intermediate Country. Link Length: 150 km Evaluation Time: (30 d).		
Bitrate	150 336 kbit/s	140 Mbit/s	64 kbit/s
EPO	$EPO = B_{\text{j}} \times (L_{\text{link}}/L_{\text{R}}) + C_{\text{j}}$ In this case $j = 1$ because $L_{\text{min}} < L_{\text{link}} < 1000$ and B_{R} is assumed equal to 1		
	G.828 and intermediate country. Table 1a in F.1668	G.826 and intermediate country. Table 2a in F.1668	
ESR	168×10^{-6}	672×10^{-6}	168×10^{-6}
Number of ES/mo	435	1741	436
SESR	84×10^{-7}	84×10^{-7}	84×10^{-7}
Number of SES/mo	22	22	22
BBER	4.2×10^{-7}	8.4×10^{-7}	–
Number of BBE/mo	8709	17 418	–

Case 2 National Portion: PDH Links in Access and Short Haul

TABLE 5.23 EPO Calculations in National Portion and PDH Links

Case 2	Links on the National Portion. Various Lengths and PDH Bitrates	
Bitrates	2 Mbit/s	34 Mbit/s
Link section/length	Access section/20 km	Short-haul section/80 km
EPO	Table 5b in Rec. F.1668 (C is assumed 0.075)	Table 4b in Rec. F.1668 (B is assumed 0.075)
ESR	$0.04\ C = 3 \times 10^{-3}$	$0.075\ B = 5.625 \times 10^{-3}$
Number of ES/mo	7776	14 580
SESR	$0.002\ C = 1.5 \times 10^{-4}$	$0.002\ B = 1.5 \times 10^{-4}$
Number of SES/mo	389	389
BBER	$2 \times 10^{-4} \times C = 1.5 \times 10^{-5}$	$2 \times 10^{-4} \times B = 1.5 \times 10^{-5}$
Number of BBE/mo	77 760	311 040

Case 3 National Portion with Combined Access and Short-Haul Sections

TABLE 5.24 EPO Calculations for Links Forming Part of the National Portion with Combined Access and Short-Haul Sections

Case 4	Link on the National Portion, Composed of a Section on the Access Network L1 (10 km) and Another Section on the Short Haul, L2 (100 km) Overall Link Length = L1 + L2 = 110 km
Bitrates	64 kbit/s < Primary rate
EPO	Tables 4b and 5b in F.1668 (Assuming $B + C = 0.16$)
ESR	$0.04\ (B + C) = 6.4 \times 10^{-3}$
Number of ES/mo	16 589
SESR	$0.002\ (B + C) = 3.2 \times 10^{-4}$
Number of SES/mo	830
BBER	–
Number of BBE/mo	–

Case 4 National Portion: SDH Links in Long Haul

TABLE 5.25 EPO Calculations in National Portion and SDH Links

Case 3	National Portion and Long-Haul Section. SDH Bitrates on a 75 km Link.			
Bitrates	STM-1 (155.52 Mbit/s). System Before March 2000		STM-1 (155.52 Mbit/s). System After March 2000	
EPO	Table 3b in Rec. F.1668 System according to G.826		Table 3a in Rec. F.1668 System according to G.828	
	Lower limit $A_l = 0.01$	Upper limit $A_l = 0.02$	Lower limit $A_l = 0.01$	Upper limit $A_l = 0.02$
ESR	$0.16\ A = 0.16\ (A_l + 0.002) \times 75/100 = 144 \times 10^{-5}$	$0.16\ A = 0.16\ (A_l + 0.002) \times 75/100 = 264 \times 10^{-5}$	$0.04\ A = 0.04\ (A_l + 0.002) \times 75/100 = 36 \times 10^{-5}$	$0.04\ A = 0.04\ (A_l + 0.002) \times 75/100 = 66 \times 10^{-5}$
Number of ES/mo	3733	6843	933	1711
SESR	$0.002\ A = 0.002 \times (A_l + 0.002) \times 75/100 = 18 \times 10^{-6}$	$0.002\ A = 0.002 \times (A_l + 0.002) \times 75/100 = 33 \times 10^{-6}$	$0.002\ A = 0.002 \times (A1 + 0.002) \times 75/100 = 18 \times 10^{-6}$	$0.002\ A = 0.002 \times (A1 + 0.002) \times 75/100 = 33 \times 10^{-6}$
Number of SES/mo	47	86	47	86
BBER	$0.0002\ A = 0.0002 \times (A_l + 0.002) \times 75/100 = 18 \times 10^{-7}$	$0.0002\ A = 0.0002 \times (A_l + 0.002) \times 75/100 = 33 \times 10^{-7}$	$0.0001\ A = 0.0001 \times (A_l + 0.002) \times 75/100 = 9 \times 10^{-7}$	$0.0001\ A = 0.0001 \times (A_l + 0.002) \times 75/100 = 165 \times 10^{-8}$
Number of BBE/mo	37 324	68 429	18 662	34 214

APPENDIX IV: CASE STUDIES FOR AVAILABILITY OBJECTIVE CALCULATIONS

RECOMMENDATION ITU-R F.1703

This section presents some application examples of the Recommendation ITU-R F.1703. The calculations assume that a year has 525 960 min.

Case 1 International Portion

TABLE 5.26 Calculation of the Parameters Related to Availability Objectives for Links Forming Part of a International Portion of Constant Bit-Rate Digital Path Element

Case 1	International Portion. Various Lengths.		
Link Length	L = 30 km (< 50 km) L_{link} = 50 km	L = 80 km (50 < L <250)	L = 1056 km (250 km < L < 2500 km)
AR	0.99985 (99.985%)	0.9983 (99.83%)	0.99873 (99.873%)
MO	18.87×10^{-3}	18.25×10^{-3}	10.28×10^{-3}
OI Events/yr	53	55	97
Unavailable min/yr	78	90	667
Mean TimeBetween Unavailability Events	9922 min (6.9 d)	9596 min (6.7 d)	5402 min (3.7 d)

Case 2 National Portions with Different Sections

TABLE 5.27 Calculation of Availability Parameters for Links Forming Part of a National Portion of Constant Bit-Rate Digital Path Element

Case 2	National Portion		
Section Type	Access Section, 30 km long L = 30 km (< 50 km) L_{link} = 50 km	Short Haul, 105 km long L = 105 km (50 < L <250)	Long Haul, 960 km long L = 9 km (250 km < L < 2500 km)
Link Length	0.9995 (99.95%)	0.9996 (99.96%)	0.9988 (99.88%)
AR	1×10^{-2}	8.34×10^{-3}	1.071×10^{-2}
MO	100	120	93
OI Events/yr	263	210	606
Unavailable min/yr	5257 min	4381 min	5627 min

Case 3 National Portions Composed of a Combination of Different Sections

Assume a link, 1095 km long, comprised of an access section (30 km), a short-haul section (105 km), and a long-haul section (960 km). The objectives for AR in this link will be obtained as the summation of all the availability objectives associated with each section of the network:

$$\begin{aligned} AR &= 1 - UR = 1 - (UR_{AN} + UR_{SH} + UR_{LH}) \\ &= 1 - (5 \times 10^{-4} + 4 \times 10^{-4} + 0) = 0.9991 \end{aligned}$$

where

UR = total unavailability ratio
UR_{AN} = unavailability ratio objective for the access portion
UR_{SH} = unavailability ratio objective for short-haul portion
UR_{LH} = unavailability ratio objective for long-haul portion

The MO objective is given by the reciprocal of the sum of OI objectives referred to the part of link belonging to each network portion:

$$MO = \frac{1}{OI_{AN} + OI_{SH} + OI_{LH}} = \frac{1}{100 + 120 + 93} = 3.19 \times 10^{-3}$$

where

MO = total mean time between outages
OI_{AN} = outage intensity objective of the access portion
OI_{SH} = outage intensity objective for short-haul portion
OI_{LH} = outage intensity objective for long-haul portion

These values correspond to an AR of 99.91% (unavailability of 473 min/yr), number of events per year OI = 313 and the mean time between unavailability events MO = 1674 min.

CHAPTER 6

LINK PATH ENGINEERING

6.1 GENERAL CONSIDERATIONS ON LINK PATH ENGINEERING

The design and planning of a microwave LOS link is usually a complex task that involves diverse interrelated aspects. In most cases, it is an iterative process that initiates with a preliminary rough design approach that is refined on various stages that lead to a final configuration that, ideally, would be technically (and economically) optimal. Most designs involve three phases:

1. Initial planning or preliminary dimensioning, where the basic parameters of the link are determined.
2. A further step in the design is detailed enough to enable system deployment. This phase includes all the link design algorithms that in many cases will imply a change in the initial set of basic parameters assumed in the first phase.
3. Test, optimization, operation, and maintenance. Once the system has been installed, and after the first test and validation measures, the system parameters might be optimized to enhance the system performance, and thus return to Phasc 2 and for a system parameters refinement.

The first step will define the basic primary parameters of the link that, at least will include the following information:

- Information related to the network nodes that will be connected by means of the link under design. The nodes can be circuit switching offices, transmission

Microwave Line of Sight Link Engineering, First Edition. Pablo Angueira and Juan Antonio Romo.

nodes, packet switching server nodes, gateways, base stations, control stations in cellular networks, user centers, etc.

- Type, volume, and priorities of the traffic that will be conveyed by the link: traffic matrix, origin-destination routes, throughput, etc.
- Physical connection topology of the network nodes.

Using these specifications as the initial reference, the capacity and frequency band will be defined. Once the band and capacity are fixed, the equipments from different manufacturers are selected. The hardware will be characterized by a list of relevant parameters as shown in Chapter 4. The configuration of the different equipment components and primary antennas will be selected depending upon the design criteria: protection type, frequency bands, and associated maximum hop lengths. In further phases of the design it might be necessary to reconsider the configuration (diversity, protection schemes) and/or the use of special antennas (e.g., in order to avoid interference degradations).

The next step in the planning process is the path engineering analysis. The path engineering analysis studies the alternatives for providing a line of sight (LOS) path between transmitting and receiving antennas, and at the same time enabling a clearance high enough to avoid diffraction even in the case of rough troposphere conditions (severe subrefraction) guaranteeing the link performance within the minimum quality and availability specifications.

In this phase, the location of each one of the nodes of the link (repeater stations), the dimensioning of each transmission site, including building and equipment room, power supply, access, towers, and antenna installation special requirements.

The absence of relevant obstacles along the path to ensure line of sight (LOS) clearance is based on geometrical calculations. These calculations require topographical databases used for path profile extraction. Profiles will be drawn in a path profile diagram along with the expected path followed by the radio signal. The diagram obtained will allow to compare both plots (profile and radio signal path) to analyze the possible diffraction or reflection effects on the terrain, specially in cases where the link profile is flat and the expected conductivity of the soil is high (links over water surfaces for example).

The link path engineering will also tackle the potential problems arising from co-channel interferences between different stations. The interference level that will be received by an antenna will be influenced by the geometrical conditions of the link, specifically by the relative position of all the repeater and nodal stations. The co-channel interference signals might arrive arriving directly through the side lobes of the antenna or by indirect scattered signal components on the terrain that will be received through the main lobe.

The engineering processes for link availability and error performance will guarantee that objectives and criteria according to a predefined environment and system operation quality requirements are fulfilled. The calculations will include the evaluation of the power link budget as well as the estimation of the time percentages associated with various propagation phenomena causing degradation. The algorithmia used for this purpose will be described in detail in Chapters 7 and 8 of this book.

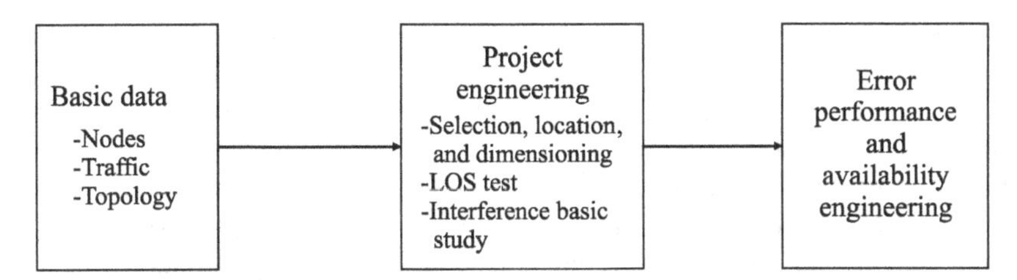

FIGURE 6.1 Planning process diagram.

Obviously, all the mentioned engineering tasks: path engineering, link performance engineering, the design of an adequate frequency plan, and the choice of equipment of the link will be closely interrelated. Figure 6.1 provides a simplified diagram that illustrates the design flow in the planning process of a microwave LOS link.

There are a few software tools that support the task of decision making in each one of the design phases of a microwave LOS link. These tools include equipment and antenna databases, as well as transmission lines and other devices necessary for a complete design. Additionally, these tools include modules to handle digital terrain databases as well as other cartography layers. In fact, according to the interface and architecture of these systems they could be regarded as another type of Geographical Information System (GIS). Finally, these software applications will include modules that perform any calculation required in the design process, ranging from algorithms related to link path engineering to all ITU-R methods associated with propagation calculations and performance engineering.

The further sections of this chapter, and Chapter 7 develop the concepts required to understand and enhance the operation of the mentioned software tools. The objective of the explanations will also be to provide the guidelines for the whole process of planning and designing a microwave LOS link. All the factors that influence the correct operation of the final design will be analyzed, in order to reach an optimal design. Moreover, during the final phase of testing, verification and equipment installation, the procedures should facilitate the identification of potential causes of failure and introduce adequate measures to solve them.

It should be always kept in mind that simulations will always be preliminary in nature, and they do not substitute the phase of inspections on the field. The inspection phase could include the verification of the exact location of the sites, checking the line of sight, confirm tower room availability for antenna installation, room for equipment in auxiliary buildings (shelters), potential interference identification, soil conditions for possible new towers and the infrastructures of the area.

6.2 SITE SELECTION CRITERIA

Whenever possible, the location of the nodal stations of the microwave LOS link will coincide with the network centers that it will connect. In the event that either technically or economically the decision can be justified, it might be recommended to concentrate the terminal stations on multiservice installations: that is, communication towers, multiservice radio sites on the surroundings of urban areas. In those cases, an additional link infrastructure is required to connect the network node and the terminal station, either with cabled or wired solutions.

In any case, the site selection process of end terminal and intermediate repeater stations should be carried out during the first phases of a preliminary design of the link, and should include, in the most general scenario, an analysis of the site characteristics, the access infrastructures for vehicles and persons, the requirements and characteristics for dimensioning and building the center and antenna tower, the energy supply infrastructure and the external radio and environmental conditions.

The size and characteristics of the shelter that will host the equipment will be mostly determined by the room requirements of the equipment, the power supply elements and the transmission line infrastructure associated with the distribution circuits that link the antenna and the transmitter/receiver units.

The site will require power supply that in modern systems has to be converted to –48 V DC. The shelter will also contain elements working on standard AC conditions, such as light, temperature conditioning equipment and others specific to each case. The shelter will contain converters or inverters that will provide energy for that equipment not working on the standard AC range value.

The characteristics of the power supply system are one of the critical elements that will have a direct influence on the availability of the microwave LOS link. In most cases, the supply system will be designed with some type of redundancy and backup subsystem, either based on uninterrupted power system units or/and batteries. If the site is being designed for the purpose of the link under construction, the supply will be dimensioned not according to the link demand but foreseeing the long-term requirements of the equipment of the link and, in many cases considering the additional radio systems that will probably share the site with the link under design.

The type and size of the tower will depend strongly on the antennas that will be installed on it. The choice will be done again with a long-term view considering the final capacity, number of antennas and height requirements of the site.

The first option will be always to use existing tower infrastructures. If this is the case, it will be necessary to carry out an interference analysis of the systems that are already using the tower. If it was necessary to build a new tower for the link, a geotechnical analysis of the soil will be required. This study will ensure that the soil type will be able to maintain the mechanical stability of the tower structure, even under the presence of strong winds and considering the worst scenario: maximum expected number of antennas and associated diameters. Mechanical stability checking should be one of the items of the maintenance task list of the site.

Last but not least, it will be of key importance to consider all the aspects related to environmental specifications as well as staff safety and equipment protection that will depend very much on the specific area where the system is being deployed. In consequence, among the design criteria of the site there will be electromagnetic compatibility requirements, electromagnetic human exposure limits, grounding, bonding, and shielding of all indoor and outdoor installations etc.

6.3 DIGITAL TERRAIN DATABASES

In general, most propagation prediction algorithms, at least those that provide acceptable accuracy for real system design, are based on detailed topographical information. The topographical information is provided by Digital Terrain Elevation (DTE) Models and Digital Elevation Models that are composed by digital terrain topographical databases and represented in the form of digital terrain maps. These models provide accurate information that will be used to find the best locations for the link stations. Also, they will be used to evaluate the path clearance and the potential loss associated with diffraction. They will be also used in the algorithmia that are required for determining the interferences to and from other stations of the link or arriving to/from different radio systems sharing the same bands.

There are a variety of digital terrain models (DTMs), with elevation data of different resolutions on a global scale that cover practically the whole surface of the Earth. Some of them are accessible though the Internet. Relevant examples are the Shuttle Radar Topography Mission (SRTM), Advanced Spaceborne Thermal Emission and Reflection Radiometer (ASTER) Global Digital Elevation Model (GDEM). Additionally, many national, regional or local administrations have their own databases that usually are commercialized to the general public with standard compatible formats. In addition, there is also quite a wide choice of companies that provide commercial digital terrain databases.

The SRTM model is the result of a joint initiative between the National Geospatial-Intelligence Agency (NGA) and the National Aeronautics and Space Administration (NASA). The digital topographic data cover the 80% of the Earth's surface between 60° North to 56° South latitudes. The latitude/longitude grid resolution of the database corresponds to an arc of 1 s (approximately 30 m/98 ft grid) with an elevation accuracy of 16 m (52 ft). At the time of writing this book, the data are available on the official web sites of NASA and NGA.

The GDEM model was created by image processing (stereo correlation techniques) a total of 1.3 million scenes of the earth available from the ASTER optical image database, that cover Earth regions between 83° North and 83° South. The GDEM has a resolution of 30 m (98 ft) and is composed of 23000 cells. The GDEM has been created by the NASA, the Ministry of Economy, Trade, and Industry of Japan. At the time of writing this book, the GDEM is available for direct download from the official NASA and Japan's Earth Remote Sensing Data Analysis Center websites.

Many applications, in addition to DTE data, require additional details and features of the terrain close to the sites where the stations will be located. The variety of information required as well as the amount of data involved, has motivated the creation of the so-called DTMs for efficient data management and storage purposes. DTM's are composed of layers (usually raster layers) that represent different features of the Earth's surface associated with the elevation data layer that will be specific for each application and geographical area. The DTM are closely related to the general term digital cartography and the tools that facilitate the handling of DTM in relation with diverse applications (a small part of which are the radio planning tools) are called GIS.

In the specific case of microwave LOS link planning, the DTM should include different soil types, vegetation layers, layers describing the existence of buildings and associated heights, etc. These DTM will be integrated with additional information required for link planning such as equipment databases as well as the data regarding other stations operating in the same area.

The evaluation and calculations of interferences and the coordination area with the rest of radio stations, is carried out with the assistance by a tool that is generically called radio environment map (REM). This application is a GIS that contains a geo-localized database of all the technical characteristics of the radio transmitters, receivers, antenna systems, and radio channel arrangements and frequency plans within the area under study. The database, in addition to position coordinates for each radio station, it also contains information such as height over average sea level, bearing angle, and radiation pattern (at least the radiation pattern envelope). The REMs will usually be one of the layers of the GIS tool (DTM) being used for link design.

There are currently powerful software packages that provide all the modules required to carry out each one of the preliminary and detailed designs of a microwave LOS link on the market. These tools contain all the algorithmia that will be described in Chapters 7 and 8, granting accurate and fast simulation results. However, it is important to remind that these tools, the associated databases and simulation results do not substitute the phases of inspections on the field, measurements and information validation in the real environment where the link will be installed.

6.4 PROFILE EXTRACTION, CLEARANCE, AND OBSTRUCTIONS

The LOS clearance analysis is based on the study terrain heights along the path profile between stations of the hop in comparison with the expected path followed by the radio signal (first Fresnel ellipsoid). This analysis will also include the study of the reflection point on the Earth's surface in those cases where this phenomenon is considered relevant. These studies will provide the required antenna heights and imply the extraction and visualization of the path profile between stations.

A radio path profile is composed by a set of terrain height samples, measured with respect to the average sea level, and organized in intervals along the path according to their distances from the transmitter (one of the stations of the hop). The first and the last samples will be each one of the hop stations. Each distance and height pair is assigned an index that increases starting from one end of the path and finishing at the opposite end. It is desirable for the sake of simplicity, even though not essential, that the distance between samples is constant.

The path profile extraction can be carried out directly from a topographical map or using a digital terrain database. Today, the first approach might seem a bit archaic, but under certain conditions, a first preliminary path analysis might be well performed using a simple map.

If the more usual case of using a digital terrain database is the choice, the specifications and characteristics of the digital data are relevant. Terrain height profiles can

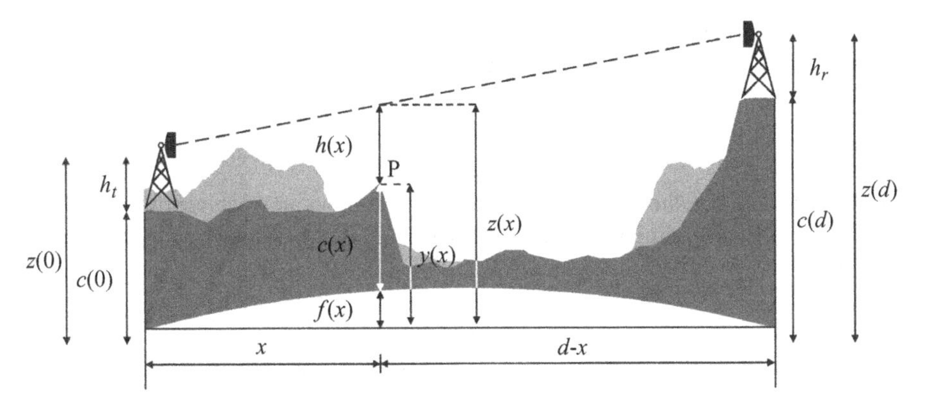

FIGURE 6.2 Example of path profile represented by the fictitious Earth model.

be represented using different Earth models that take into account the refraction (and thus the curvature of the signal path) according to different models:

- *Real Earth radius*: The Earth's real curvature is represented using a baseline or "zero height curve," parabolic in nature. The signal path is represented by plotting the real path travelled by the radio signal.
- *Flat Earth*: The Earth's surface is represented as a straight line, and signal paths will be plotted taking into account their own curvature caused by refraction as well as a correction to include the fact that Earth is considered flat.
- *Fictitious Earth*: In this case the signal path is represented by a straight line. In consequence, Earth is drawn using a parabolic baseline (zero height curve) that includes the real Earth's curvature and the factor related to refraction, to compensate for having plotted the propagation as a straight line. The resulting Earth radius is kR_o and it is referred to as equivalent Earth radius.

The model widely used is the fictitious Earth model. Path profiles will be represented adding to the real Earth's curvature, the variations originated by refraction (k factor). The effective height at each one of the path profile samples will be then calculated as the height above average sea level plus the addition of the Earth's bulge associated with a specific k factor. Figure 6.2 contains a path profile represented by the fictitious Earth model.

The Earth's bulge $f(x)$ associated with a generic location P on the path profile, at a distance x from the transmitter station and at $(d - x)$ from the receiver station, is calculated using equation (6.1), where all the distances are expressed in m. R_0 is the real value of Earth's radius:

$$f(x) = \frac{x(d-x)}{2\mathrm{k}R_0} \tag{6.1}$$

The effective terrain height, $Z(x)$, at P, is the geographical elevation, plus the Earth's bulge at that location P, that is, $f(x)$:

$$Z(x) = C(x) + f(x) \tag{6.2}$$

Following this model, the straight line that links transmitter and receiver (end stations of the hop), T-R, according to this model (Figure 6.2) is represented by the following equation:

$$Y(x) - Z(0) = \frac{Z(d) - Z(0)}{d} x \tag{6.3}$$

where d is the horizontal distance between the antennas of the transmitter and receiver stations, $Z(0)$ and $Z(d)$ the heights over average sea level of these stations:

$$Z(0) = Y(0) = C(0) + h_t \tag{6.4}$$

$$Z(d) = Y(d) = C(d) + h_r \tag{6.5}$$

where h_t and h_r are the heights of the antenna installation positions on the tower above ground. $C(0)$ and $C(d)$ are the terrain heights above average sea level at the transmitter and receiver antenna sites, respectively.

The clearance $h(x)$, of an arbitrary location P on the profile with an abscise x is the difference between the ordinate value of the threshold report line at location P and the effective terrain height at this location, as shown in equation (6.6). The clearance represents the distance between the signal path and the terrain obstacles at that location P:

$$h(x) = Z(x) - Y(x) \tag{6.6}$$

The sign of the clearance might be either positive or negative. The clearance is negative is the signal path is above the effective terrain height, and otherwise negative.

The theory indicates that diffraction losses due to on an obstacle occur if the signal path (T-R line) has a margin above the obstacle that is equal or lower than 0.577 times the radius of the first Fresnel ellipsoid. This condition is equivalent to a clearance $h \geq -0.577\, R_1$. The algorithmia to calculate the diffraction loss caused by different types, number and obstacles configurations along the path between transmitter and receiver will be described in Chapter 7.

Figure 6.3 shows an example of profiles on a fictitious Earth, with k = 4/3 where the clearance is indicated by h.

The Earth's bulk is a function of the k factor. If k factor varies according to the refractivity gradient, the Earth's bulk will also vary accordingly and, in consequence, the clearance will also with refraction. Referring to clearance and bulk variations

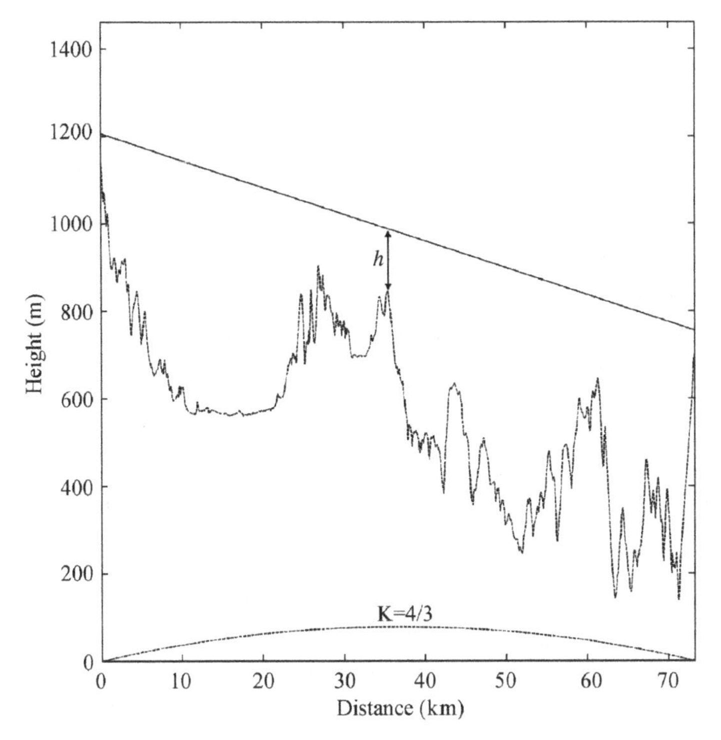

FIGURE 6.3 Profile example and clearance for K = 4/3.

as Δh and Δf respectively, their values are obtained by equation (6.7):

$$\Delta h\,(x) = \Delta f\,(x) = \frac{x\,(d-x)}{2R_0}\left(\frac{1}{k_2} - \frac{1}{k_1}\right) \tag{6.7}$$

where

Δh = clearance variation
Δf = bulk variation
d = path distance
x = path location
R_0 = Earth's radius (real value)
k_1 and k_2 = k factor, situations 1 and 2

If k diminishes, the bulk will increase, and thus, the absolute value of the clearance will decrease (assuming obstacles below the signal path). This effect is illustrated by Figure 6.4.

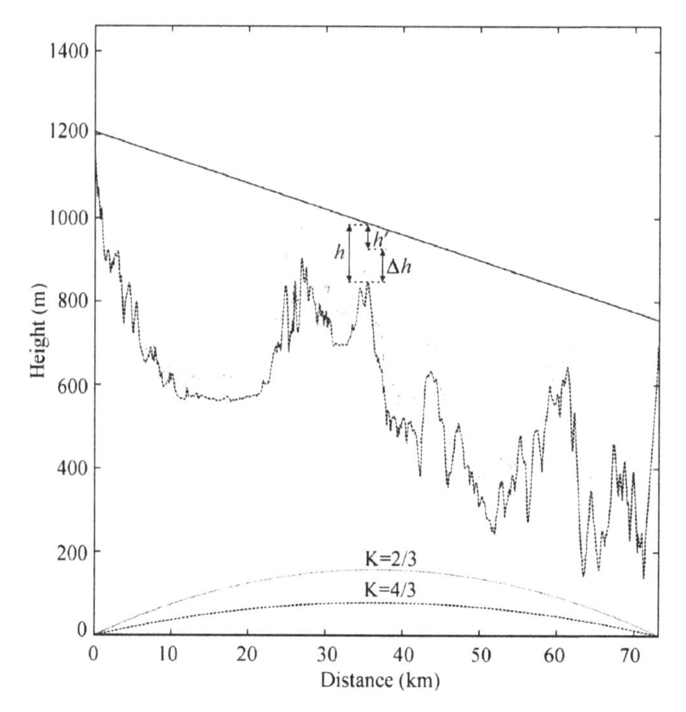

FIGURE 6.4 Clearance variation as a function of the different values of k.

6.5 OPTIMUM CHOICE OF ANTENNA HEIGHTS

The heights where the antennas of a microwave LOS link are installed will be calculated as a compromise between a series of factors that include propagation aspects as well as practical criteria. The most important ones follow:

1. Guarantee the absence of obstacles that block the line of sight minimizing the possible diffraction losses under anomalous refraction conditions. The usual practice is to follow the criteria and clearance objectives recommended by ITU-R in Recommendation ITU-R P.530.
2. Minimize the presence of multipath propagation fading phenomena. Multipath fading can vary significantly as a function of the antenna heights and especially on the link path inclination. The criteria provided by ITU-R P.530 only account for diffraction, and it might happen that the fulfillment of those clearance criteria maximizes the presence of multipath. The ITU-R does not currently include general criteria for a compromise solution that accounts both for diffraction and multipath. The experience shows that paths with low inclination are more likely to suffer multipath fading that those with relevant inclination angles. Thus, taking advantage of the terrain profile and/or increasing the antenna height imbalance seems adequate if multipath propagation is expected.

Nevertheless, the usual procedure is based on protecting the link with the required protection margin against multipath (as described in Chapter 8) considering that the antennas will be calculated only according to clearance criteria. In special cases, during the system tests it might be possible to analyze the behavior of the link in multipath propagation conditions, considering different antenna heights and/or diversity configurations that might solve a potential degradation caused by severe multipath.

3. Minimize the impact of reflections on the Earth's surface. The usual procedure is again based on the antenna heights calculated according to clearance criteria. Once these criteria have been applied, the reflection points are identified. Then, by varying the initial antenna height proposal, the translation of the reflection point to a less critical reflection area is studied. This study will be only carried out in cases where the reflection point is located on special zones, where the reflection is remarkable. Those cases will be usually links over bodies of water, and in general, flat profiles with good conductivity soil.
4. Minimize interferences caused by external emissions in the vicinity of the installation. The analysis will be based on estimating the C/I ratios associated with other radiating systems in the same or nearby towers, as described in Chapter 8. In consequence, in order to protect the system against interference, the antenna heights might be changed and/or the radiation pattern choice (antenna model) could be reconsidered.

In summary, antenna heights are mostly calculated according to clearance criteria, avoiding the blockage of the Fresnel ellipsoid under standard conditions and minimizing the diffraction losses under anomalous refraction periods. The antennas will be installed on the closest available position on the tower to the calculated values, and should reflection be a problem those heights might be changed in order to translate the reflection spot to a more favorable scenario.

In order to protect the system against multipath fading occurrences that exceed the link margin, diversity configurations will be analyzed. The usual diversity scheme is spatial diversity, where two or more antennas arranged vertically and separated a distance that is a function of the frequency and the probability associated with a maximum fading depth. Each antenna should also fulfill the clearance criteria.

If special measures would be required against external interferences, the first approach will be based on modifying the antenna choice, picking up a model with better directivity patterns and higher front-to-back and front-to-side lobe ratios. If necessary, the antenna heights might be changed. If this was the case, the clearance, reflection, and/or multipath related criteria should be applied recursively in order to find a position that also accounts for all factors.

6.5.1 Obstacle Clearance Objectives and Criteria

In order to ensure a certain degree of clearance over the obstacles along the path, the antenna heights of the stations at both ends of a hop must be maintained over a minimum value that will be calculated according to predefined criteria.

The variability of the refractivity, and in consequence the variability of the radio path, will also imply a variable clearance that will require a statistical analysis related to the statistics of refractivity. The clearance objectives will be then described by the time percentages defined thresholds are exceed. The clearance thresholds will in turn be related to statistics of the k factor and the maximum acceptable diffraction losses.

The clearance objectives have evolved during the past two decades. Initially the criteria were very conservative, and antenna heights were selected to avoid any diffraction losses even during extreme anomalous refraction (very small k values). Currently the criteria imply short duration outages (very small time percentages) based on more accurate statistics of the k factor.

6.5.1.1 Non-diversity Antenna Configurations Even though there is not any specific diffraction method for small time percentages, the following calculation procedure, described in Recommendation ITU-R P.530, can be used to calculate the antenna heights in non-diversity installations.

The antenna heights will be the maximum values obtained in the following two situations:

1. The antenna heights will be calculated to provide a clearance over the highest obstacle equal to the first Fresnel ellipsoid R_1 assuming the median value of the k factor (if the median is not available the default k = 3/4 will be supposed). This condition applies both in temperate and tropical climates.
2. For the effective k factor value, k_e, exceeded the 99.9% of the time, and for the specific hop length, the antenna heights will be such that the following conditions are fulfilled:

Temperate climate	Tropical climate
Clearance: 0.0 R_1 (i.e., grazing) if there is a single isolated path obstruction	Clearance: 0.6 R_1 for path lengths greater than about 30 km
Clearance: 0.3 R_1 if the path obstruction is extended along a portion of the path	

The k values can be obtained from refractivity statistics of the zone where the link is being deployed. If local data are not available, the ITU-R provides a curve to obtain the effective k factor, k_e, exceeded approximately during 99.9% of the most unfavorable month, for continental temperate climate. The curve provides values as a function of the hop distance as shown in Figure 6.5.

If there were doubts about the type of climate in order to apply the previous criteria, the worst-case situation should be assumed. In certain cases, other rules based on average clearance values of temperate and tropical climates could be also applied, leading to less pessimistic results.

Specific conditions might apply for frequencies below 2 GHz, where it might be necessary to consider lower percentages of R_1 clearance. The reason is to avoid the

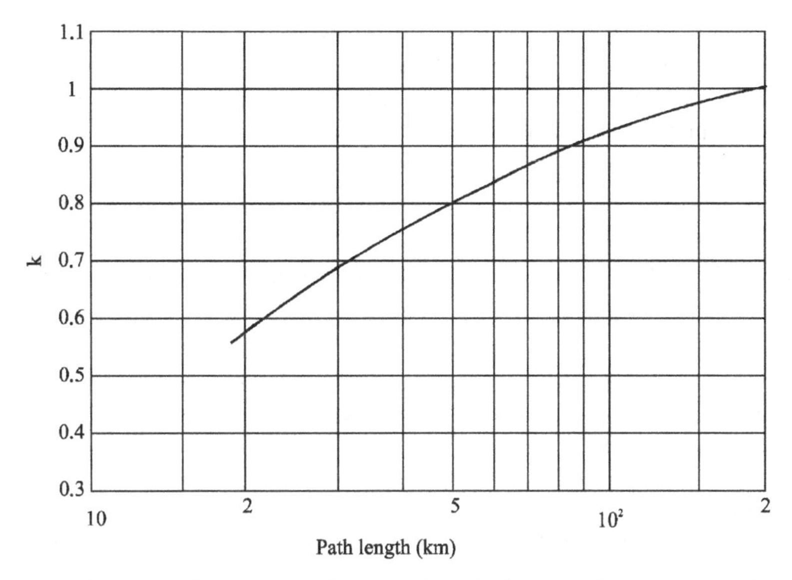

FIGURE 6.5 Value of k_e exceeded for approximately 99.99% of the worst month (continental temperature climate). (*Figure courtesy of ITU-R.*)

installation of antennas with remarkable diameter values on the highest part of the communication towers.

Also, in the case of frequencies over 13 GHz, the accuracy resulting from the estimation of the obstacle height, might be equal to the radius of the first Fresnel zone. The accuracy (error) associated with the obstacle heights should be added to the clearance distance in order to avoid clearance underestimation derived from the calculation process.

If the path profile would be such that the previous conditions are fulfilled for any antenna height (even zero) the installation height will be chosen according to economical and other project engineering criteria associated with the specifics of the transmission site. Usually, among the criteria applied in those cases are minimum antenna heights over close buildings and vegetation for security and protection of the installation. In those cases, increase the height of the antennas does not provide any further enhancement on the median received signal and might increase the probability of multipath fading.

6.5.1.2 Two and Three Antenna Space Diversity Configurations Microwave LOS links below 10 GHz, in geographical areas where the refractivity statistics are such that multipath is a relevant perturbation, will require diversity configurations in order to fulfill the performance specifications. In these cases, the antenna heights will be calculated according to the specific procedure given in ITU-R P.530.

First in this method, the antenna that will be installed at the highest position will be calculated using the procedures and criteria for single antenna installations and explained in previous sections. The height of the antenna in the lowest position will be obtained applying the following clearance criteria:

- $0.6 \times R_1$ to $0.3 \times R_1$ if the path obstruction extends along a portion of that path (round obstacles).
- $0.3 \times R_1$ to $0.0 \times R_1$ if the obstruction is associated with one or two isolated obstacles on the path profile.

The value of k used in the calculations will be the median value of the point k factor (if no other value is available 4/3 will be used). These clearance criteria have been established in order to permit a maximum diffraction loss that ranges from 3 to 6 dB as well as to reduce simultaneously the occurrence probability of surface multipath fading.

Once the antenna heights have been obtained, the distance between the antennas should satisfy the diversity requirements under multipath fading conditions. If that was not the case, the heights should be modified accordingly and the diversity requirements and the clearance criteria would not be compatible, a different technique against multipath, apart from spatial diversity should be sought.

If the channel plan of the link is below 2 GHz and in situations where the maximum available height on the tower limits the clearance values, the lower values of the clearance ranges mentioned earlier will be chosen.

Alternatively, the clearance of the lower antenna may be chosen to give about 6 dB of diffraction loss during normal refractivity conditions (i.e., during the middle of the day), or some other loss appropriate to the fade margin of the system. These diffraction losses might be determined either by calculations or test measurements. If measurements should be carried out, they need to be performed on several different days to avoid anomalous refractivity conditions. The calculation of the diffraction losses can be done according to the methods described in Chapter 7.

In the case of paths where the multipath is associated with reflections on stable surfaces (i.e., water or flat terrain), it might be necessary to obtain first the height of the antenna that will be installed on the highest position according to the criteria provided by single antenna cases. Once the maximum value is determined, the position of the other antenna (or antennas) will be obtained calculating the minimum antenna diversity separation distance according to multipath protection requirements.

In extreme situations (i.e., over long paths over wide water surfaces) it might be necessary to employ diversity configurations with three antennas. In those cases, the lowest antenna position will be obtained using the minimum height criteria for the lowest antenna described at the beginning of this section. The highest position will be obtained according to general clearance criteria for single antenna cases, and finally the third antenna that will be in the middle, will be positioned according to the minimum distance that provides enough protection against multipath fading.

6.5.2 Relevant Reflection Reduction Procedures for Non-diversity Configurations

In cases where there are relevant surface reflection components, the optimal antenna position will not only account for diffraction. The antenna heights will be also calculated considering the possibility of eliminating the specular reflection or at least trying to transform those specular components into low amplitude scattered components.

This technique minimizes the effect of surface multipath propagation without the need for diversity configurations. The first step in the reflection reduction process, the location of the reflection point on the path should be identified. If the reflection point has special features, as will be described in the next subsection, a specific study will be carried out in order to reduce the reflection using different techniques, most of them interrelated.

6.5.2.1 Areas Causing Relevant Reflection Significant reflections are produced in areas where the effective reflection coefficient of the soil is higher Earth's surface. The effective reflection coefficient can be either measured or calculated from soil conductivity and permittivity values according to the methods provided in Chapter 7.

Critical reflections are produced on large bodies of water such as sea, lakes, marshes or water reservoirs. Desserts, flat plateaus, flat building roofs, rice fields, or any steppes without vegetation can also be good reflection surfaces. On the contrary, areas with irregular topography and zones covered with vegetation are less likely to produce relevant reflection components. Additionally, it should be taken into account that the reflection coefficient is also a function of the frequency. The effective reflection area will also depend on the frequency, the smaller the frequency, the wider the reflection area. Figure 6.6 illustrates conceptually a hop with relevant reflections, that is, a hop over the sea.

6.5.2.2 Shielding of the Reflection Point This technique consists of taking advantage of the obstacles along the path profile (hills, mountains, buildings etc) in order to shield the antennas from the surfaces that might cause relevant reflections along the path (see Figure 6.7). In order to find an obstacle that shields adequately

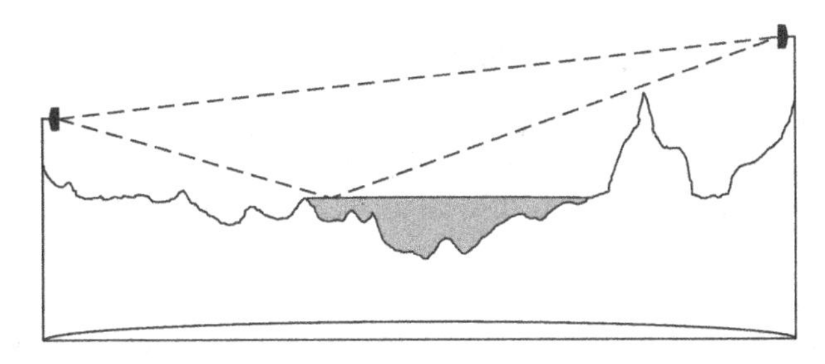

FIGURE 6.6 Hop with reflection point on the surface of a water body.

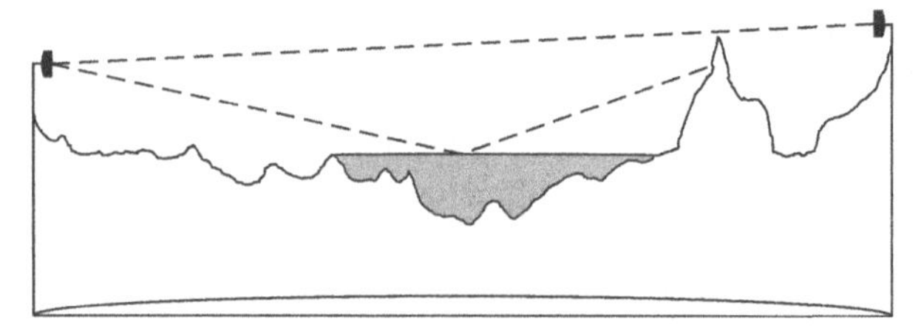

FIGURE 6.7 Shielding of the reflection point.

the system, and thus blocking the propagation path of the reflected component, it is necessary to analyze different paths of the signal associated with a range of effective k values that vary from k_e (99.9%) (or any other minimum value) to any high value. The reflection component should be blocked or at least conveniently shielded, not only for average values but also for high values of the effective k factor.

The ideal shielding situation will be based on hills covered with vegetation, in order to reduce the possible diffraction of the reflected path. Also, the shielding will be easier in profiles with reduced clearance over the path obstacles.

The benefit of shielding the reflection point using one of the obstacles of the path might be lost if one or more reflected components are super refracted over the obstacles because reflection multipath and distortion are more probable under those conditions. Additionally, any changes in the antenna heights in order to block the reflection components should be done maintaining the clearance criteria described in previous sections.

6.5.2.3 *Translating the Reflection Point* This technique consists of adjusting the antenna height at one or both ends of the hop in order to place the reflection point on a surface with vegetation or an irregular area. In the case of mixed water–land paths, it would be desirable that the reflection point happens on the land portion and not on the water body, or even better, if the land where the reflection is located is covered with trees or any other vegetation layer. As a rule of thumb for preliminary designs, diminishing the antenna height will get the reflection point closer to that antenna. The translation of the reflection point will start analyzing the potential reflection areas as described in the next subsection.

6.5.3 Calculation of Potential Reflection Areas

The position of potential reflection areas long the path profile, in order to optimize the antenna heights for diversity and non-diversity installation can be determined using any of the two following methods, known as Geometrical and Analytical Method respectively.

6.5.3.1 *Geometrical Method* An approximate method that is very useful in real practice for its simplicity is the location of the reflection point as the path points

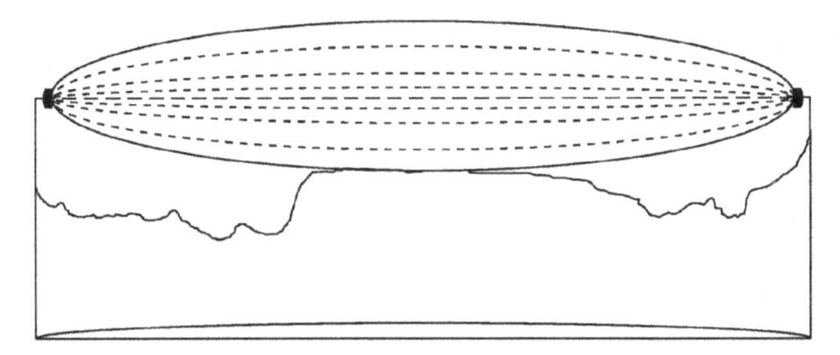

FIGURE 6.8 Geometrical method for obtaining the reflection location.

that are tangent to any Fresnel ellipsoid. Those locations will fulfill the condition of having equal incidence and reflection angles. In consequence, the reflection point will be found increasing the radius of the Fresnel ellipsoid until it touches the path profile. If the tangent of the ellipse is parallel to the terrain profile, then that reflection point can be considered relevant. Figure 6.8 illustrates schematically this geometrical approach.

This method can also be used to determine the optimum antenna separation in cases where space diversity schemes are required to protect the system against multipath fading (Figure 6.9). In those cases, once the reflection point is determined, the Fresnel radius is incremented $\lambda/2$. Then, the ellipsoid is displaced again to be tangent to the terrain and the new antenna height is recorded. The difference between the antenna heights after and before increasing the radius of the ellipsoid will be the vertical separation of the antennas that compose the diversity system.

6.5.3.2 Analytic Method If a higher accuracy of the situation of the reflection point is required an analytic method described in ITU-R P.530 might be used. This

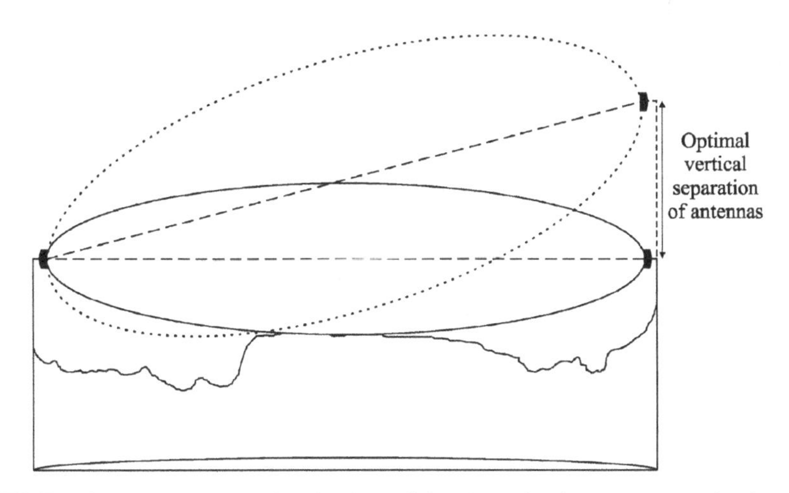

FIGURE 6.9 Geometrical method for determining the optimal antenna separation in spatial diversity configurations.

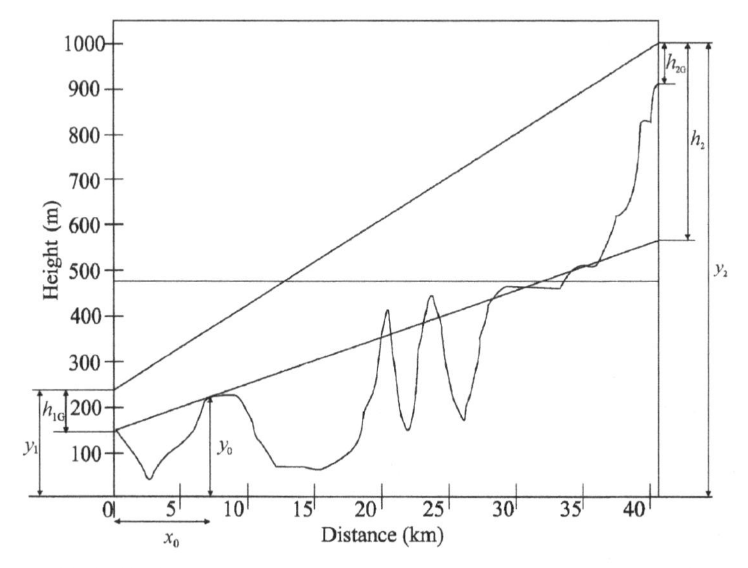

FIGURE 6.10 Geometrical representation of the analytical method.

method assumes that the antenna heights at both ends of the hop have been already determined according to the clearance criteria for non-diversity configurations. The method is applied iteratively and it is composed of two steps as shown in Figure 6.10.

The first step consists of the calculation of the heights of the transmitting and upper receiving antennas above possible specular reflection areas on or near the path profile. Such areas may or may not be horizontal depending on the path profile, and there may be more than one of them. While some areas can be determined from maps, others may require a detailed inspection of the terrain along and in the close vicinity of the path. The heights h_1 and h_2 of the antennas above a reflection area of inclination angle ν are calculated using equations (6.8) and (6.9):

$$h_1 = h_{1G} + y_1 - y_0 + x_0 (\tan \nu)\, 10^3 \tag{6.8}$$

$$h_2 = h_{2G} + y_2 - y_0 - (d - x_0)(\tan \nu)\, 10^3 \tag{6.9}$$

where

y_1 = altitude above sea level at one of the hop ends (site 1)(m)
y_2 = altitude above sea level at the other hop end (site 2)(m)
h_{1G} = antenna height above ground at site 1
h_{2G} = antenna height above ground at site 2
y_0 = altitude of mid-point of reflection area above sea level (m)
x_0 = distance of mid-point of reflection area from site 1 (km)

Once the antenna heights have been calculated, the distance from the potential reflection areas to the sites 1 and 2 are calculated (d_1 and d_2 in km) using equations (6.10) and (6.11). The calculation is carried out iteratively using a range of k values from k_e (99.9%) to 10^9. On each iteration, the values of h_1 and h_2 will be adjusted to move the reflection point away from the specular reflection area:

$$d_1 = \frac{d(1+b)}{2} \tag{6.10}$$

$$d_2 = \frac{d(1+b)}{2} \tag{6.11}$$

where

$$b = 2\sqrt{\frac{m+1}{3m}}\cos\left[\frac{\pi}{3} + \frac{1}{3}\arccos\left(\frac{3c}{2}\sqrt{\frac{3m}{(m+1)^3}}\right)\right] \tag{6.12}$$

$$m = \frac{d^2}{4a_e(h_1+h_2)}10^3 \tag{6.13}$$

$$c = \frac{(h_1-h_2)}{(h_1+h_2)} \tag{6.14}$$

In equation (6.13), a_e represents the effective Earth's radius associated with a k factor value ($a_e = ka$ where $a = 6375$ km, the actual Earth's radius). The distance d is expressed in km and the heights h_1 and h_2 in m.

Sometimes, it will not be possible to avoid specular reflections. In those cases, the length difference between the direct and reflected paths is calculated for the same range of k values as before: k_e (99.9%) to 10^9 using the following expression:

$$\tau = \frac{2f}{0.3d}\left[h_1 - \frac{d_1^2}{12.74k}\right]\left[h_2 - \frac{d_2^2}{12.74k}\right]10^{-3} \tag{6.15}$$

Each time that the number of wavelengths τ is an integer, with k variable, the received signal level has a minimum. This condition should be avoided whenever possible. The greater the number of integer values of $\tau_{max} - \tau_{min}$ as k varies over its range, the more likely is the performance to be degraded and diversity will be required. The ITU-R provides the following criteria to evaluate the problem and solve the situation with diversity configurations:

- If $\tau_{máx} - \tau_{mín} < 1$ while k varies over its range, diversity configurations will not probably be required. In this case, the antenna heights (one or both of them) should be adjusted so $\tau \approx 0.5$ for the median value of k.
- If $\tau_{máx} - \tau_{mín} \geq 1$, the depth of surface multipath fades and whether some kind of diversity might be necessary depends on how well the signal is reflected.

BIBLIOGRAPHY

ITU-R Rec. P. 453: The radio refractive index: its formula and refractivity data. International Telecommunication Union. Radiocommunication Sector. ITU-R. Geneva. 2003.

ITU-R Rec. P. 530: Propagation data and prediction methods required for the design of terrestrial line-of-sight systems. International Telecommunication Union. Radiocommunication Sector. ITU-R. Geneva. 2009.

CHAPTER 7

PROPAGATION CALCULATION METHODS ACCORDING TO ITU-R P SERIES RECOMMENDATIONS

7.1 GENERAL CONSIDERATIONS ON PROPAGATION CALCULATION METHODS

This chapter aims at providing a clear and detailed description of the procedures for calculating the parameters that characterize fading that are associated with different propagation effects. The sections will contain information related only to the propagation phenomena relevant to the design and operation of radio-relay links.

The algorithms, methods and expressions presented here are mostly covered by ITU-R P.530 Recommendation that specifically refers to the "propagation data and prediction methods required for the design of terrestrial line-of-sight systems". ITU-R Recommendation P.530 provides prediction methods for the propagation effects that should be taken into account in the design of digital fixed line-of-sight links, both under clear air and rainfall conditions. The propagation phenomena that are included in the prediction procedures of ITU-R Rec. P.530 are:

- Attenuation due to hydrometeors.
- Attenuation caused by atmospheric gases.
- Diffraction fading due to obstruction of the path by terrain obstacles under abnormal refraction periods.
- Fading due to multipath reflection on Earth's surface.
- Frequency selective fading and delay during multipath propagation.

Microwave Line of Sight Link Engineering, First Edition. Pablo Angueira and Juan Antonio Romo.

- Fading due to atmospheric multipath or beam spreading associated with abnormal refraction.
- Variation of the angle-of-arrival at the receiver terminal and angle-of-launch at the transmitter terminal due abnormal refraction conditions.
- Cross-polarization discrimination (XPD) reduction in multipath or rain conditions.

During the last years, a significant amount of changes have been proposed to the algorithms used to evaluate different propagation phenomena. These changes have led to a relevant number of versions of ITU-R Rec. P.530.

In addition to this reference document, the Recommendation ITU-R P.452 provides methods for evaluating the propagation phenomena (in this case associated with interference) between stations on the surface of the Earth, accounting for both clear air and hydrometeor scattering interference mechanisms. The set of propagation models included in ITU-R Rec. P452 ensure that the predictions take into account all the significant interference propagation mechanisms that can arise. Methods for analyzing the radioelectric, meteorological, and topographical features of the path are provided for any practical propagation path up to a distance limit of 10 000 km

Table 7.1 summarizes both ITU-R Recommendations that refer to propagation in point-to-point LOS systems. The objective of the table is to provide a look-up table that can be used as a fast index to identify the method, application conditions and limitations for preliminary studies without requiring exhaustive consultation of all the ITU-R material.

These recommendations usually refer to geophysical data represented by global maps. Table 7.2 summarizes a description of the geophysical parameters of the maps as well as their names and recommendations where they can be found.

7.2 FADING DEFINITION

Propagation in radio-relay links is mainly affected by atmospheric conditions in such a way that received signal level is constantly fluctuating over time. These signal variations can be caused by any of the propagation mechanisms that degrade received signal and that have been described in Chapter 2.

In radiocommunication system design, the nominal received power is a concept associated with the median received power during a long enough observation time. The power decrease from the nominal value is usually called fading. The difference between nominal level and received level under fading conditions is called instantaneous fading or fade depth and is expressed in dB. The time interval between decrease and recovery of nominal level is called fade duration. Figure 7.1 illustrates an example of typical record of received signal levels over time that shows fading and provides an example of fading parameters.

Fast small amplitude variations of the received signal around a nominal or median value of P_0 (dBm) will always be present under normal operation conditions. These variations cause, known as scintillation are caused by small-scale refraction effects.

TABLE 7.1 ITU-R Methodology for Calculating Propagation in Point-to-Point Line-of-Sight Communications Systems

	ITU-R Method	
	P.452	P.530
Application	Services employing stations on the surface of the Earth; interference	Line-of-sight fixed links
Output	Path loss	Path loss Outages due to selective fading Outages due to flat fading XPD outage Diversity enhancements (clear air conditions)
Frequency	100 MHz–50 GHz	Approximately 150 MHz–40 GHz
Distance	Up to and beyond the radio horizon	Up to 200 km if line-of-sight
% time	0.001–50 / Average year and worst month	All percentages of time in clear-air conditions 1 – 0.001 in precipitation conditions
Terminal height	No limits specified	Enough to ensure specified path clearance
Input data	Frequency Time percentage Transmitter coordinates Receiver coordinates Receiver antenna height Transmitter antenna height Path profile data Meteorological data	Distance Frequency Transmitter height Receiver height Percentage time Path obstruction data Terrain information Climate data

As illustrated in Figure 7.1, occasionally, a severe fading situation will start at t_0. At a later t_1, the received power value is P_1 (dBm) and the fading depth is F_1, equal to $P_0 - P_1$ (dB). After t_1, the signal level continues to decrease until thc minimum value P_2 is reached (fading depth $F_2 = P_0 - P_2$). Signal level then increases again, reaching P_1 at t_2. Signal recovers the nominal value level, $P_{0,}$ at the t_3 time. The fade duration thus described by this example is $t_3 - t_0$ and the fade depth is considered to be F_1 during the interval:

$$\tau_1 = t_2 - t_1 \tag{7.1}$$

During the fade occurrence, the received median power is P_f (dBm) that is obviously lower than P_0.

The difference between the median received value and the median value during the fading occurrence is referred to as *depression of the median* or Pearson decrease

TABLE 7.2 Geophysical Parameters Used in Prediction Methods

Rec. ITU-R	Description	Grid Resolution	Spatial Interpolation Required	Interpolation in Probability	Interpolation of the Variable
P.839	Mean annual 0°C isotherm height (km) (zero deg)	1.5° × 1.5°	Bilinear	N.A.	N.A.
P.837	Rain rate exceedance probability (%)(rain rate)	1.125° × 1.125°	Bilinear	N.A.	N.A.
P.1511	Topographic altitude (a.m.s.l.) (km)(altitude)	0.5° × 0.5°	Bicubic	N.A.	N.A.
P.836	Total columnar water vapor exceedance probability (%) (IWVC)	1.125° × 1.125°	Bilinear (1)	Logarithmic	Linear
P.836	Surface water vapor density exceedance probability (%) (Rho)	1.125° × 1.125°	Bilinear (1)	Logarithmic	Linear
P.836	Water vapor scale height	1.125° × 1.125°	Bilinear	Logarithmic	Linear
P.1510	Mean annual surface temperature (temperature)	1.5° × 1.5°	Bilinear	N.A.	N.A.
P.453	Median value of the wet term of the refractivity (Nwet)	1.5° × 1.5°	Bilinear	N.A.	N.A.
P.840	Columnar CLW exceedance probability (%)	1.125° × 1.125°	Bilinear	Logarithmic	Linear
P.840	Statistical distribution of total CLW content	1.125° × 1.125°	Bilinear	N.A.	N.A.

($P_0 - P_f$ in dB). Alternatively, a quantitative expression of fading can be written as a function of the signal envelope voltage. If that voltage is r:

$$F_1 = P_0 - P_1 = 20 \log \frac{r_0}{r_1} \tag{7.2}$$

7.3 REFLECTION ON EARTH'S SURFACE

7.3.1 Effective Surface Reflection Coefficient

The effective reflection coefficient of a surface depends on parameters that are difficult to estimate such as ground conductivity, surface roughness, degree of humidity within

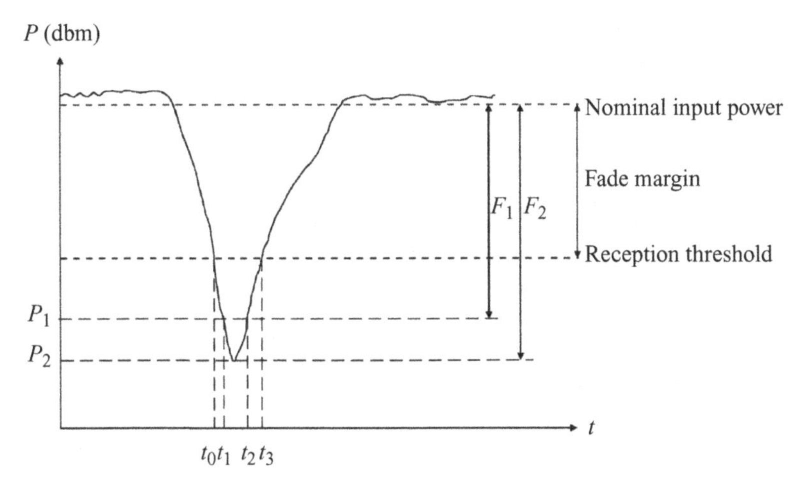

FIGURE 7.1 Fading and associated parameters.

the soil in the area, etc. For this reason, a precise determination might be complex and subjected to uncertainties, particularly at high frequencies. Furthermore, we have to take into account that reflection effect does not only happen at the geometric point of reflection. It is necessary to consider the entire surface formed by Fresnel zones along the geometric point of reflection.

Nevertheless, for practical purposes a rough estimate procedure defined in ITU-R P.530 Recommendation can be followed. This procedure is based on the principles of optical geometry that are valid as long as the grazing angle of the incident ray at the reflection point, φ, given in milliradians, is above a threshold value that is equal to:

$$\varphi = \left[\frac{2.1}{f}\right]^{1/3} \tag{7.3}$$

where f is the frequency in GHz.

For narrower angles, where diffraction phenomena dominate, diffraction models over spherical Earth. Figure 7.2 shows geometry of the reflection profile over spherical Earth.

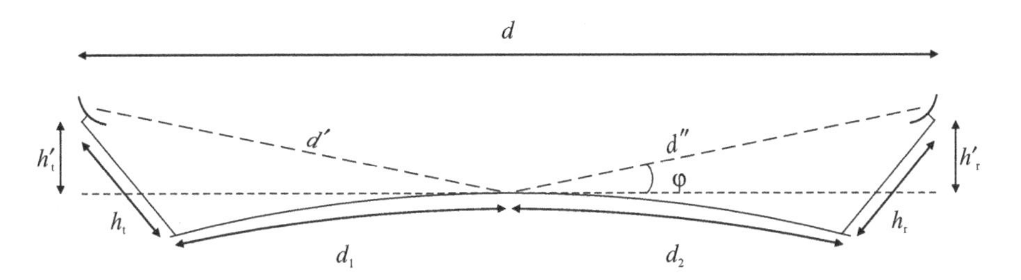

FIGURE 7.2 Reflection on the Earth's surface (spherical Earth model).

The effective reflection coefficient, ρ_{eff}, in the most general case, can be obtained by the following equation:

$$\rho_{\text{eff}} = \rho \cdot D \cdot R_s \cdot R_r \tag{7.4}$$

where

ρ = specular reflection coefficient of the flat surface
D = divergence factor of the Earth's surface
R_s = specular reflection factor
R_r = surface roughness factor

The specular reflection coefficient of a flat surface, ρ, can be calculated by the following equation:

$$\rho = \left| \frac{\sin\varphi - \sqrt{C}}{\sin\varphi + \sqrt{C}} \right| \tag{7.5}$$

Depending on the polarization:

$$\text{Horizontal} \quad C = \eta - \cos^2\varphi \tag{7.6}$$

$$\text{Vertical} \quad C = \frac{\eta - \cos^2\varphi}{\eta^2} \tag{7.7}$$

The parameter η is the complex dielectric constant of the Earth's surface near the surface reflection area that can be obtained from equation (7.8):

$$\eta = \varepsilon_r - j18\frac{\sigma}{f} \tag{7.8}$$

ε_r is the relative dielectric constant and σ is the conductivity (S/m) that can be estimated according to the information available in ITU-R P.527 Recommendation.

The reflection coefficient of equation (7.5) depends on the angle of incidence of the reflected signal. This angle, φ, depends on k (equivalent Earth radius) and can be obtained from equation (7.9):

$$\varphi = \frac{h_1 + h_2}{d}\left[1 - m\left(1 + b^2\right)\right] \tag{7.9}$$

$$m = \frac{d^2}{4a_e\left(h_1 + h_2\right)} \cdot 10^2 \tag{7.10}$$

$$b = 2\sqrt{\frac{m+1}{3m}}\cos\left[\frac{\pi}{3} + \frac{1}{3}\arccos\left(\frac{3c}{2}\sqrt{\frac{3m}{(m+1)^3}}\right)\right] \tag{7.11}$$

$$c = \frac{h_1 - h_2}{h_1 + h_2} \tag{7.12}$$

where

h_1 and h_2 = transmitter and receiver antennas heights (m)
d = distance between antennas (km)
a_e = effective Earth radius for a specific k, $a_e = k\ a$ (km), $a =$ 6375 km

Distances d_1 and d_2 to each possible reflection surface from the transmitter and receiver (end points of the hop), are given by equations (7.13) and (7.14):

$$d_1 = \frac{d\,(1+b)}{2} \tag{7.13}$$

$$d_2 = \frac{d\,(1-b)}{2} \tag{7.14}$$

In order to take into account the effect of specular reflection on a spherical surface, we add a divergence factor of the Earth's surface, D, to the effective coefficient equation. D is calculated by the following equation:

$$D = \sqrt{\frac{1-m\left(1+b^2\right)}{1+m\left(1+b^2\right)}} \tag{7.15}$$

If the reflection surface is completely specular or smooth across the first Fresnel zone, the effective reflection coefficient is given by equation (7.16), and reflection is called specular:

$$\rho_{\text{eff}} = \rho \cdot D \tag{7.16}$$

If this is not the case, we need to add the reduction factors associated with the type of reflection surface.

Evaluating the reflection area requires a previous calculation of the intersection surface of the first Fresnel ellipsoid with the Earth's surface in the reflection area. The major axis of the Fresnel ellipsoid will intersect the surface exactly on the reflection point. The ellipse length (major axis), L_1 (parallel to the signal path), and width (minor axis), W_1, perpendicular, can be calculated through the following equations in kilometers:

$$L_1 = d\sqrt{1+\frac{4f\ h_1 h_2 10^{-2}}{3\ d^2}}\cdot\left[\frac{f\,(h_1+h_2)^2\ 10^{-2}}{3\ d^2}\right]^{-1} \tag{7.17}$$

$$W_1 = \sqrt{\frac{3\cdot 10^{-4} d}{f}} \tag{7.18}$$

When the surface is rough, the reflected signal has two components: a specular component that is consistent with the incident signal and another diffuse scattering component that fluctuates in amplitude and phase with a Rayleigh distribution. The roughness of an area can be estimated by the Rayleigh roughness criterion that regards a surface specular if the parameter g is lower than 0.3. The parameter g is calculated from equation (7.19):

$$g = \frac{40\pi f \sigma_{\mathrm{h}} \sin\varphi}{3} \tag{7.19}$$

where

f = frequency (MHz)
σ_{h} = standard deviation of the relative height values within the first Fresnel ellipsoid intersection area. The relative height values are calculated with respect of the mean value of the height within the ellipsoid (m)
φ = grazing angle measured with respect the horizontal (tangent to the Earth's surface at the reflection point)

If the surface within the first Fresnel zone ellipse intersection is relatively rugged, with a g value equal or more that 0.3, the estimation of a surface roughness factor can be based on the following equation:

$$R_{\mathrm{r}} = \sqrt{\frac{1+\left(\frac{g^2}{2}\right)}{1+2.35\left(\frac{g^2}{2}\right)+2\pi\left(\frac{g^2}{2}\right)^2}} \tag{7.20}$$

If there is clearly a portion, or portions, of the total first Fresnel zone ellipse that will be associated with specular reflection, a specular reflection reduction factor, R_{s}, must be used. Otherwise, R_{s} will be 1. The R_{s} factor is calculated from equation (7.21):

$$R_{\mathrm{s}} = \sqrt{\frac{f\,(h_1+h_2)^4\,(\Delta x)^2\,10^{-2}}{3\,h_1 h_2 d^2}} \tag{7.21}$$

where

Δx = length of the portion of the Fresnel ellipse with specular reflection (km)
h_1 and h_2 = transmitter and receiver antenna heights (m)

7.3.2 Fading Associated with Specular Reflections

If the path between two stations of the hop has an effective reflection coefficient, and if the hop length is moderate, the reflection on Earth's surface might become the dominant effect behind multipath reflection fading. This situation is typical in paths over bodies of water such as seas, lakes or marshes.

In these cases, an estimation of the fading produced by the vector sum of the components associated with the direct and reflected paths can be carried out with a two-ray combination model. The transfer function of a two-ray model is expressed in terms of the relative amplitudes and delays of both components as shown by equation (7.22):

$$H(w) = 1 + b\exp[-j(w\tau + \beta)] \tag{7.22}$$

where

b = relative amplitude of the reflected path with respect to the LOS path
τ = relative delay of the reflected path with respect to the LOS component
β = reflection phase rotation (phase of the reflection coefficient)

The values of b (GHz) and τ (ns) can be calculated using the expressions (see previous section) for the reflection coefficient. The relative amplitude of the reflected path is described by equation (7.23):

$$b = |R|\,D\,g_{\mathrm{TR}}\,g_{\mathrm{RR}} \tag{7.23}$$

$$\tau = \frac{\Delta}{2\pi f} \tag{7.24}$$

where

$|R|$ = reflection coefficient module of the soil on the Fresnel ellipsoid area
D = divergence factor for terrain roughness within the reflection area
$g_{\mathrm{TR}}\,g_{\mathrm{RR}}$ = gain values of the transmitting and receiving antennas on the launch and arrival directions of the reflection component
Δ = path difference between the LOS and reflected components (m)
f = frequency (GHz)

The squared module of the transfer function of the channel will be a periodic function with maxima $(1 + b)$ and minima $(1 - b)$ is provided by equation (7.25) as well as the fade depth in equation (7.26):

$$|H(w)|^2 = 1 + b^2 + 2b\cos(w\tau + \beta) \tag{7.25}$$

$$F_{\mathrm{R}}(w) = -10\log\left[|H(w)|^2\right] = -10\log\left[1 + b^2 + 2b\cos(w\tau + \beta)\right] \tag{7.26}$$

The fade depth $F_{\mathrm{R}}(w)$ depends on the link distance (by b, ι y β), the frequency, the antenna height and the effective Earth's radius. The minima of equation (7.26) are in frequencies that can be calculated using (7.27):

$$2\pi f + \beta = (2n + 1)\pi \qquad \text{with } n = 1, 2, 3 \ldots \tag{7.27}$$

The maximum depth at these frequencies will then be:

$$F_{\mathrm{R}}(w) = -20\log(1 - b) \tag{7.28}$$

Previous expressions are in fact the same expressions of a generic two-ray composition that was also the model used for analyzing multipath fading in Chapter 2. This case also depends on the frequency and in consequence it can also be considered an example of selective fading.

As an illustrative example lets calculate the frequency associated with a maximum fade depth that is closer to the radio channel center frequency, assumed β and τ are known. The calculation is equivalent to find the value of n associated with the closest frequency, f_0, to the center frequency f_c. The relative difference between two consecutive minima is $1/\tau$ and thus, the following condition must be fulfilled:

$$|f_c - f_0| \leq \frac{1}{2\tau} \tag{7.29}$$

In addition the index of the f_0 will be:

$$2\pi f_0 + \beta = (2n_0 + 1)\pi \tag{7.30}$$

If we consider that $\beta = c\pi$, it is immediate to deduce the value of n_0:

$$n_0 = E\left[f_c\tau + \frac{c}{2}\right] \tag{7.31}$$

where E represents the whole part of the calculated value between brackets and finally f_0 will be calculated with equation (7.30).

7.4 ATTENUATION DUE TO ATMOSPHERIC GASES

Attenuation due to absorption of oxygen and water vapor along a d (km) long hop is given by equation (7.32):

$$A = \gamma_a\, d = (\gamma_a + \gamma_w)\, d \tag{7.32}$$

where

γ_a = total specific attenuation due to atmospheric gases (dB/km)
γ_o = specific attenuation due to oxygen (dB/km)
γ_w = specific attenuation due to water vapor (dB/km)

A simplified method for calculating the specific attenuation coefficients is provided by the Appendix 2 of the ITU-R P.676 Recommendation. The values provided there are only valid for the frequency range from 1 – 350 GHz. that covers all the frequencies allocated to microwave LOS links where this attenuation is relevant.

Figure 7.3 shows a graphical representation of the equations provided by the mentioned recommendation.

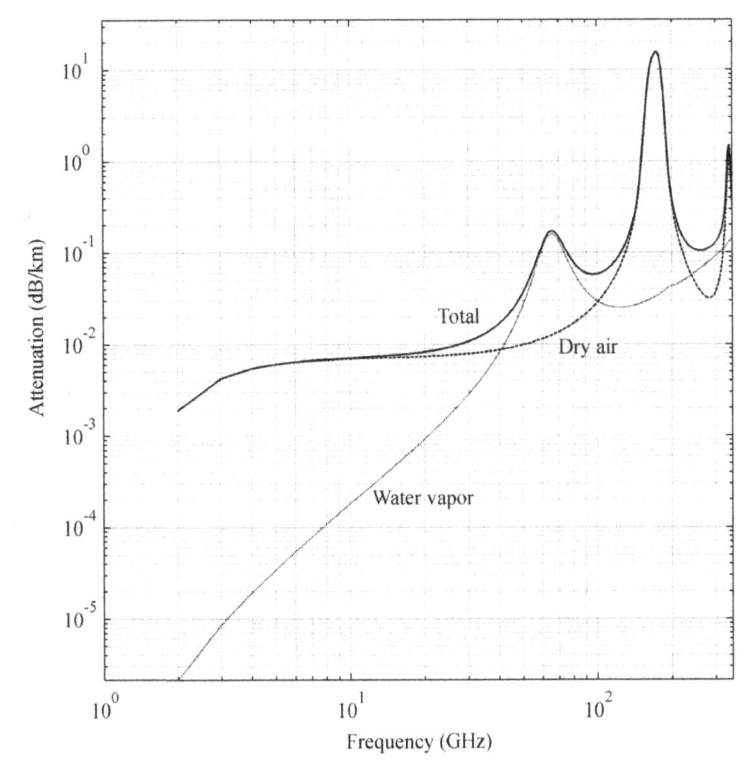

Attenuation in dB/km for pressure of 1013 hpa, temperature of 15°
and water vapor density of 7.5 g/cm3

FIGURE 7.3 Specific attenuation of oxygen and water vapor as a function on frequency.

The curves are specific attenuation at sea level for dry air and water vapor with a density of 7.5 g/m^3. This water vapor value is the practical reference for estimating the specific attenuation due to gases. The model assumes that the density of oxygen is constant.

7.5 DIFFRACTION FADING

Losses due to diffraction on an obstacle happen when the first Fresnel ellipsoid is blocked by any terrain irregularity or by any kind of artificial construction. Diffraction fading fundamentally affects to the determination of antenna heights in each hop, so its study is one of the key tasks of link engineering procedures (see Chapter 6).

Diffraction losses are time-varying because of refractivity gradient variations. Reliable local statistics in the lower atmosphere, approximately up to 100 meters above ground level, should be used to quantify this variability, as long as they are available. If local statistics are not available, ITU-R P.453 Recommendation provides global predictions of the refractivity gradient in the first 100 m of the troposphere.

Diffraction losses are evaluated through the parameter known as clearance and generally referred to as h. Clearance is positive when the obstacle is above the line of sight or negative otherwise. Diffraction losses on an obstacle happen when clearance takes values such as $h > -0.577\, R_1$.

Different models are used to calculate the attenuation caused by diffraction. The calculation model choice will depend on:

- Existence of an isolated obstacle or several obstacles along the profile.
- In the case of several obstacles, the existence of a dominant obstacle.
- Geometric model to be applied to the obstacle: sharp edge, rounded obstacle, or wedge with finite conductivity.

An obstacle is considered isolated if there is no interaction between with the surrounding area, and the clearance is at least $0.577\, R_1$ on both sides of the obstacle. Under this simplified condition, diffraction attenuation will be only caused by the obstacle and the surrounding area will not contribute to this attenuation.

For calculation purposes, isolated obstacles are modeled either as edges with negligible thickness and curvature radius close to zero or as thick and smooth edges, with a well defined curvature radius in its highest part. Additionally, there will be different methods for the cases where there is a single isolated obstacle and for those where several irregularities are relevant in the path profile.

Figure 7.4 shows a general organization chart on the procedure to evaluate diffraction.

7.5.1 Simplified Calculation from ITU-R Recommendation P.530

For design cases that only require an approximate calculation of diffraction losses caused by a specific obstacle, the formula that appears in ITU-R P.530 Recommendation can be applied. This expression is suitable for losses of more than about 15 dB:

$$A_{\mathrm{d}} = -20\frac{h}{R_1} + 10 \tag{7.33}$$

where

h = clearance of the most important obstacle in route
R_1 = radius of the first Fresnel ellipsoid

If detailed evaluation of diffraction losses was required, we can use the methods described in ITU-R P.526 Recommendation that are appropriate for cases of spherical Earth's surface and uneven ground with different types of obstacles. The following sections summarize the ones that could be used in certain for radio-relay link designs.

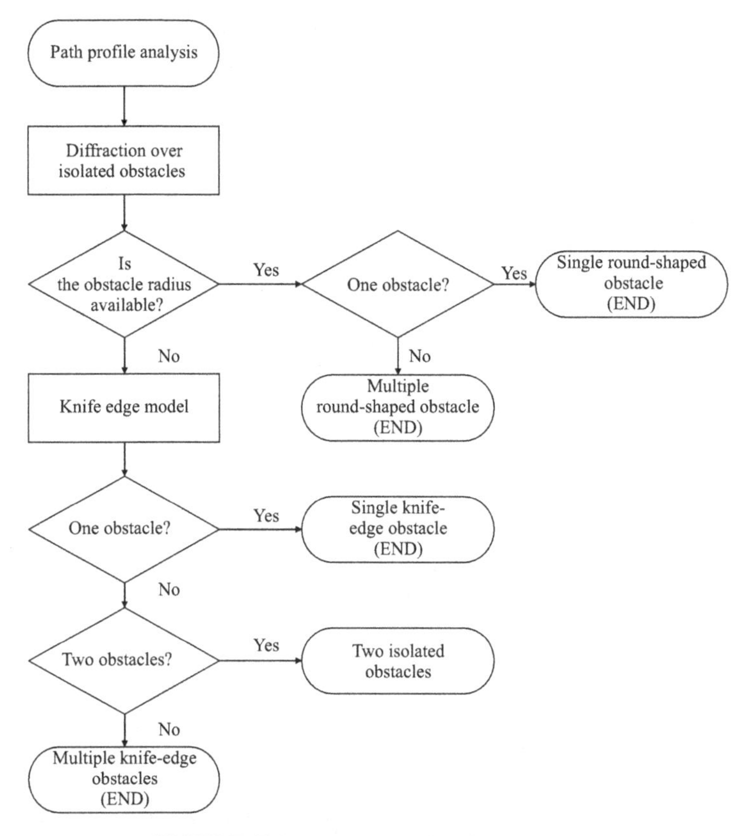

FIGURE 7.4 Evaluation of diffraction losses.

7.5.2 Diffraction on a Knife-Edge

Diffraction on a "knife-edge" isolated obstacle is based on the geometric model shown in Figure 7.5. As in the other methods, losses associated with this diffraction component depend on clearance h.

Diffraction losses are calculated through a parameter that depends on the geometry shown in Figure 7.5. This parameter is dimensionless and is usually represented by ν. It is obviously directly related to the Fresnel ellipsoid radius:

$$\upsilon = h\sqrt{\frac{2}{\lambda}\left(\frac{1}{d_1} + \frac{1}{d_2}\right)} \tag{7.34}$$

where

λ = wavelength (m)
h = clearance (m)
d_1 = distance to transmitter (km)
d_2 = distance to receiver (km)

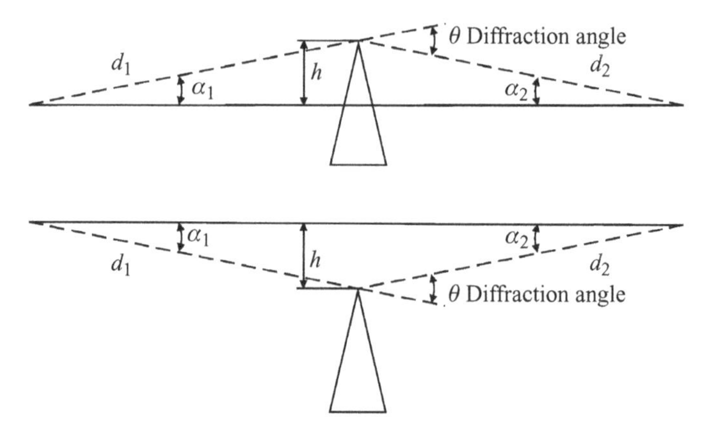

FIGURE 7.5 Model for diffraction losses on a knife-edge.

It is usual to work with an equivalent equation in function of the frequency (in MHz), with distances given in kilometers:

$$\nu = 2.58 \cdot 10^{-3} h \sqrt{\left(\frac{fd}{d_1 d_2}\right)} \tag{7.35}$$

Attenuation due to diffraction is calculated by the general equation:

$$J(\nu) = -20 \log \left(\frac{\sqrt{[1 - C(\nu) - S(\nu)]^2 + [C(\nu) - S(\nu)]^2}}{2} \right) \tag{7.36}$$

where $C(\nu)$ and $S(\nu)$ are the Fresnel integrals of the argument ν. In practice, we work with graphical representations of equation (7.36) such as the one in Figure 7.6 or with simplifications for certain ranges of ν. Particularly, for $\nu > 0.7$ it is usual to use equation (7.37):

$$J(\nu) = 6.9 + 20 \log \left(\sqrt{(\nu - 0.1)^2 + 1} + \nu - 0.1 \right) \tag{7.37}$$

This equation provides losses only related to effects in the first Fresnel ellipsoid. It is worth mentioning that $J(-0.78) \approx 0$, and this value $\nu = -0.78$ define the lower limit to which this approach must be used. $J(\nu)$ is set to zero for $\nu < -0.78$.

7.5.3 Single Round Shaped Obstacle

Figure 7.7 summarizes the model for calculating diffraction on a single round shaped obstacle.

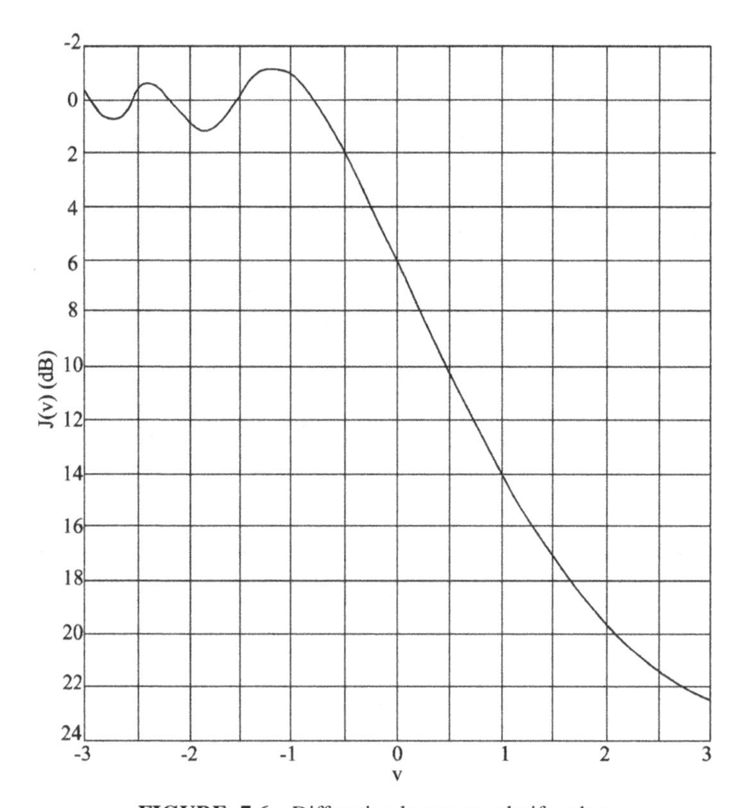

FIGURE 7.6 Diffraction losses on a knife-edge.

Distances d_1 and d_2 and height h above the base line in Figure 7.7 (line of sight), are measured with respect to the apex formed by the intersection of rays projection on the obstacle. The curvature radius of the obstacle, R, is the curvature radius of the vertex of a parabola fitted to the obstacle profile near its top. When the parabola is fitted, the maximum vertical distance from vertex to use in this procedure must be of the order of the first Fresnel zone radius at the path profile location of

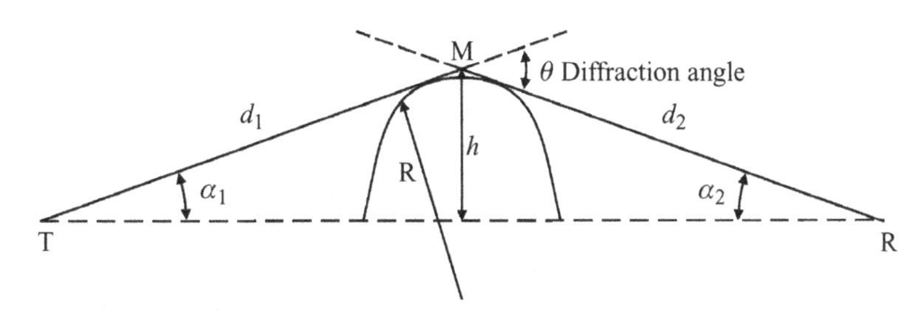

FIGURE 7.7 Geometry of diffraction on an isolated round shaped obstacle.

the obstacle. Diffraction loss associated with this geometry can be calculated from equation (7.38):

$$A = J(\nu) + T(m, n) \tag{7.38}$$

In equation (7.38), $J(\nu)$ is the Fresnel–Kirchoff loss due to the equivalent knife-edge obstacle with a maximum height that equals the vertex of the parabola. The dimensionless parameter ν can be evaluated by any of the equations described in the previous section. $T(m,n)$ is the additional attenuation due to the curvature of the obstacle and can be calculated by equation (7.39):

$$T\,(m, n) = 7.2\,m^{1/2} - (2 - 12.5\,n)\,m + 3.6\,m^{3/2} - 0.8\,m^2 \tag{7.39}$$

$$T(m, n) = -6 - 20\log(mn) + 7.2\,m^{1/2} - (2 - 17n)\,m + 3.6\,m^{3/2} - 0.8\,m^2 \tag{7.40}$$

Equation (7.39) is used for mn ≤ 4 while equation (7.40) is used for mn > 4. Values of m and n are calculated from equations (7.41) and (7.42):

$$m = \frac{R\left[\dfrac{d_1 + d_2}{d_1 d_2}\right]}{\left[\dfrac{\pi R}{\lambda}\right]^{\frac{1}{3}}} \tag{7.41}$$

$$n = \frac{h\left[\dfrac{\pi R}{\lambda}\right]^{2/3}}{R} \tag{7.42}$$

where

$R =$ obstacle radius from Figure 7.7 (m)
$d_1, d_2 =$ distances to transmitter and receiver (m)
$h =$ obstacle height (m)
$\lambda =$ wavelength (m)

When R tends to zero, $T(m,n)$ also tends to zero and equation (7.38) equals the diffraction on a knife-edge for a cylinder with zero radius.

7.5.4 Two Isolated Obstacles

When there are two perfectly identified isolated obstacles in the route, either knife-edge or rounded, methods derived from the theory of diffraction on isolated obstacles described in previous sections can be applied.

First, the clearance at each obstacle will be calculated and normalized to the radius of the first Fresnel zone: h/R_1. This step will allow identifying the relevant irregularities.

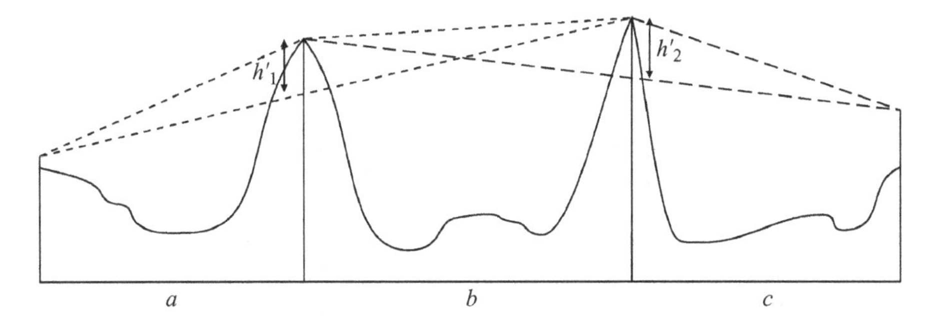

FIGURE 7.8 Profile with two isolated obstacles with similar clearance.

Two different methods can be used to estimate diffraction losses depending on the whether the clearance values of both obstacles are similar or significantly different. Both methods are regardless of the shape of the obstacle, either knife-edge or rounded.

7.5.4.1 Two Obstacles with Similar Clearance In this case the h/R_1 ratios are similar in both obstacles and therefore produce similar losses. Figure 7.8 shows an example of terrain profile for this case.

The method consists of successively applying the isolated obstacle diffraction theory on both obstacles. The top of the first obstacle acts as a diffraction source on the second one. The first diffraction part, defined by distances a and b and height h_1', causes a L_1 (dB) loss, and the second one, defined by distances b and c and height h_2', a L_2 (dB) loss. Total diffraction loss is then:

$$L = L_1 + L_2 + L_c \tag{7.43}$$

L_1 and L_2 are calculated using the formulae of knife-edge or round isolated obstacle, depending on the shape of each obstacle. A correction term L_c (dB) must be added in for taking b spacing between both obstacles into account. We can estimate L_c by the following formula:

$$L_c = 10 \log \left[\frac{(a+b)(b+c)}{b(a+b+c)} \right] \tag{7.44}$$

Equation may be considered valid when both L_1 and L_2 are higher than about 15 dB.

7.5.4.2 Two Obstacles with Different Clearance This second method is applied when h/R_1 ratios are significantly different in both obstacles and therefore they cause different losses. Figure 7.9 shows an example of terrain profile for this case. The obstacle with higher h/R_1 ratio is called predominant or main (obstacle M in Figure 7.9). The other obstacle is called secondary (obstacle S in Figure 7.9).

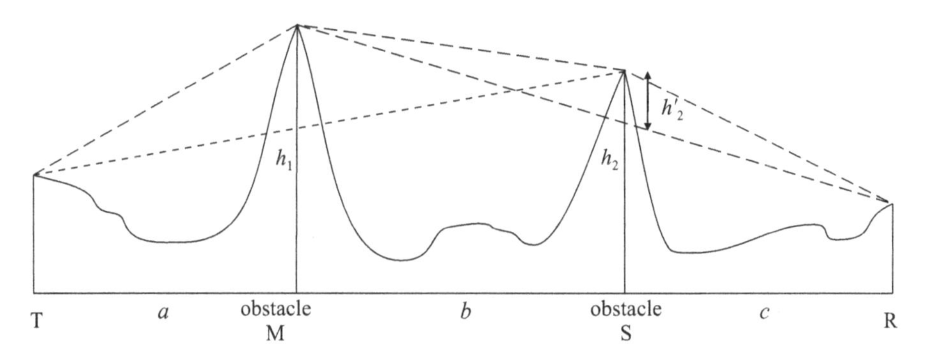

FIGURE 7.9 Profile with two isolated obstacles with different clearance.

This method consists of successively applying the diffraction theory to both obstacles. Total diffraction losses are given by the equation:

$$L = L_1 + L_2 - T_c \tag{7.45}$$

The L_1 term represents diffraction losses in the whole route if there were just that obstacle on it. In order to calculate these losses, we use the equations for an isolated knife-edge of rounded obstacle, depending on how it is, with an h_1 clearance.

The L_2 term represents diffraction losses in the MSR subroute, with a clearance h'_2, with respect to the secondary obstacle. In this case, just like in the previous case, we use the calculation formulas of an isolated obstacle for knife-edge or rounded obstacle. If the dominant obstacle is located closer to the receiver than the secondary obstacle, the subroute to calculate the L_2 attenuation component is the one formed by TSM. The correction factor T_c (dB) is used to consider the spacing between both obstacles as well as their height. T_c (dB) can be obtained from equation (7.46).

$$T_c = \left[12 - 20 \log \left(\frac{2}{1 - \frac{\alpha}{\pi}}\right)\right] \left(\frac{q}{p}\right)^{2p} \tag{7.46}$$

$$p = \left[\frac{2}{\lambda} \frac{(a+b+c)}{(b+c)a}\right]^{1/2} h_1 \tag{7.47}$$

$$q = \left[\frac{2}{\lambda} \frac{(a+b+c)}{(a+b)c}\right]^{1/2} h_2 \tag{7.48}$$

$$\text{tg}\, \alpha = \left[\frac{b(a+b+c)}{ac}\right]^{1/2} \tag{7.49}$$

where h_1 and h_2 are clearance of each separately obstacle in the direct transmitter–receiver route.

When the obstacle that causes diffraction can be obviously identified as a flat roof building, it has been demonstrated that an approximation valid enough for preliminary estimations consists of considering a diffraction model of a double knife-edge obstacle, without taking the additional reflection component caused by the roof surface into account.

7.5.5 Multiple Isolated Obstacles

When there are multiple obstacles in the route we can use cascaded obstacle calculation methods. ITU-R P.526 Recommendation describes two methods: cascaded cylinder method, for which detailed vertical profiles of the obstacles are required and cascaded knife-edge method, whose application does not require detailed information of each obstacle.

For planning three obstacle routes we can use the cascaded knife-edge procedure that is based on the Deygout method. This method assumes the existence of one main and two secondary edges. In this case, the path terrain profile heights are required.

It is worth reminding here that if resulting design from route study is correctly done, multiple diffraction situations should not happen during important time percentages (longer than 1% of the worst month).

7.6 MULTIPATH FLAT FADING

This section describes the calculation methods that provide the probability of fading occurrence associated with multipath. There are some restrictions that apply to the procedures described in this section. First of all, it is assumed that the systems under analysis will be narrow band. This condition implies in fact that the width of the fading in frequency is similar or wider than the radio channel (or channels) being affected. The effect is then considered flat fading. Additionally, the method is applicable under clear sky conditions and it is assumed uncorrelated with hydrometeors.

The effect is associated with multipath propagation and also related to mechanisms such as beam spreading and variations in the angle of launch and arrival.

The flat fading probability calculation methods assume very deep fades that can be characterized by a Rayleigh probability distribution function. In consequence, the probability of fading being deeper than F_1, is written as:

$$P(F > F_1) = P_0 \cdot 10^{-F_1/10} \tag{7.50}$$

P_0 is a parameter called multipath fading occurrence factor that depends on radio link length, frequency, terrain roughness, and weather factors. The time percentage when fading exceeds an A_1 attenuation value can be written in percentage as equation (7.51).

$$p(A > A_1) = P_0 \cdot 10^{-A_1/10}.100 \tag{7.51}$$

All calculation methods for probability of flat fading exceeding an F depth, use equations derived from the previous formula and differ from each other in the different ways of calculating P_0. Flat fading has historically been calculated by the Barnett and Vigants formula, but the methods suggested by ITU-R in Recommendation P.530 are nowadays more frequently used.

This recommendation has continuously been updated in recent years, as the influence of climatic and terrain factors have been validated by further tests. The version described in this section corresponds to ITU-R P.530 version 13 released in 2009. The obtained results apply to the average worst month anywhere in the world.

It is interesting the fact that the method does not require the path profile heights and can be used not only for initial planning but also for the detailed design phases. The method consists in fact of two complementary methods: a first method valid for small time percentages, in other words, for deep fading occurrences, and a second method that extents the first one to any value of depth and time percentage.

The recommendation provides a practical limit as a criterion to differentiate a deep and non-deep fade situation. A cut-of or transition attenuation A_t can be obtained by the following empirical interpolation procedure:

$$A_t = 25 + 1.2 \log p_0 \tag{7.52}$$

where p_0 is the multipath propagation occurrence factor expressed as a time percentage during which a 0 dB attenuation is exceeded. In each case, the time percentage, p_w, during which fading depth A (dB) is exceeded in the average worst month is obtained from the general equation:

$$p_w = p_0 \times 10^{-A/10} \tag{7.53}$$

7.6.1 Prediction for Small Percentages of Time

The first method of the ITU-R P.530 recommendation is used for very deep fades. Two calculation procedures are described: a less detailed one valid for the initial planning of the system and a wider second one to be used during the detailed planning stage of the system.

7.6.1.1 Initial Planning Stage The time percentage, p_w, when fading depth A (dB) is exceeded in the average worst month is calculated in the preliminary planning stage from the following equation:

$$p_w = p_0\, 10^{-\frac{A}{10}} = K d^{3.1} \left(1 + |\varepsilon_p|\right)^{-1.29} f^{0.8} 10^{-0.00089 h_L - A/10} \tag{7.54}$$

where

K = geoclimatic factor
d = hop distance (km)
f = frequency (GHz)
$|\varepsilon_p|$ = path inclination (mrad)
p_0 = multipath propagation occurrence factor (%)

7.6.1.1.1 Calculation of the Geoclimatic Factor For a fast calculation of K on initial planning applications, a quite precise estimation can be obtained from the formula:

$$K = 10^{-4.6-0.0027dN_1} \tag{7.55}$$

dN_1 is the refractivy gradient not exceeded for 1% of the average year in the lower 65 m of the troposphere. If local data of the area is not available, the ones provided by ITU-R P.453 recommendation can be used, using bilinear interpolation of the four nearest points of the data grid. Data can be obtained from the Radiocommunication Bureau (BR), on the web site of the ITU-R Study Group 3.

7.6.1.1.2 Path Inclination Calculation Path inclination magnitude $|\varepsilon_p|$ (mrad) is calculated from the following equation:

$$|\varepsilon_p| = \frac{|h_r - h_e|}{d} \tag{7.56}$$

where h_e and h_r (m) are the antenna heights above the sea level and d is the path length (km).

7.6.1.2 Detailed Planning Stage When a detailed planning or design of the radio-relay link is required, the time percentage, p_w, when the fading depth A (dB) is exceeded in the average worst month is calculated from the following equation:

$$p_w = p_0 10^{-\frac{A}{10}} = Kd^{3.4}\left(1 + |\varepsilon_p|\right)^{-1.03} f^{0.8} 10^{-0.00076h_L - A/10} \tag{7.57}$$

where

K = geoclimatic factor
d = hop distance (km)
f = frequency (GHz)
$|\varepsilon_p|$ = path inclination (mrad)
p_0 = multipath propagation occurrence factor (%)

Equations that allow p_w estimation are valid for frequencies between f_{min} and f_{max}, being:

$$f_{min} = \frac{15}{d}; \quad f_{max} = 45 \tag{7.58}$$

with d in kilometers and frequency in GHz.

7.6.1.2.1 Geoclimatic Factor Calculation When we have data from different routes in a similar climate and terrain zone, or form several frequencies in a single route, it is recommended to obtain a mean geoclimatic factor by averaging the values of $\log K$. If we do not have measured values for K, it is recommended to estimate the geoclimatic factor for the average worst month from equation:

$$K = 10^{-4.4-0.0027dN_1}(10 + s_a)^{-0.46} \tag{7.59}$$

where

dN_1 = refractivity gradient not exceeded for 1% of the average year in the lower 65 m of the troposphere

s_a = roughness of terrain in the area, defined as the standard deviation of terrain heights (m) in a 110 km × 110 km area with resolution of 30 s

dN_1 is calculated in the same way as in the initial planning method, through the maps of ITU-R P.453 recommendation. The parameter s_a is calculated using a database of terrain heights, In this case, the 110 × 110 km area must be aligned with the terrestrial length, in such a way that both equal halves of it are located on each side of the length that crosses the path center.

7.6.1.2.2 Path Inclination Calculation Path inclination magnitude $|\varepsilon_p|$ (mrad) shall be calculated with the same procedure as in the initial planning method.

7.6.2 Prediction for All Percentages of Time

This method is appropriate for any fading depth because it combines procedure for very deep fading that is explained in the previous point with an empirical interpolation procedure for shallow fading down to 0 dB. It can also be used for calculations in the initial planning or detailed design stages according to the method in the previous section.

If the required fading depth, A, is equal or greater than cut-off or transition attenuation A_t, we use the calculation formulas stated for small time percentages. If it is lower than cut-off value, the time percentage, p_w, when fading depth, A (dB), is exceeded in the average worst month, can be expressed by equation (7.60):

$$p_w = 100\left[1 - \exp\left(-10^{-q_a \frac{A}{20}}\right)\right] \tag{7.60}$$

The q_a factor can be calculated by the following expression:

$$q_a = 2 + 10^{-0.016A}\left[1 + 0.310^{-A/20}\right]\left[q_t + 4.3\left(10^{-A/20} + A/800\right)\right] \tag{7.61}$$

Being q_t a parameter that can be obtained from the following equation:

$$q_t = (q'_a - 2)\Big/\left[10^{-0.016A_t}\left(1 + 0.310^{-A_t/20}\right)\right] - 4.3\left(10^{-A_t/20} + A_t/800\right) \tag{7.62}$$

where A_t is the transition fading depth and q'_a can be calculated from the following equation:

$$q'_a = -20\log\left[-\ln\left(\frac{100 - p_t}{100}\right)\right]\Big/A_t \tag{7.63}$$

Where A_t is the transition fading depth, and p_t is the transition time percentage. This percentage is the percentage of time when A_t is exceeded in the average worst month. This percentage p_t can be calculated by equation (7.64), being p_0 the multipath propagation occurrence factor.

$$p_t = p_0\ 10^{-A_t/10} \tag{7.64}$$

As a summary of both suggested methods in ITU-R P.530 Recommendation, the first step is calculating p_0, with the equations and considerations that are taken into account in the method for small time percentages for the initial or detailed planning stages. Next, transition attenuation value is calculated from equation (7.52). If the excess attenuation value for which the time percentage is being calculated is equal or higher than the transition attenuation value, the method for small percentages of time will be used and if this was not the case, the second method for all percentages of time will be the best option. Figure 7.10 provides a group of curves that graphically represent the method with p_0 as a parameter.

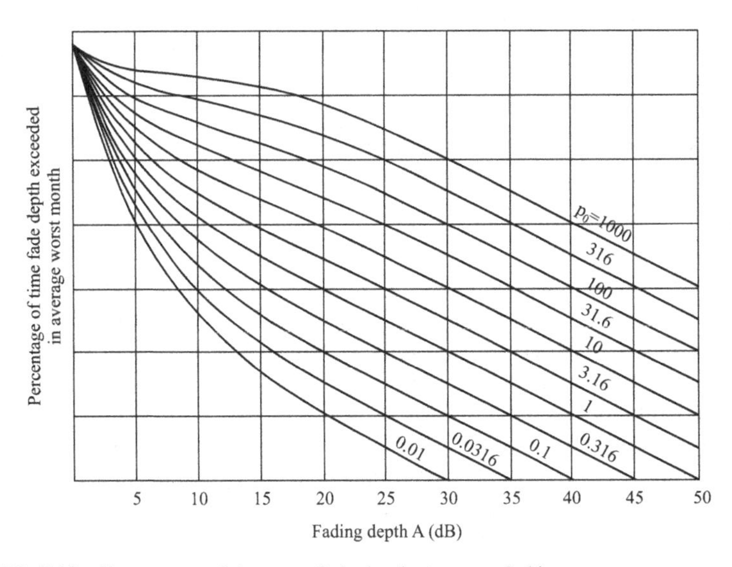

FIGURE 7.10 Percentage of time, p_w, fade depth, A, exceeded in average worst month, with p_0. *(Figure Courtesy of ITU-R)*.

The ITU-R P.530 Recommendation also provides models for converting prediction of exceeded percentages for the average worst month into predictions of either the average year or shorter periods of time.

7.7 DISTORTION DUE TO MULTIPATH PROPAGATION EFFECTS UNDER CLEAR SKY

Distortion in line of sight radio links using the ultra high-frequency and super high-frequency bands is mainly caused by the frequency dependence of both amplitude and group delay, under multipath propagation and clear sky conditions.

From the point of view of system design, this effect can be mitigated by an increase of the gross margin (margin above the system threshold) only in analog systems. It is important here to stress that using a wider fading margin does not mitigate frequency selective fading.

The degradation associated with frequency selective effects during multipath propagation depends on:

- Properties of the radioelectric system (modulation scheme, capacity, used bandwidth, etc.) and its susceptibility to dispersion effects.
- Intensity of frequency selective along a specific radioelectric route.
- Intensity of amplitude distortions associated with frequency selective fading.

It is important to stress that unlike predictions for flat fading that provide the statistics of the fading phenomena, prediction models for selective fading provide the system outage probability. These methods depend, therefore, on the specific equipment. Three approaches have traditionally been considered for predicting outages that affect radio link error performance:

- Methods that use linear amplitude dispersion (LAD).
- Net-fade margin against fading.
- Methods based on Signature curves.

The description of a method based on the LAD calculation can be found in the Handbook on Digital Radio-Relay Systems from ITU-R. The method is based on empirical results from observations and measurements carried out in an experiment at 140 Mbit/s by the Research Laboratories of Telstra (formerly, Telecom Australia) under real conditions. The field test was done during the period between November 1982 and April 1984. As in other methods for outage prediction, the LAD method predicts the total time when system BER threshold (10^{-3} generally) is exceeded.

ITU-R P.530 Recommendation has not included a description and recommendation guidelines for applying this method so far.

The net-fade margin method consists performing the selective outage calculation using a similar model as the one of flat fading, despite the effect and behavior of flat and selective fading will not be the same. This method directly depends on radio link design parameters and it is consequently included in Chapter 8, within the Section dedicated to radio link design according to error performance criteria.

According to ITU-R, there is not enough data to take an optimal decision for the use of each option. Nevertheless, the body of ITU-R P.530 Recommendation contains only a description of the Signatures method.

7.7.1 Signature

The signature is the curve that represents the sensitivity of a specific receiver against distortion caused by multipath. We need these curves because the different receivers' response to different radio channel arrangements is different. In this way, they allow to include goodness (or defects) of receiver features into the radio link design according to error performance criteria. Signatures are experimental curves that manufacturers measure in the laboratory using propagation channel simulators that will reproduce the multipath effect at the receiver input. The two-ray model is the most frequently used model for obtaining the signature curve. Figure 7.11 shows a graphical representation of the frequency response.

Equation (7.65) describes the transfer function:

$$H(\omega) = a\left[1 - b\exp\left(-j(\omega - \omega_0)\,\tau\right]\right. \tag{7.65}$$

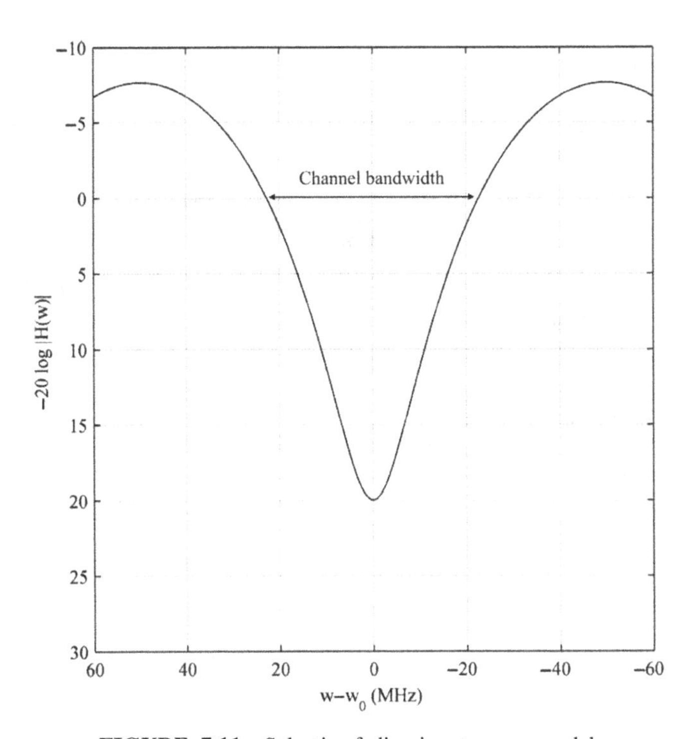

FIGURE 7.11 Selective fading in a two-ray model.

where

a = amplitude factor

b = relative amplitude of the second path with respect to the first one, that will be assumed normalized to 1

f_0 = distance between radio channel center and the maximum depth frequency ($w_0 = 2\pi f_0$)

τ = relative path delay

$20\log\lambda$ = Fading depth ($\lambda = 1 - b$)

The signature is the curve of critic value B_c, related to a specific bit error rate (BER). It indicates the selective fading depth at a specific frequency $f - f_0$ that causes BER degradation equal to the BER_0 condition it has been measured for. Figure 7.12 shows the signature curve provided by an equipment manufacturer for high capacity radio links and associated with a BER of 10^{-3}. Two signature curves are usually given, one for each of the minimum (MP) and non-minimum (NMP) phase situations described in Chapter 2.

The signature width $W(f_0)$ remains almost constant for any delay and it is a function of the modulation and equalization schemes. The most usual value for τ is 6.3 ns, providing a single curve for each receiver, although it is worth outlining that, from a theoretical standpoint, there is a signature curve for each delay value. Some outage calculation methods assume that τ is a continuous random variable. So, those

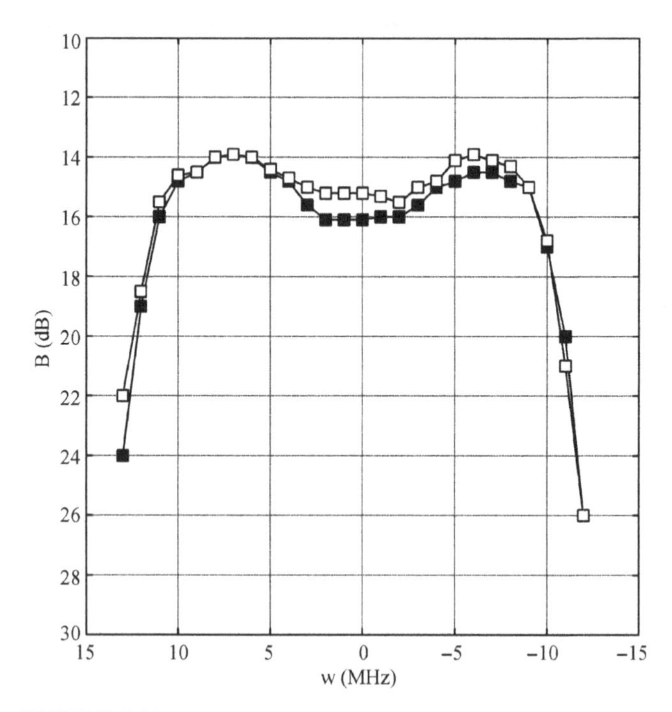

FIGURE 7.12 Example of signature of a radio-relay receiver unit.

cases need proportionality rules for estimating $b_c(\tau)$ variation with τ. For example, ITU-R mentions the linear rule that is only valid for small delays where it establishes that height in λ is proportional to delay τ.

In addition, the height of signature curve depends on τ, modulation scheme and on the existence of receiver equalization units and their complexity. From the radio link operation point of view, signature curves should be as narrow and low as possible because the outage probability under multipath conditions will depend on the integration of its function.

In reality it is usual to work with rectangular shape approximations to the real signature curves, so calculation of the multipath degradation functions use only the height and width of the curve. These W and B_c are considered through a standardized system parameter called standardized signature constant K_n. This parameter is calculated from equation (7.66):

$$K_n = \frac{T^2 W \lambda}{\tau} \tag{7.66}$$

where

T = symbol period (ns)
W = signature width (GHz)
λ = average value of the signature (linear) $\lambda(f) = 1 - b_c(f) = B_c$
τ = reference delay of the signature curve for λ (ns)

7.7.2 Outage Probability Calculation Based on Signatures

Steps for calculating the probability of exceeding a specific error rate due to multipath are detailed in the following text. This method is described in ITU-R P.530 and it is based on the general expression for multipath fading probability:

$$P_s = p\,(o/\eta)\,\eta \quad 100(\%) \tag{7.67}$$

where

η = multipath activity factor
$p(o|\eta)$ = conditional outage probability due to multipath:

$$p\,(o/\eta) = C\,P_b(1)k < \tau^2 > \frac{1}{T^2} \tag{7.68}$$

where

C = constant factor
$P_b(1)$ = probability of b (normalized echo amplitude in the two-ray model)
k = standardized signature constant
$<\tau^2>$ = second-order momentum of the echo delay distribution that is $2\tau_m$ in the case of negative exponential distributions, τ_m being the average delay.
T = symbol period (ns)

The method starts with the estimation of the average time delay, by using the empirical expression of equation (7.69) in function of the radio link distance in kilometers (d):

$$\tau_m = 0.7 \left(\frac{d}{50}\right)^{1,3} \tag{7.69}$$

Once τ_m is known, we calculate the time percentage P_0 when 0 dB fading is exceeded in the average worst month from one of both methods of ITU-R P.530 recommendation described in the previous section and shown in equations (7.70) and (7.71) for initial and detailed planning respectively. The result $P_0 = p_w/100$ is the multipath occurrence factor:

$$p_w = Kd^{3.1}\left(1 + |\varepsilon_p|\right)^{-1.29} f^{0.8} 10^{-0.00089 h_L} \tag{7.70}$$

$$p_w = Kd^{3.4}\left(1 + |\varepsilon_p|\right)^{-1.03} f^{0.8} 10^{-0.00076 h_L} \tag{7.71}$$

where

K = geoclimatic factor
d = hop distance (km)
ε_p = path inclination (mrad)
f = frequency (GHz)
h_L = height of the lowest antenna (minimum of both antennas)

The multipath activity factor η is calculated from P_0 from equation (7.72):

$$\eta = 1 - e^{-0.2(P_0)^{0,75}} \tag{7.72}$$

and once the multipath activity factor is known, we can obtain the selective outage probability of transmission from equation (7.73):

$$P_s = 2.15\eta \left(W_M \times 10^{-B_M/20} \frac{\tau_m^2}{|\tau_{r,M}|} + W_{NM} \times 10^{-B_{NM}/20} \frac{\tau_m^2}{|\tau_{r,NM}|}\right) \tag{7.73}$$

where

η = multipath activity factor
$W_{M/NM}$ = signature width (GHz)
$B_{M/NM}$ = signature depth (dB)
$\tau_{r,M/NM}$ = reference delay (ns) used to obtain the signature
M and NM = subscripts refer to minimum and nonminimum phase situations respectively.

If we only have the standardized K_n system parameter, we can calculate the selective outage probability as follows:

$$P_s = 2.15\eta\ (K_{n,M} + K_{n,NM})\ \frac{\tau_m^2}{T^2} \tag{7.74}$$

where

T = symbol period (ns)

$K_{n,M/NM}$ = standardized signature constant for the minimum phase fading (M) or the nonminimum phase fading (NM)

7.8 ATTENUATION DUE TO HYDROMETEORS

Absorption and dispersion due to hydrometeors such as rain, snow, hail, and fog produce attenuation in a radio link route that can be calculated by integrating the specific attenuation along the whole route if the hydrometeor variation along route is known.

There are numerous methods for predicting rain attenuation statistics in the literature that are based on empirical cumulative distributions of the average rainfall at the location of interest. The method of ITU-R P.530 Recommendation described here, defines an equivalent path length along which we can suppose that the rain intensity will remain constant for each rain occurrence. This method provides long-term statistic results for both rain and wet snow.

It is important to note that the influence of other type of hydrometeors different from rain, as fog, can be noticeable for fixed radio links at high frequencies, in EHF frequencies for the fixed service around 100 GHz. In this case, the only reference available today at ITU-R is Recommendation P.840 that contains information for predicting attenuation due to fog and clouds. The methods of P.840 should be handled with care, as the main objective of the Recommendation is the design of Earth-to-space paths, with elevation angles higher than 5°.

7.8.1 Prediction of Attenuation Due to Rain

Attenuation due to rain is characterized through the A_p parameter that represents the exceeded attenuation during a p time percentage. The values of this attenuation have been calculated empirically, as a function of the geographical latitude from equations (7.75) and (7.76):

$$A_p = A_{0.01} 0.12 p^{-(0.546+0.043 \log_{10} p)} \text{ Latitude N/S} \geq 30^\circ \tag{7.75}$$

$$A_p = A_{0.01} 0.07 p^{-(0.855+0.139 \log_{10} p)} \text{ Latitude N/S} < 30^\circ \tag{7.76}$$

where $A_{0.01}$ is the exceeded attenuation during 0.01% of the time (dB). The previous equations are valid for a probability range from 0.001% – 1%. Exceeded attenuation

TABLE 7.3 $A_p/A_{0.01}$ Ratios for Relevant Time Percentages

	$A_p/A_{0.01}$ Ratio	
Time Percentage	N or S Latitude Higher or Equal to 30°	N or S Latitude Lower to 30°
1%	0.12	0.07
0.1%	0.39	0.36
0.01%	1.00	1.00
0.001%	2.14	1.44

value during 0.01% of time is calculated from equation (7.77):

$$A_{0.01} = \gamma_R d_{\text{eff}} = \gamma_R d r \tag{7.77}$$

where

γ_R = specific attenuation (dB/km)
d_{eff} = effective length of the radio link (km)

The specific attenuation is calculated using ITU-R P.838 Recommendation. There will be a value for frequency, polarization and rainfall intensity $R_{0.01}$, exceeded during 0.01% of the time, with a 1 min integration time.

The rainfall intensity $R_{0.01}$ should be obtained from long-term local data. In case this information was not available, reference data can be obtained in ITU-R P.837 Recommendation, although the accuracy of the latter will not be comparable to empirical data gathered locally.

The effective path length of the radio link, d_{eff}, is calculated by multiplying the length of the real path, d, by a distance factor, r that can be estimated from equation (7.78):

$$r = \frac{1}{1 + d/d_0} \tag{7.78}$$

$$d_0 = 35\exp(-0.015 R_{0.01}) \tag{7.79}$$

This estimation is valid for $R_{0.01} \leq 100$ mm/h. In the case of rain rates exceeding the 100 mm/h intensity value, the calculations will be done with $R_{0.01}$ equal to 100 mm/h. Table 7.3 shows the values for the most significant time percentages.

The previous prediction procedure is valid worldwide, for frequencies up to 40 GHz and path lengths up to 60 km.

If we want to obtain the worst month statistics, p_w, in function of the calculated percentages of annual time, p, they can be obtained by the information about weather that is specified in ITU-R P.841 Recommendation.

7.8.1.1 Specific Attenuation Calculation The specific attenuation γ_R (dB/km) can be calculated by following the method described by ITU-R in

ITU-R P.838 recommendation. The specific attenuation is obtained from rainfall intensity R (mm/h) from equation (7.80):

$$\gamma_R = kR^{\alpha} \tag{7.80}$$

k and α coefficient values are determined as a function of frequency, f (GHz), in the range from 1 – 1000 GHz from equations:

$$\log_{10} k = \sum_{j=1}^{4} a_j \exp\left[-\left(\frac{\log_{10} f - b_j}{c_j}\right)^2\right] + m_k \log_{10} f + c_k \tag{7.81}$$

$$\alpha = \sum_{j=1}^{5} a_j \exp\left[-\left(\frac{\log_{10} f - b_j}{c_j}\right)^2\right] + m_\alpha \log_{10} f + c_\alpha \tag{7.82}$$

where

f = frequency (GHz)
k = constant that depends on frequency and polarization (k_V or k_H)
α = constant that depends on frequency and polarization (α_V or α_H)

Appendix I to this chapter includes the data required for are attached in the. These contain the coefficients for calculating k_H, k_V, α_H, and α_H (Tables 7.4 and 7.5). Coefficients for the frequency range from 1 – 1000 GHz have been also calculated and compiled in Table 7.6 of Appendix I.

7.8.1.2 Method for Rain Intensity Excess Value for a Certain Probability Target

Specific rain attenuation along a certain path depends on rain intensity R and on k and α coefficients. Having described the calculation procedure for k and α coefficients it will be necessary to decide the most adequate source values for rain intensity R.

ITU-R P.837 Recommendation provides an equation for calculating rain intensity, exceeded during $p\%$ of the average year, at any location with an integration time of 1 min. The model takes as input several empirical parameters that depend on latitude and longitude (in a range between +90° N and – 90° S) and are available in digital terrain maps that can be downloaded from Radiocommunications Study Group 3 web at ITU-R's site. Parameters from this database are given to a precision of 1.125°, and a bilinear interpolation is recommended (ITU-R P.1144). The method will provide rain intensity R_p (mm/h) exceeded during $p\%$ of the time.

The same recommendation provides global maps with rain intensity excess values for 0.01% of the time, values that, as described in the previous section, are required for the calculation of γ_R. Figure 7.13 shows one of the maps in of recommendation ITU-R P.837, in particular, the figure shows the overall map for all areas of the planet. In addition to this global map, ITU-R Rec. P.837 contains more precise values

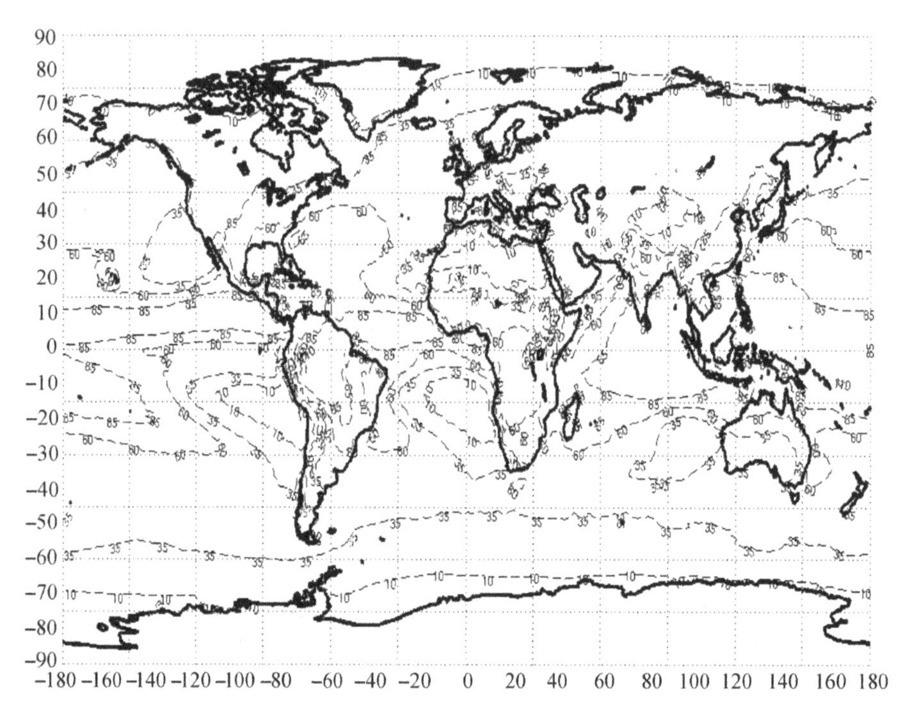

FIGURE 7.13 Global map of the excess rainfall during 0.01% of the time in an average year.

in detailed maps for North and Central America, South America, Europe, Africa and Middle East, Asia and Oceania.

As stated previously, whenever reliable local data are available, they will provide more accurate than ITU-R maps for obvious granularity reasons.

7.8.2 Combined Method for Rain and Wet Snow

The presence of rain combined with snow may cause significant attenuation in routes at high altitudes or latitudes, due to the effect of melted ice particles or wet snow at the melting layer. Melting layer conditions only have an additional effect to rain in radio link attenuation when the following is met:

$$h_{\text{link}} < h_{\text{rain}} - 3600 \tag{7.83}$$

Being h_{link} the height of the radio link center that can be calculated by the following equation:

$$h_{\text{link}} = 0.5\,(h_1 + h_2) - \left(\frac{D^2}{17}\right) \tag{7.84}$$

where

$h_{1,2}$ = heights of radio link terminals (m) above average sea level.
D = path length
h_{rain} = median value of rain height above the average sea level that can be obtained from ITU-R P.839 recommendation

Calculation of the combined attenuation due to rain and wet snow is based on determining an F coefficient. The attenuation associated with rain and wet snow will then be expressed as the multiplication of the attenuation of rain by this coefficient:

$$A_{\text{rs}} = A_{\text{p}} F \tag{7.85}$$

The F factor can be obtained based on an iterative process that compares the radio link height related to rain, by a series of wet snow probabilities associated with each height value. The calculation procedure consists of 49 iterations, using an i index that progressively varies from de $0 - 48$. For $i = 0$ the F factor is 0. For progressive indexes the F factor is the factor of the previous index increased by a factor, ΔF that corresponds with the new i index. Thus for each new i index, up to $i = 48$, we calculate:

$$F = F + \Delta F \tag{7.86}$$

The value of each ΔF can be obtained in function of a $\Gamma(\Delta h)$ multiplicative factor that takes into account the different specific attenuations according to rain height, Δh, and of probability of the radio link being at the rain height, Δh:

$$\Delta F = \Gamma(\Delta h) P_{\text{i}} \tag{7.87}$$

The value of the radio link height related to rain height can be obtained for each i iteration index by equation (7.88):

$$\Delta h = h_{\text{link}} - h_{\text{rain}} \tag{7.88}$$

were the rain height, h_{rain}, is adjusted in each iteration from the average height value calculated according to ITU-R P.839 Recommendation (both in meters above the average sea level):

$$h_{\text{rain}} = h_{\text{average\_rain}} - 2400 + 100i \tag{7.89}$$

The $\Gamma(\Delta h)$ multiplicative factor can be obtained by equation (7.90):

$$\Gamma(\Delta h) = \begin{cases} 0 & 0 < \Delta h \\ \dfrac{4\left(1 = \mathrm{e}^{\Delta h/70}\right)^2}{\left(1 + \left(1 - \mathrm{e}^{-(\Delta h/600)^2}\right)^2 \left(4\left(1 - \mathrm{e}^{\Delta h/70}\right)^2 - 1\right)\right)} & -1\,200 \leq \Delta h \leq 0 \\ 1 & 0 < \Delta h - 1\,200 \end{cases} \tag{7.90}$$

The P_i probability of the radio link being in Δh, is shown in Table 7.7 of Appendix II to this chapter.

7.8.3 Frequency Extrapolation

When the attenuation data at one frequency is known, equation (7.91) can be used to extrapolate statistics in the range between 7 and 50 GHz for the same hop length and in the same climatic region, using a long-term distribution obtained from measurements. If we consider A_1 the attenuation at f_1 frequency and A_2 the attenuation at f_2 frequency, the equations are:

$$A_2 = A_1 \left(\frac{\Phi_2}{\Phi^1} \right)^{1-H(\Phi_1, \Phi_2, A_1)} \tag{7.91}$$

$$\Phi(f) = \frac{f^2}{1 + 10^{-4} f^2} \tag{7.92}$$

$$H(\Phi_1, \Phi_2, A_1) = 1.12\ 10^{-3} \left(\frac{\Phi_2}{\Phi_1} \right)^{0.5} (\Phi_1 A_1)^{0.55} \tag{7.93}$$

Attenuations are in dB and frequencies in GHz.

7.8.4 Polarization Extrapolation

If long-term statistics of attenuation with a specific polarization (vertical (V) or horizontal (H)) for a given radio link are available, attenuation can be extrapolated for the cross polarization in the same radio link. The results will apply to links within the same path length and frequency range of the attenuation prediction method, by the following equations:

$$A_V = \frac{300 A_H}{335 + A_H} \tag{7.94}$$

$$A_H = \frac{335 A_V}{300 - A_V} \tag{7.95}$$

7.8.5 Attenuation Due to Clouds and Fog

Attenuation due to clouds and fog has not traditionally been considered in radio link design. Its effect in the most used frequency bands (below 40 GHz) is considered irrelevant compared with other phenomena such as rain or fading due to atmospheric multipath.

With the saturation of bands below 40 GHz and as the radio links become more common in frequencies close to 100 GHz, the problem cannot always be simplified dismissing the attenuation due to clouds and fog. In places where these phenomena are usual they might degrade the performance significantly. Models and data already accepted internationally for radiocommunication systems planning are applied to satellite links.

ITU-R P.840 Recommendation provides the method for predicting attenuation due to clouds and fog on Earth-space paths.

Clouds and for are totally formed by minute drops, generally smaller that 0.01 cm. These dimensions allow approximating, for frequencies lower than 200 GHz, calculation methods for specific attenuation by Rayleigh dispersion. The γ_c (dB/km) specific attenuation inside a cloud of fog can be written in terms of the total liquid water containing per volume unit.

$$\gamma_c = K_1 M \tag{7.96}$$

where

M = liquid water density inside the cloud or fog (g/m^3)
K_1 = specific attenuation coefficient ((dB/km)/(g/m^3))

Liquid water density in fog is typically about 0.05 g/m^3 in moderate fog (visibility about 300 m) and 0.5 g/m3 in thick fog (visibility about 50 m). In order to calculate K_1 value, we can use a mathematical model based on Rayleigh dispersion that uses a double Debye model for dielectric permittivity $\varepsilon(f)$ of water and is valid for frequencies up to 1000 GHz:

$$K_l = \frac{0.819 f}{\varepsilon''(1+\eta^2)} \tag{7.97}$$

where f is frequency (GHz), and:

$$\eta = \frac{2+\varepsilon'}{\varepsilon''} \tag{7.98}$$

The complex dielectric permittivity of water is:

$$\varepsilon''(f) = \frac{f(\varepsilon_0-\varepsilon_1)}{f_p\left[1+(f/f_p)^2\right]} + \frac{f(\varepsilon_1-\varepsilon_2)}{f_s\left[1+(f/f_s)^2\right]} \tag{7.99}$$

$$\varepsilon'(f) = \frac{\varepsilon_0-\varepsilon_1}{\left[1+(f/f_p)^2\right]} + \frac{\varepsilon_1-\varepsilon_2}{\left[1+(f/f_s)^2\right]} + \varepsilon_2 \tag{7.100}$$

where

$\varepsilon_0 = 77.6 + 103.3\,(\theta - 1)$
$\varepsilon_1 = 5.48$
$\varepsilon_2 = 3.51$
$\theta = 300/T$ (temperature in °K)
f_p = principal relaxation frequency
f_s = secondary relaxation frequency

Principal and secondary relaxation frequencies can be obtained from equations (7.101) and (7.102):

$$f_{\mathrm{p}} = 20.09 - 142\,(\theta - 1) + 294\,(\theta - 1)^2 \tag{7.101}$$

$$f_{\mathrm{s}} = 590 - 1500\,(\theta - 1) \tag{7.102}$$

both in GHz.

In order to determine the attenuation of a specific radio link, statistics of the path portion affected by for would be necessary. Once this parameter is known, it would be immediate to infer total attenuation in the radio link. This information is not currently available in ITU-R recommendations. Nevertheless, as they are critic frequencies near or above 100 GHz, path lengths are thought to be of very few kilometers or even less than one kilometer. Under these conditions, the fact of supposing the whole profile affected by clouds or fog does not seem to be a significant source of error.

7.9 REDUCTION OF CROSS-POLAR DISCRIMINATION (XPD)

The parameter XPD mainly affects to frequency arrangements of radio channel plans and frequency sharing in different hops and links (Chapters 1, 2, and 3).

XPD may deteriorate enough as for causing co-channel interference and, to a lesser extent, adjacent channel interference. XPD reduction must be taken into account under both clear sky and rain conditions.

7.9.1 XPD Outage Under Clear-Air Conditions

Under clear weather conditions, multipath propagation is the main cause of polarization changes of the transmitted signal. As expected, if the target calculation is the quality affection in one hop of a specific radio link, the cross-polar radiation pattern of transmitter and receiver antennas of that hop will be equally important.

Polarization changes under clear sky conditions usually happen during small percentages of time. This phenomenon is in the line with the characteristics of multipath propagation occurrences. In fact, the methods for estimating changes in XPD under these conditions will be derived from the ones used for determining multipath selective fading.

The method for estimating XPD is described next and it is part of the contents of ITU-R Rec. P.530 recommendation. As other methods of this recommendation, the calculation is described as a series of steps:

1. Calculate reference XPD_0. Calculation depends on the minimum XPD for the axis of the main lobe of the antenna. The minimum is referred to transmission

or reception:

$$\mathrm{XPD_0} = \begin{cases} \mathrm{XPD_g} + 5 & \text{for} \quad \mathrm{XPD_g} \leq 35 \\ 40 & \text{for} \quad \mathrm{XPD_g} > 35 \end{cases} \tag{7.103}$$

2. Evaluate the multipath activity factor following the equation used in previous calculations, with P_0 the multipath occurrence factor in percentage and calculated according to the methods seen in equation (7.54) or (7.57):

$$\eta = 1 - \mathrm{e}^{-0.2(P_0)^{0.75}} \tag{7.104}$$

3. Calculate the k_{XP} factor as a function of the number of antennas in transmission with equation (7.105). The two transmitter antennas case is referred to two transmissions of orthogonal polarization of two different antennas spaced s_t(m) from each other, being λ (m) the wave length:

$$k_{\mathrm{XP}} = \begin{cases} 0.7 & \text{single transmitting antenna} \\ 1 - 0.3\exp\left[-4 \times 10^{-6}\left(\frac{s_t}{\lambda}\right)^2\right] & \text{two transmitting antennas} \end{cases} \tag{7.105}$$

4. Calculate the Q factor from equation (7.106):

$$Q = -10\log\left(\frac{k_{\mathrm{XP}}\eta}{P_0}\right) \tag{7.106}$$

5. Calculate the factor C:

$$C = \mathrm{XPD_0} + Q \tag{7.107}$$

6. Calculate the equivalent margin due to cross-polar discrimination for a defined BER value from equation (7.108):

$$M_{\mathrm{XPD}} = \begin{cases} C - \frac{\mathrm{C_0}}{\mathrm{I}} & \text{without XPIC} \\ C - \frac{\mathrm{C_0}}{\mathrm{I}} + \mathrm{XPIF} & \text{with XPIC} \end{cases} \tag{7.108}$$

   In equation (7.108) $\mathrm{C_0/I}$ is the carrier to interference ratio for a specific BER value. XPIF is an improvement factor due to the use of cross-polar interference cancellation circuits. The typical value is 20 dB.

7. Calculate the outage probability due to cross polarization through equation (7.109):

$$P_{\mathrm{XP}} = P_0 \times 10^{-\frac{M_{\mathrm{XPD}}}{10}} \tag{7.109}$$

7.9.2 XPD Outage Caused by Rain

Cross-polar discrimination decrease caused by rain by that rain is the dominant depolarization factor in bands above 10 GHz, especially if rain is intense. Estimation methods for XPD decrease due to rain are based on the calculation of rain attenuation that is usually called co-polar attenuation (CPA) in this context. Equation (7.110) shows the XPD value as a function of CPA:

$$\mathrm{XPD} = U - V(f)\log \mathrm{CPA} \tag{7.110}$$

In equation (7.110), U and V values are empirical factors that depend on frequency, path elevation angle and polarization. These coefficients can be approximated by the following equations:

$$U = U_0 + 30\log f \tag{7.111}$$

$$\begin{aligned} V(f) &= 12.8 f^{0.19} & 8 \le f \le 20\ \text{GHz} \\ V(f) &= 22.6 & 20 < f \le 35\ \text{GHz} \end{aligned} \tag{7.112}$$

Empirical evidence concerning U_0 suggests values between 9 and 15 dB for attenuations caused by rain above 15 dB. In the same way, polarization is not relevant for CPA calculation in order to obtain XPD values, so ITU-R recommends using circular polarization for CPA if its value is going to be used in XPD calculation.

As far as frequency variation is concerned, empirical values have been obtained for relating discrimination in two different frequencies (XPD_1, XPD_2) for the same time percentage and polarization as far as they are in the range between 4 and 30 GHz:

$$\mathrm{XPD}_2 = \mathrm{XPD}_1 - 20\log(f_2/f_1) \qquad 4 \le f_1, f_2 \le 30\ \text{GHz} \tag{7.113}$$

A step-by-step procedure is next presented for calculating outages due to polarization losses caused by rain. Calculation requires of the previous knowledge of rain attenuation exceeded 0.01% of the time (already covered by in previous sections). Calculation steps are summarized as follows:

1. Calculate the attenuation caused by rain exceeded 0.01% of the time $A_{0.01}$.
2. Obtain the equivalent path attenuation Ap (dB) from equation (7.114). This attenuation depends on the empirical XPD factors (U and V) as well as on the use of XPIC circuit. Furthermore, this attenuation, A_p, depends on the minimum carrier to interference ratio C_0/I (dB) specified for the radio link:

$$A_p = 10^{((U - C_0/I + XPIF)/V)} \tag{7.114}$$

3. Calculate m and n parameters using equations (7.115) and (7.116):

$$m = \begin{cases} 23.26 \log \left[A_p/0.12A_{0.01}\right] & m \leq 40 \\ 40 & \text{any other case} \end{cases} \quad (7.115)$$

$$n = \left(-12.7 + \sqrt{161.23 - 4m}\right)/2 \quad (7.116)$$

4. Finally, the outage probability is:

$$P_{\text{XPR}} = 10^{(n-2)} \quad (7.117)$$

Parameter n has a valid range of values from -3 to 0. Just in case of using XPD cancellation circuits we can obtain values lower than -3.

7.10 ITU-R DATABASES FOR TROPOSPHERIC PROPAGATION STUDIES

For the use of global prediction models of propagation in microwave radio link design, it is necessary to have geophysical and radio-meteorological statistical data.

In any case, those obtained by measurement campaigns in the study area will be more reliable and accurate. Nevertheless, in most cases, these data will not exist and it will be necessary to search in more general databases (and therefore less detailed), such as the ones provided by the Study Group 3 of the ITU.

The variables of ITU-R databases in Figure 7.14 are: Rho: Surface water vapor density exceedance probability, cloud liquid water (CLW) data, zero deg: altitude statistics of the zero-degree isotherm, Nwet: statistics of the wet term of radio refractivity at surface level, statistics of integrated water vapor content (IWVC).

Local data of each parameter in the area can be found either though general maps in P series Recommendations or through digital files available on the web site of ITU-R Study Group 3.

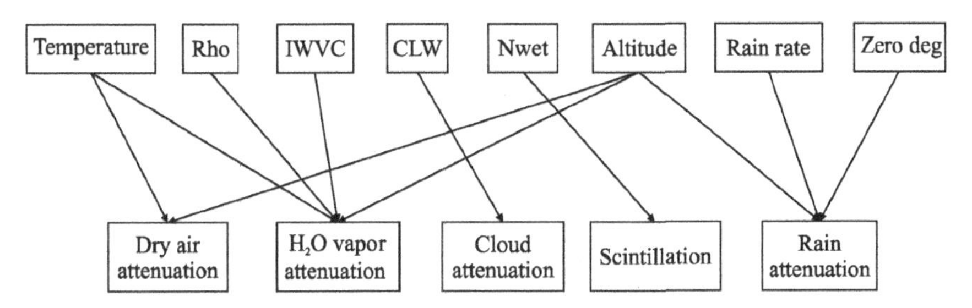

FIGURE 7.14 Geophysical and radio meteorological data as described by ITU-R. (Figure courtesy of ITU-R).

BIBLIOGRAPHY

ITU-R Rec. P.453: The radio refractive index: its formula and refractivity data. International Telecommunication Union. Radiocommunication Sector. ITU-R. Geneva. 2003.

ITU-R Rec. P.526: Propagation by diffraction. International Telecommunication Union. Radiocommunication Sector. ITU-R. Geneva. 2009.

ITU-R Rec. P. 530–13: Propagation data and prediction methods required for the design of terrestrial line-of-sight systems. International Telecommunication Union. Radiocommunication Sector. ITU-R. Geneva. 2009.

ITU-R Rec. P.676: Attenuation by atmospheric gases. International Telecommunication Union. Radiocommunication Sector. ITU-R. Geneva. 2009.

ITU-R Rec. P.837: Characteristics of precipitation for propagation modeling. International Telecommunication Union. Radiocommunication Sector. ITU-R. Geneva. 2007.

ITU-R Rec. P.838: Specific attenuation model for rain for use in prediction methods. International Telecommunication Union. Radiocommunication Sector. ITU-R. Geneva. 2005.

ITU-R Rec. P.839: Rain height model for prediction methods. International Telecommunication Union. Radiocommunication Sector. ITU-R. Geneva. 2001.

ITU-R Rec. P.840: Attenuation due to clouds and fog. International Telecommunication Union. Radiocommunication Sector. ITU-R. Geneva. 2009.

Multipath Propagation at 4, 6 and 11 GHz. W.T. Barnett. Bell System Technical Journal. 1972.

Space Diversity Engineering. A. Vigants. Bell System Technical Journal. 1973.

APPENDIX I: DATA FOR SPECIFIC ATTENUATION CALCULATION γ_R

TABLE 7.4 Coefficients for Calculating the Constant k

Coefficients for k_H					
j	a_j	b_j	c_j	m_k	c_k
1	−5.33980	−0.10008	1.13098	−0.18961	0.71147
2	−0.35351	1.26970	0.45400		
3	−0.23789	0.86036	0.15354		
4	−0.94158	0.64552	0.16817		
Coefficients for k_V					
j	a_j	b_j	c_j	m_k	c_k
1	−3.80595	0.56934	0.81061	−0.16398	0.63297
2	−3.44965	−0.22911	0.51059		
3	−0.39902	0.73042	0.11899		
4	0.50167	1.07319	0.27195		

TABLE 7.5 Coefficients for Calculating the Constant α

		Coefficients for α_H			
j	a_j	b_j	c_j	m_α	c_α
1	−0.14318	1.82442	−0.55187	0.67849	−1.95537
2	0.29591	0.77564	0.19822		
3	0.32177	0.63773	0.13164		
4	−5.37610	−0.96230	1.47828		
5	16.1721	−3.29980	3.43990		
		Coefficients for α_V			
j	a_j	b_j	c_j	m_α	c_α
1	−0.07771	2.33840	−0.76284	−0.053739	0.83433
2	0.56727	0.95545	0.54039		
3	−0.20238	1.14520	0.26809		
4	−48.2991	0.791669	0.116226		
5	48.5833	0.791459	0.116479		

TABLE 7.6 k and α as a Function of Polarization and Frequency

Frequency (GHz)	k_H	α_H	k_V	α_V
1	0.0000259	0.9691	0.0000308	0.8592
1.5	0.0000443	1.0185	0.0000574	0.8957
2	0.0000847	1.0664	0.0000998	0.9490
2.5	0.0001321	1.1209	0.0001464	1.0085
3	0.0001390	1.2322	0.0001942	1.0688
3.5	0.0001155	1.4189	0.0002346	1.1387
4	0.0001071	1.6009	0.0002461	1.2476
4.5	0.0001340	1.6948	0.0002347	1.3987
5	0.0002162	1.6969	0.0002428	1.5317
5.5	0.0003909	1.6499	0.0003115	1.5882
6	0.0007056	1.5900	0.0004878	1.5728
7	0.001915	1.4810	0.001425	1.4745
8	0.004115	1.3905	0.003450	1.3797
9	0.007535	1.3155	0.006691	1.2895
10	0.01217	1.2571	0.01129	1.2156
11	0.01772	1.2140	0.01731	1.1617
12	0.02386	1.1825	0.02455	1.1216
13	0.03041	1.1586	0.03266	1.0901
14	0.03738	1.1396	0.04126	1.0646
15	0.04481	1.1233	0.05008	1.0440
16	0.05282	1.1086	0.05899	1.0273
17	0.06146	1.0949	0.06797	1.0137
18	0.07078	1.0818	0.07708	1.0025
19	0.08084	1.0691	0.08642	0.9930
20	0.09164	1.0568	0.09611	0.9847

(continued)

TABLE 7.6 ***(Continued)***

Frequency (GHz)	k_H	α_H	k_V	α_V
21	0.1032	1.0447	0.1063	0.9771
22	0.1155	1.0329	0.1170	0.9700
23	0.1286	1.0214	0.1284	0.9630
24	0.1425	1.0101	0.1404	0.9561
25	0.1571	0.9991	0.1533	0.9491
26	0.1724	0.9884	0.1669	0.9421
27	0.1884	0.9780	0.1813	0.9349
28	0.2051	0.9679	0.1964	0.9277
29	0.2224	0.9580	0.2124	0.9203
30	0.2403	0.9485	0.2291	0.9129
31	0.2588	0.9392	0.2465	0.9055
32	0.2778	0.9302	0.2646	0.8981
33	0.2972	0.9214	0.2833	0.8907
34	0.3171	0.9129	0.3026	0.8834
35	0.3374	0.9047	0.3224	0.8761
36	0.3580	0.8967	0.3427	0.8690
37	0.3789	0.8890	0.3633	0.8621
38	0.4001	0.8816	0.3844	0.8552
39	0.4215	0.8743	0.4058	0.8486
40	0.4431	0.8673	0.4274	0.8421
41	0.4647	0.8605	0.4492	0.8357
42	0.4865	0.8539	0.4712	0.8296
43	0.5084	0.8476	0.4932	0.8236
44	0.5302	0.8414	0.5153	0.8179
45	0.5521	0.8355	0.5375	0.8123
46	0.5738	0.8297	0.5596	0.8069
47	0.5956	0.8241	0.5817	0.8017
48	0.6172	0.8187	0.6037	0.7967
49	0.6386	0.8134	0.6255	0.7918
50	0.6600	0.8084	0.6472	0.7871
51	0.6811	0.8034	0.6687	0.7826
52	0.7020	0.7987	0.6901	0.7783
53	0.7228	0.7941	0.7112	0.7741
54	0.7433	0.7896	0.7321	0.7700
55	0.7635	0.7853	0.7527	0.7661
56	0.7835	0.7811	0.7730	0.7623
57	0.8032	0.7771	0.7931	0.7587
58	0.8226	0.7731	0.8129	0.7552
59	0.8418	0.7693	0.8324	0.7518
60	0.8606	0.7656	0.8515	0.7486
61	0.8791	0.7621	0.8704	0.7454
62	0.8974	0.7586	0.8889	0.7424
63	0.9153	0.7552	0.9071	0.7395
64	0.9328	0.7520	0.9250	0.7366
65	0.9501	0.7488	0.9425	0.7339

TABLE 7.6 (*Continued*)

Frequency (GHz)	k_H	α_H	k_V	α_V
66	0.9670	0.7458	0.9598	0.7313
67	0.9836	0.7428	0.9767	0.7287
68	0.9999	0.7400	0.9932	0.7262
69	1.0159	0.7372	1.0094	0.7238
70	1.0315	0.7345	1.0253	0.7215
71	1.0468	0.7318	1.0409	0.7193
72	1.0618	0.7293	1.0561	0.7171
73	1.0764	0.7268	1.0711	0.7150
74	1.0908	0.7244	1.0857	0.7130
75	1.1048	0.7221	1.1000	0.7110
76	1.1185	0.7199	1.1139	0.7091
77	1.1320	0.7177	1.1276	0.7073
78	1.1451	0.7156	1.1410	0.7055
79	1.1579	0.7135	1.1541	0.7038
80	1.1704	0.7115	1.1668	0.7021
81	1.1827	0.7096	1.1793	0.7004
82	1.1946	0.7077	1.1915	0.6988
83	1.2063	0.7058	1.2034	0.6973
84	1.2177	0.7040	1.2151	0.6958
85	1.2289	0.7023	1.2265	0.6943
86	1.2398	0.7006	1.2376	0.6929
87	1.2504	0.6990	1.2484	0.6915
88	1.2607	0.6974	1.2590	0.6902
89	1.2708	0.6959	1.2694	0.6889
90	1.2807	0.6944	1.2795	0.6876
91	1.2903	0.6929	1.2893	0.6864
92	1.2997	0.6915	1.2989	0.6852
93	1.3089	0.6901	1.3083	0.6840
94	1.3179	0.6888	1.3175	0.6828
95	1.3266	0.6875	1.3265	0.6817
96	1.3351	0.6862	1.3352	0.6806
97	1.3434	0.6850	1.3437	0.6796
98	1.3515	0.6838	1.3520	0.6785
99	1.3594	0.6826	1.3601	0.6775
100	1.3671	0.6815	1.3680	0.6765
120	1.4866	0.6640	1.4911	0.6609
150	1.5823	0.6494	1.5896	0.6466
200	1.6378	0.6382	1.6443	0.6343
300	1.6286	0.6296	1.6286	0.6262
400	1.5860	0.6262	1.5820	0.6256
500	1.5418	0.6253	1.5366	0.6272
600	1.5013	0.6262	1.4967	0.6293
700	1.4654	0.6284	1.4622	0.6315
800	1.4335	0.6315	1.4321	0.6334
900	1.4050	0.6353	1.4056	0.6351
1000	1.3795	0.6396	1.3822	0.6365

APPENDIX II: DATA FOR CALCULATING ATTENUATION DUE TO WET SNOW

TABLE 7.7 Probability of the Height Related to Rain Height for Each *i* Index

Index "i"		Probability P_i
0	48	0.000555
1	47	0.000802
2	46	0.001139
3	45	0.001594
4	44	0.002196
5	43	0.002978
6	42	0.003976
7	41	0.005227
8	40	0.006764
9	39	0.008617
10	38	0.010808
11	37	0.013346
12	36	0.016225
13	35	0.019419
14	34	0.022881
15	33	0.026542
16	32	0.030312
17	31	0.034081
18	30	0.037724
19	29	0.041110
20	28	0.044104
21	27	0.046583
22	26	0.048439
23	25	0.049588
24		0.049977

CHAPTER 8

LINK ENGINEERING ACCORDING TO AVAILABILITY AND ERROR PERFORMANCE CRITERIA

8.1 INTRODUCTION

As described already in previous chapters, the planning of a microwave LOS link involves several design and dimensioning stages. There is an important dimensioning step associated with the system capacity, bit rates, number, and arrangement of radio channel that was covered by Chapter 3. Additionally, there is a set of procedures intended to obtain the adequate number of hops, antenna heights, and antenna separation criteria in diversity configurations against multipath and reflection. These methods were described in Chapter 6.

This chapter focuses on the design according to error performance and unavailability objectives. The function of the objectives is two fold. On one hand, they will provide the most relevant performance thresholds for dimensioning parameters such as radiated power, antenna gain, sensitivity, etc. On the other hand, when the system has been already designed and installed, they will be the reference to test the system and once it has become operational, they will serve as the reference for monitoring its operation.

From a general perspective, quality problems or error performance degradations in a radio-relay link are usually due to different propagation phenomena. These phenomena are mainly related to refraction and hydrometeors. Unavailability, in contrast, may be due to two causes; equipment failure and propagation phenomena, particularly rain.

Microwave Line of Sight Link Engineering, First Edition. Pablo Angueira and Juan Antonio Romo.

Before the design process is explained, it is worth having a brief discussion about the terminology used for defining the correct operation of a radio-relay link because there is a lack of agreement about specific terms for the general concept of "quality" in literature and different design guideline reports. The following words referring to the operation of a radio-relay link in different sources can be found: outage probability, fidelity, error performance, availability, and reliability.

Outage probability usually refers to the statistics of interruptions of the correct operation of the radio-relay link without specifying its impact or relevance and duration. The term outage is normally linked to the cause that creates malfunctioning events. So, we usually talk about outage due to precipitation, outage due to abnormal refraction, etc.

Fidelity refers to degraded operation of the radio-relay link that affects the service quality, but assumes that the link is available. Error performance is also used for referring to this aspect of quality.

Unavailability is a parameter defined by ITU and evaluates either the percentage of time the radio-relay link is out of service due to breakdowns or periods of more than 10 consecutive seconds when problems originated by propagation create severe quality degradations, serious enough to consider the link useless.

Finally, reliability is a parameter used in any engineering field as a measurement of the error free operation of the equipment. It could be regarded in most cases equivalent to unavailability associated with equipment. Nevertheless, we should bear in mind that in certain regions (the United States, for example) it is equivalent to outage probability, regardless of the cause and no matter the duration (i.e., short or long interruptions associated with error performance or availability respectively).

8.2 DESIGN ACCORDING TO AVAILABILITY AND ERROR PERFORMANCE

8.2.1 Initial Specifications

The general design of a radio-relay link is done according to three major groups of specifications: capacity, error performance and availability.

The capacity requirements of a microwave LOS link will have a direct impact on the arrangements of the frequency plan, namely, the number and bandwidth of radio channels. In addition to the number of radio channels, capacity will be associated with the frame structures, multiplexation methods and information overhead used to organize base band information that will also depend on the network where the link will be installed (plesiochronous digital hierarchy, synchronous digital hierarchy, etc.).

Error performance and availability are directly related to link budget dimensioning introduced in Chapter 1. This step of the design process will provide adequate margins to fight signal variations associated with some of the propagation phenomena, and thus minimize their impact of the system. The link budget calculation process will validate and provide additional criteria for the choice of radio equipment (modulators, transmitters, radio frequency reception units and demodulators) characteristics, branching elements, and antennas.

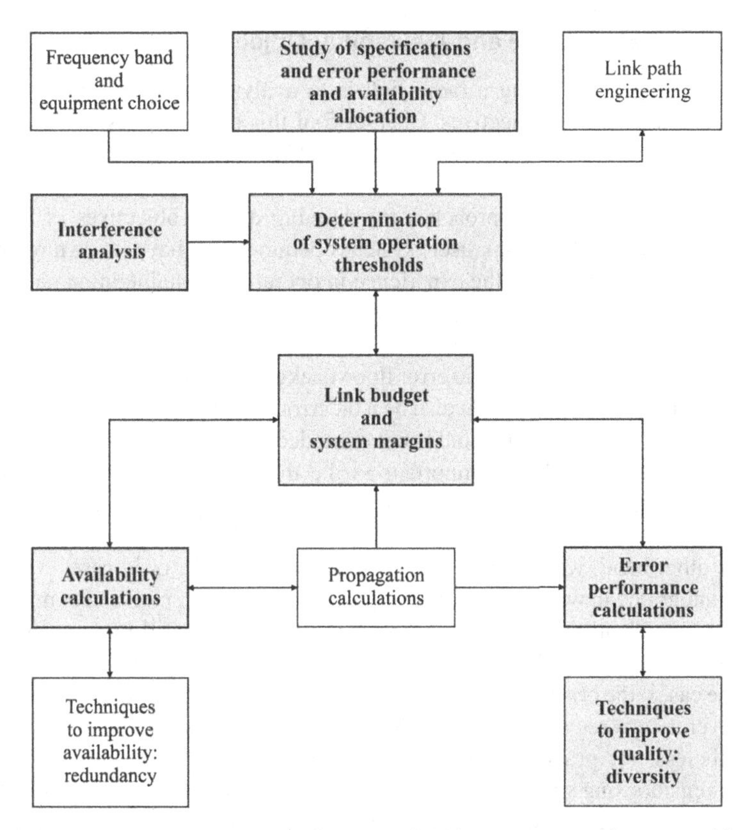

FIGURE 8.1 Design process according to availability and error performance objectives.

Finally, if the system margins calculated through the link budget are not enough to protect the system in situations of intense rain, or in cases of severe multipath propagation, other design tools will be used, that is, diversity techniques.

8.2.2 Microwave LOS Link Design Diagram

Figure 8.1 shows a simplified picture of the diverse design procedures required for the overall design of a microwave LOS link system. Most of the blocks of the Figure 8.1 should be already familiar to the reader of this book. The link path engineering block was covered by Chapters 6. The frequency planning and equipment selection criteria were described in detail in Chapter 3 and 4. The propagation calculations, with emphasis on the methods provided by ITU-R recommendations were included in Chapter 7, while the propagation foundations where these methods are based were covered by Chapter 2. Finally, redundancy and system protection arrangements are part of Chapter 4.

The figure contains additional blocks (shadowed in grey) that represent the design procedures covered by this chapter.

8.2.3 Error Performance and Availability Objectives

The first step when designing a radio link is to analyze the applicable error performance and availability objectives. Chapter 5 of this book has described in depth the parameters involved as well as the relevant ITU recommendations related to this matter.

There are two potential approaches for defining quality objectives of a radio-relay link. On one hand, the link user (system operator) may have its own reference operation values according on the experience in operating or maintaining radio links over a long time period. In these cases, the specifications can be described in terms of statistics relating to usual parameters in digital networks—errored seconds (ES), severely errored seconds (SES), and error floor-background block error ratio (BBER). Sometimes, the specification might also be a bit error rate (BER) value and maximum time percentages where the threshold is not exceeded, both fixed by the operator. The associated statistics will be as important as the threshold criterion. It is usual to work with statistics of either operation objectives for the worst month or annual statistics.

On the other hand, when practical values are not available, or in cases where the system is integrated in wider networks that require overall link, path and communication quality specifications, it is advisable to use the F Series ITU-R recommendations (F.1703 for availability and F.1668 for EPO) as a the reference tools.

In those cases, the criteria will be based on statistical values of errored second ratio (ESR), severely errored second rate (SESR), and BBER already mentioned before.

Nevertheless, for practical reasons, ESR, SESR and BBER cannot be directly applied when choosing specific equipment for the design of the radio-rely link. Manufacturers do not include these parameters in the specifications of the equipment operation thresholds. This apparent contradiction relays on the fact that these parameters (ESR, SESR, BBER) are intended for link operation monitoring (in-service monitoring) once the system has been deployed. If the system performance was done based on BER values, its monitoring would imply the interruption of the service for BER evaluation. On the contrary, ESR, SESR, and BBER can be monitored by measuring the errored blocks. These errored blocks (an errored group of bits) will be easily identified by error correction decoding algorithms in the receiver, and this error condition will be usually reflected on signaling fields of the frame.

8.2.3.1 Availability and Bit Error Rate

The BER value available in ITU-R Recommendation G.821 associated with the definition of SESs is still a fundamental reference for availability studies. This recommendation relates the SESR to bit a of $10^{-3,}$ using the usual 10 s hysteresis period for the entry and exit to/from an unavailability period.

In addition, it will be also necessary to determine the time percentage when the radio-relay link performance will be above the threshold. The ITU-R Recommendation F.1703 will provide the required reference data for each particular case. As already described in detail in Chapter 5, recommendation is based on the parameters AR and OI .

TABLE 8.1 Bit Error Rate Values and Error Performance Parameters

Parameter	Associated BER
SESR	$1.7 \cdot 10^{-5}$
BBER	10^{-12}

8.2.3.2 Error Performance Criteria and Bit Error Rate

Historically, error performance objectives (EPOs) were not an important factor in radio-relay link design, and, therefore, they were not taken into account for system dimensioning. Currently, both for design and maintenance purposes, error performance is regarded as important as availability.

During the years when ITU-R G.821 Recommendation was the main reference for digital transmission systems, the designs were based on the SESR definition in G.821 and the equivalent BBER threshold of 10^{-3}. Later, when ITU-T G.826 Recommendation was released, obtaining the BER value associated with the SESR and to the BBER was not so immediate. It must be taken into account that SESR and BBER are defined from errored block percentages and blocks will have different lengths depending on the transmission standard. The BER associated with an errored block rate basically depends on two parameters: the block length in bits and the bit error distribution function. At the beginning of the 90's some studies provided reference values based on simulations which were progressively incorporated to the first versions of microwave LOS link design recommendations such as the ITU-R P.530. Table 8.1 shows a summary of the proposed BER and error performance parameters SESR and BBER.

Later, in Version 8 of the ITU-R P.530 Recommendation (1999) BERs were described in association to different rates and block. Table 8.2 summarizes the values.

TABLE 8.2 BER_{SES} for Different SDH Routes and Sections

Route	Bit Rate (Mbit/s)	BER_{SES} [a,b]	Blocks/s, n[a]	Bits/Block, *NB*[a]
VC-11	1.5	$5.4 \times 10^{-4}\ \alpha$	2000	832
VC-12	2	$4.0 \times 10^{-4}\ \alpha$	2000	1120
VC-2	6	$1.3 \times 10^{-4}\ \alpha$	2000	3424
VC-3	34	$6.5 \times 10^{-5}\ \alpha$	8000	6120
VC-4	140	$2.1 \times 10^{-5}\ \alpha$	8000	18 792
STM-1	155	$2.3 \times 10^{-5}\ \alpha$	8000/92 000	19 940/801
		$1.3 \times 10^{-5}\ \alpha + 2.2 \times 10^{-4}$		

[a] $\alpha = 1$ indicates a Poisson distribution of errors.

[b] Blocks per second are defined in ITU-T G.826 recommendation for the SDH routes and in ITU-T G.829 for the SDH sections. Some SMT-1 equipment can be designed with 8000 blocks/s (19 940 bits/block), but ITU-T G.829 recommendation defines block rate and block length as 192 000 blocks/s and 801 bits/block, respectively.

TABLE 8.3 Technical Specifications of a Digital Radio-Relay Link Unit

Frequency (GHz)	6	7/8	11	13	15	18
Receiver Sensitivity (BER = 10^{-6}) 32 QAM	−74	−74	−74	−74	−74	−74
All frequencies						
Receiver overload (BER = 10^{-6})	Better than −20 dB					
Unfaded BER	Less than 10^{-13}					

The aforementioned version of the recommendation gives methods for inferring BER values associated with BBER and ESR. The current version of the recommendation (ITU-R P.530-13) has omitted references to specific values of the BER.

Other bibliographic sources suggest a calculation method of BER assuming Poisson distributions for the distribution of errored bits per block. Results do not significantly differ from the ones shown in Table 8.2.

Nowadays, we can continue assuming $1.7 \cdot 10^{-5}$ the reference threshold for SESR EPOs and 10^{-11} for BBER. In practice, the equipment manufacturers will be the ones setting these values accurately. The equipment sensitivity or minimum SNR values will be defined in catalogues as a function of BER values for different propagation conditions. A clear example would be equipment signatures intended for system behavior under strong multipath conditions that will be provided as a function of the BER. Table 8.3 shows the specifications of real link equipment as an example.

Analyzing the data provided by this manufacturer, we could consider a BER of 10^{-6} as an error performance threshold related to SESR. The bitrate in absence of propagation impairments (unfaded BER) is 10^{-13} that could be associated with BBER and constitutes the "error floor" of the system.

8.2.4 Link Budget and Thresholds

The link budget of a radio-relay link and associated parameters were described in detail in Chapter 1. The link budget is the equation that relates the available power at the receiver input as a function of the transmitter power and the different loss and gains found along the route from the transmitter to the receiver. The link budget can be written using equation (8.1):

$$P_{rx} = P_{tx} - L_{tt} + G_t - L_b + G_r - L_{tr} \tag{8.1}$$

where

P_{rx} (dBm) = available power at the receiver input
P_{tx} (dBm) = power delivered by the transmitter to the antenna circuit
L_{tt} (dB) = losses on the antenna distribution and coupling circuits
L_{tr} (dB) = losses on the distribution and coupling circuits connecting the antenna and the receiver

G_t, G_r = gain of transmitter and receiver antennas
L_b (dB) = basic propagation loss (between isotropic antennas), function of the distance, frequency and propagation mechanism

In the process of designing a radio-relay link, the link budget will be the tool to distribute adequately the gain and loss sources by an adequate choice of equipment. There are two design approaches at this point.

The first scenario is when there is not a previous restriction associated with the radio-relay link equipment, ranging from radio-frequency units to antennas. In this case, it is possible to work out desired system gains (transmitted power, line losses, antenna gains and sometimes required sensitivity) through the choice of different units and models.

The second scenario is the one where the equipment manufacturer or even the equipment models are already specified previously. There could be a wide range of reasons for this: commercial reasons, maintenance issues, capacity restrictions, frequency band limitations and other specifications. These cases usually allow limited action on transmission gain values and powers.

In either case, the link budget is used to determine system margins according to availability and EPOs. In order to obtain the margins, the first step is to calculate power thresholds associated with error events (actually associated with BER) by following the criteria recommended in Chapters 1, 4, and 5. Thresholds are usually provided by the manufacturer. If this was not the case, as they depend on the modulation scheme and the internal noise of the receiver units, it will be necessary to obtain them from theoretical calculations for each modulation and coding scheme plus a penalization associated with practical implementation. In consequence, we need to take into account considerations on BERs related to the error events summarized in Table 8.4.

Once thresholds are determined, we need to calculate the margin of the radio link. The gross margin, also called thermal noise margin or flat fading margin, is the difference between the level received in nominal conditions and the threshold power level. If we designate gross margin as M_3 associated with a BER of 10^{-3} and being

TABLE 8.4 Error Parameters, BER Values and Power Threshold Calculations

	Criterion	ITU Recs.	BER	Associated Power Threshold
SESR	Availability	ITU-T: G.821, G.827 ITU-R F.1703	10^{-3}	T_{h3}
SESR	Error performance	ITU-T: G.826, G.828,G.829 ITU-R F.1668	10^{-6}, $1{,}7{\cdot}10^{-5}$	T_{h6}, T_{h5}
BBER	Error performance	ITU-T: G.826, G.828,G.829 ITU-R F.1668	10^{-12}	T_{h12}

C(dB) the received power, we obtain:

$$M_3 = C - T_{h3} \tag{8.2}$$

Similarly, for BER of 10^{-6}, BER of $1.7 \cdot 10^{-5}$, and BER of 10^{-12} just changing subscript:

$$\begin{aligned} M_5 &= C - T_{h5} \\ M_6 &= C - T_{h6} \\ M_{12} &= C - T_{h12} \end{aligned} \tag{8.3}$$

8.2.5 Impact of Propagation on Availability and Error Performance Calculations

Determining the origin and type of fading, losses and other perturbations is a key task of the link planning engineering process. This task requires access to weather data, refraction statistics and other map databases (i.e., vegetation, buildings, etc). Table 8.5 summarizes the different loss and fading sources associated with various propagation phenomena that were covered in detail in Chapters 1 and 2.

Identifying the impact of propagation phenomena on availability and error performance is as important as calculating the exact attenuation values caused by these phenomena in the geographical area of operation. The analysis require access to basic data of the radio-relay link frequency plan, at least the center frequency of the radio channel plan and if the link will be a low, medium or high capacity

TABLE 8.5 Losses and Fading Associated to Different Propagation Phenomena

Basic propagation loss	L_{fs}	Free space loss
Basic fixed losses in excess to the free space loss L_{bexc}	L_{gas}	Gas (O_2) and water vapor (H_2O) absorption
	$L_{vegetation}$	Vegetation attenuation.
Basic variable losses in excess to the free space loss L_{bexv}	$L_{diffraction}$	Obstacles diffraction in (caused by anomalous refraction, see Chapter 2)
	$L_{scintillation}$	Tropospheric scintillation fading
	$L_{hydrometeors}$	Fading caused by rain and other hydrometeors
	$L_{multipath}$	Multipath fading, including reflection effects on earth surface and multipath originated due to various refraction phenomena in higher troposphere layers.
	L_{XPD}	Depolarization losses (associated with hydrometeors and anomalous refraction)
	$L_{misalignment}$	Associated with anomalous refraction conditions.
	$L_{beam\ spreading}$	

system. The following recommendations are common practice for designing radio-relay links in reality:

- Above 10 GHz, fading is mainly caused by rain and other hydrometeors. The attenuation created by hydrometeors is a flat fading effect within the frequency range of the radio-relay link channel plan. This implies that all radio channels of the arrangement will suffer the same attenuation.
- Below 10 GHz, the main cause of degradation is multipath associated with anomalous refraction conditions (usually super-refractive troposphere conditions). If the bandwidth of the radio-relay link frequency plan is wider than the propagation channel coherence bandwidth, selective fading will dominate. In practice, we consider that only medium and high capacity radio links (above 10 Mbps) will suffer fading selective effects associated with multipath. The effect of multipath in narrowband systems (low capacity links) will be flat fading.
- Diffraction losses are excluded from the link budget. This effect will be taken into consideration in the link path engineering process: the number of hops and tower height will be calculated according to clearance criteria (by usually following methods in ITU-R Rec. P.530).
- Hydrometeors and multipath are studied independently and calculated assuming that they are not correlated. There are currently no reliable data about their joint likelihood, and consequently until these data are known they will be treated independently. Moreover, in current design procedures, they are considered as mutually exclusive phenomena. So, when we refer to rain conditions, we assume that the main disturbance suffered by the signal is due to hydrometeors, fundamentally absorption and depolarization. When we refer to design for clear sky conditions, we assume that the most important disturbances are related to anomalous refraction conditions (flat or selective fading, misalignment or depolarization).
- In the case of frequencies where both effects of anomalous refraction and hydrometeors could be considered important (5–10 GHz) ITU-R advises to work with total fading through the sum of probabilities associated with the exceeded values due to each cause.

Table 8.6 summarizes practical criteria for associating propagation phenomena to availability and EPO calculations in a radio-relay link design.

The ITU-R recommends in P.530 that 5 GHz should be the lower limit for considering rain a relevant planning criterion in a radio link design. This value is relatively conservative for most climatic regions of the world. In temperate climates, hydrometeors are not usually considered in design below 10 GHz. It is important to discriminate different consequences in system performance associated with hydrometeors. Fading caused by rain is only taken into account for availability calculations. Rain events are statistically assumed to last more than 10 consecutive seconds. If rain is not intense enough, the margin against flat fading (gross margin) will absorb the effects of that fading. Moreover, as they are not rapid variation effects, the automatic gain control

TABLE 8.6 Practical Criteria for Propagation Impact Allocation on Availability and Error Performance

Frequency	Propagation Phenomenon	Disturbance in the Radio Link	Influenced Objective
$F_0 > 10$ GHz	Anomalous refraction[a]	Medium and hig capacity links: Distortion/selective fading	Error performance
		Low capacity links: Flat fading	Availability and error performance[b]
$F_0 < 10$ GHz	Hydrometeors	Flat fading	Availability

[a]We assume that the route is designed in such a way there are no diffraction losses associated with sub-refraction.
[b]We do not have data for predicting duration of anomalous refraction effects. In hops of typical length below 10 GHz, we assume that availability objectives are complied if error performance objectives are complied.

(AGC) circuits in receiver will be able to absorb the impact of attenuations due to moderate intensity rain.

The case of fading due to anomalous refraction is more complex. First, there are not reliable statistics on fading duration. This is an obstacle to discriminate accurately where their impact should be included: availability or error performance.

As mentioned previously, radio-relay links can be divided into three categories in practice: low, medium and high capacity radio links, with capacities in the ranges of 0–34 Mbps, 13–140 Mbps and higher that 140 Mbps. This division aims at relating radio link bandwidths with the spectral fading width. Hence, the effect of beam spreading and multipath will be considered as flat fading for medium and low capacity radio links, whereas in high capacity radio links beam spreading will cause also flat fading but multipath will cause selective fading.

As far as the link budget design is concerned, under nominal conditions, an appropriate choice of system powers and gains, will enable operating with a received power level some decibels above the availability and error performance thresholds (gross margin). Nominal conditions will be considered the absence of variable fading (no matter the perturbation source). If variable fading exists, the gross margin will absorb, to a great extent the variable propagation effects.

Summarizing, the objective of the link budget calculations will be obtaining the gross margin value under nominal propagation conditions. The minimum power threshold will be the one corresponding to Availability (BER not exceeded over an observation period of time). It will be assumed that the gross margin will absorb flat fading that creates unavailability and error performance degradation.

Finally, it should be remarked that the gross margin will not protect the system against selective fading caused by multipath and other countermeasures might be necessary if this phenomena is relevant on the area where the link is being deployed.

8.3 MICROWAVE LOS LINK DESIGN ACCORDING TO AVAILABILITY CRITERIA

8.3.1 General

This section is develops the design methods associated with unavailability in a radio-relay link design. As it has been described in previous chapters, unavailability (U_{link}) has two causes. The first one is related to propagation events that cause serious disturbances in the radio link with a duration of more than 10 s and it is usually referred to as U_{P}. The second cause is failure of radio equipment, which is called U_{E}. Both causes are uncorrelated in nature and independently considered within the radio link design process. In consequence, total availability corresponds to the equation (8.4):

$$U_{\text{link}} = U_{\text{P}} + U_{\text{E}} \tag{8.4}$$

Electric power supply outages can be also considered a source of unavailability events in some parts of the World. This third factor is complex to describe by statistical functions for a general case. Depending on how critic the radio link is, it can be mitigated by uninterruptible power supplies or through autonomous power supplies.

In the case of unavailability associated with propagation, the first thing to do is to establish the causes that drive the link bit rate higher than availability threshold (usually 10^{-3}). This step depends mostly on the part of the world and on the frequency band. Despite this case-by-case analysis requirement, two general scenarios of unavailability caused by propagation can be identified:

- Flat and show fading originated by anomalous refraction (frequency plans below 10 GHz).
- Fading caused by rain and other hydrometeors such as snow or wet snow. (frequency plans above 10 GHz).

As it will be also covered by the section on error performance, it is usual to consider one dominating phenomenon and carry out the design according to that effect. In any case, if it was necessary to consider both causes, they would be considered statistically independent and therefore, the total unavailability statistics would be the sum of individual statistics associated with each effect.

Below 10 GHz, there is no available accurate method to predict the distribution and duration of anomalous refraction phenomena that may drive the system into unavailability periods. The ITU-R considers that for typical length radio links (without specifying) below 10 GHz not subjected to intense ducts propagation, if fidelity criteria are fulfilled, availability criteria will also regarded as completed.

In the case of unavailability associated with equipment failure, this depends on the reliability of equipment (mean time between failures (MTBF)), and on the required time to repair or replace defective units mean time to repair (MTTR).

Each link design will be based on a single availability objective value that is obtained from ITU-R recommendations, generally using criteria from ITU-R F.1703. This and other recommendations do not provide any guideline to distribute the overall objective into U_{E} and U_{P}. The optimal distribution is a task of the link planning engineer that should take decisions according to the characteristics on the geographical area were the radio-relay link is located and at best, using long-term local propagation statistics, equipment availability parameters (MTBF and MTTR) and, in certain cases also power supply reliability data.

The following subsections describe the method for calculating gross margin values of the radio link for counteracting fading due to rain and equipment failures. Measures that can be taken in the design to improve radio link reliability are also covered.

8.3.2 Unavailability Caused by Rain

Rain is the main cause of unavailability of links in most parts of the world above 5 or 10 GHz. The starting frequency from which this factor has a relevant effect depends on the reference bibliographic source being consulted. Despite of the variety of values 10 GHz in template climatic zones and 5 GHz in zones of intense convective rainfall can be recommended as general practice values.

Fading due to rain is flat across the whole frequency plan of any radio link. On the other hand, the duration statistics of these fading phenomena vary within a range of several seconds or minutes, so they can be considered to be relatively slow fading occurrences (specially compared with multipath phenomena related to anomalous refraction).

ITU-R does not provide information about the statistical distributions of fading duration that could be generally applicable. ITU-R Recommendation P.530 proposes an approximate empirical formulation for calculating rain attenuation excess A, during a specified period of time. In order to evaluate accurate availability statistics associated with rain, the fading value exceeded during at least 10 s should be calculated as well as the number of such occurrences over an observation period. The ITU-R provides an approximate equation to calculate the number of fade events exceeding attenuation A for 10 s or longer:

$$N_{10\mathrm{s}}(A) = aA^{b} \tag{8.5}$$

where coefficients a and b depend on frequency, weather, and route length.

Unfortunately, ITU-R does not have significant statistical data for a and b for different frequencies, hop lengths and climatic zones. Data from a radio link in the north of Europe is provided as an example of reference values, where between 95% and 100% of all rain events with attenuation above about 15 dB caused unavailability.

In the absence of other criteria, authors recommend to consider hydrometeors causing only unavailability (not error performance) and account for hydrometeor fading (basically rain) by means of the gross margin calculated through the link budget.

Gross margin (M_3) was defined in previous sections as the difference between the received level and a determined operation threshold (T_{h3}) for availability objectives. Once operation threshold is known, the following steps summarize the design procedure:

1. Availability objectives of the radio-relay link are obtained from recommendation ITU-R F.1703. We call this target value U_{TOTAL}.
2. Objectives will be distributed into equipment failures, propagation and power plant (if necessary). The objective related to propagation will be called U_P.
3. The link budget is evaluated, either analytically as in equations (8.6) and (8.7) or by means of a hypsogram (power level diagram):

$$T_{h3} - M_3 = P_{tx} - L_{tt} + G_t - L_b + G_r - L_{tr} \tag{8.6}$$

$$L_b = L_{fs} + L_{vegetation} + L_{hydrometeors} \tag{8.7}$$

4. We will calculate the attenuation caused by hydrometeors associated with a probability value of U_P. Specific methods for each hydrometeor type will be required for this purpose. The ITU-R provides methods for rain and wet snow in recommendations ITU-R P.838 and P.530 (see Chapter 7). In the case of rain, and according to the method and parameters developed in Chapter 7, we will calculate the parameter A_p, attenuation not exceeded during a p percentage of time. For high latitude regions or radio links located at relevant heights, the effect of hail or snow particles will also be included. The impact of snow will be determined by a corrector factor applied to the rain attenuation A_p, that will depends on the height of the 0°C isotherm. In those specific cases when, because of special climatic conditions of the hops, it would also be necessary considering attenuation due to clouds and its statistics, methods provided by ITU-R P.840 recommendation can be used.
5. Once the attenuation associated with a probability of U_P is known, that value is brought into equation (8.6) and M_3 will be obtained. This way, we have the rain attenuation value that occurring with a probability of U_P will be equal to the gross margin, which from another perspective can be stated as: the gross margin that exception made for an U_P of the time will protect the system against rain attenuation.

8.3.3 Rain Attenuation in Multiple Hop Links

In radio-relay links with several hops in tandem, the global effect of rain attenuation depends on the spatial correlation of rain phenomena across the geographical area where the radio-relay link is installed. If rain occurrences are spatially uncorrelated the overall probability of fading can be calculated as the sum of fading probabilities in each one of the hops. Otherwise, if there is correlation between nearby hops, total probability equation can be approximated as:

$$P_{\mathrm{T}} = K \sum_{\mathrm{i}=1}^{n} P_{\mathrm{i}} \tag{8.8}$$

where K is an empirical factor that includes the total effect of the spatial correlation of the rain.

Certain information about K parameter values can be found in ITU-R 530 recommendation through curves in function of the route of the weather fronts, number of hops of the radio-relay link, hop lengths and exceeded probabilities associated with each hop. The ITU-R acknowledges a lack of information on this topic. In those radio links where rain is a critic problem of availability, route diversity is an alternative solution. The key factor for determining the diversity gain obtained with this solution relies on the geographical distance between both routes. Once again, empirical data are scarce and usually only applicable to the geographical area where they were obtained, with little generalization value.

8.3.4 Equipment Reliability and Unavailability

The characterization of reliability of equipment for calculating system availability requires a more complex procedure than the statistical calculation of fading associated with hydrometeors. Equipment availability is identified with the concept of reliability of the group of hardware elements that compose the radio-relay link. The calculation procedure will require distinguishing if the radio link has backup (reserve) units as well as the architecture of the different switching sections where reserve units (and radio channels) become active in case of breakdown.

Reliability of a radio-relay link is the probability of proper operation of the equipment, under certain conditions, during a specified period of time. The statistics will consider failure probabilities, as well as the time required to repair those failures. Under normal operation conditions and for theoretical calculation purposes, if the radio link is working properly, equipment failure frequency should be independent of the total time that has passed from the installation of this hardware and should only be related to the statistics of the average time between failures. This average time between failures will be empirical parameters provided by the manufacturers and will be usually described by a negative exponential distribution from equation (8.9):

$$p(t) = \frac{1}{\tau} \exp\left(\frac{-t}{\tau}\right) \tag{8.9}$$

where τ is the average time between failures, which is usually written as MTBF. The physical interpretation of this equation is, of course, that the probability of having a failure will increase as the amount of time since the last failure increases.

The MTBF value of equipment will be a characteristic of its reliability and will be one of the parameter values contained in the specification documentation provided by the manufacturer. It is usually to have values in the order of magnitude of 10^5.

Nevertheless, one of the key aspects of an appropriate reliability design will be determining whether the real environmental operation conditions fit the test conditions used by the manufacturer to obtain the MTBF specification.

In the case of systems with neither diversity nor protection, different modules that form a radio-relay link hop, and at a higher level hops that form a complete radio-relay link can be considered as series connected elements. In consequence, the equation of the average time between failures of two series units A and B:

$$(\mathrm{MTBF})_{\mathrm{Total}}^{-1} = (\mathrm{MTBF})_{\mathrm{A}}^{-1} + (\mathrm{MTBF})_{\mathrm{B}}^{-1} \tag{8.10}$$

This equation, expanded for the N units or modules that form the radio-relay link, is used for calculating the total average time between failures.

In order to calculate radio link availability, we also need to take into account that when a failure happens, if there is no reserve equipment, it is necessary to repair or replace the unit damaged. This time is called MTTR and depends on the equipment unit, fault diagnosis and fault management characteristics of the radio link operator. Depending on the location, the MTTR can be limited to 2 hours in urban areas or to a significantly longer time in remote areas.

Equipment unavailability (in the case of equipment without protection) is given by equation (8.11):

$$\mathrm{U}_{\mathrm{Equipment}}(\%) = \frac{\mathrm{MTTR}}{(\mathrm{MTBF} + \mathrm{MTTR})} 100 \tag{8.11}$$

As generally, MTBF >> MTTR:

$$\mathrm{U}_{\mathrm{Equipment}}(\%) = \frac{\mathrm{MTTR}}{\mathrm{MTBF}} 100 \tag{8.12}$$

Equation (8.12) has an important consequence. If the operator achieves similar average repairing times for any of the units and modules that form the radio-relay link, unavailability of the series connection of the units that form it is the sum of unavailability values associated with each module.

8.4 MICROWAVE LOS LINK DESIGN ACCORDING TO ERROR PERFORMANCE OBJECTIVES

8.4.1 General

Once design is analyzed according to availability objectives criteria, this section describes the procedures to account for error performance criteria.

A multitude of different causes can degrade BER (or SESR, BBER) of an information stream transported by a radio-relay link. On the contrary to the availability studies, it will be assumed that equipment does not have any further impact on the error performance degradations exception made for determining the error floor or irreducible BER (BBER according to ITU-R nomenclature). This irreducible threshold is usually in the range of 10^{-11} to 10^{-13}. Thus, propagation phenomena will be the main cause of error performance degradation. Among these, the following ones can be mentioned:

- Tropospheric scintillation.
- Depolarization under both clear sky and rain conditions.
- Fading due to hydrometeors that do not exceed availability limit.
- Multipath propagation due to ground, water, and reflection on other surfaces; or very negative values of the refraction gradient at ground level or combinations of reflections on the ground and tropospheric multipath conditions. In all these cases (situations with very negative refraction gradients) frequency selective fading will degrade the system error performance, especially for low frequency bands.
- Fading due to beam spreading and variations of the angle of launch and arrival under anomalous refraction associated with stratified troposphere conditions and ducts. These situations will present a combination of slow and fast fading.

Generally, in terms of design, tropospheric scintillation is negligible in amplitude, below 40 GHz. The gross margin of the radio-relay link completely eliminates the possible effect on error performance. On the other hand, flat and selective fading caused by anomalous refraction effects have a significant effect at frequencies below 10 GHz.

Depolarization effects must be taken into account in two design steps: the intra-system interference analysis for radio channel arrangement procedures (based on minimum cross-polar discrimination (XPD) values), and inter-system interference evaluation phase in the case of evaluating interference among different radio links. Additionally, depolarization can be also a cause of outage due to XPD degradation, as it will be explained in later sections.

The effect of hydrometeors effect on error performance (mainly rain or wet snow) should be considered when attenuation produces a received signal below the objective error performance threshold (T_{h5} or T_{h6}) and at the same time above unavailability threshold (T_{h3}). This calculation is not usually included in design procedures related to error performance.

Summarizing, radio-relay link design for a certain EPO is limited, in most cases, to selective fading, flat fading, and depolarization. Next sections describe the application of different methods for applying flat and selective fading estimation methods for dimensioning a radio-relay link according to BER (SESR and BBER) objectives.

8.4.2 Design for Operation in Flat Fading Conditions

Flat fading appears in a radio link as a more or less abrupt fall in received level with respect to the median value observed during a period of time long enough. Receiver equipment has a certain margin for adapting to varying input power values. These

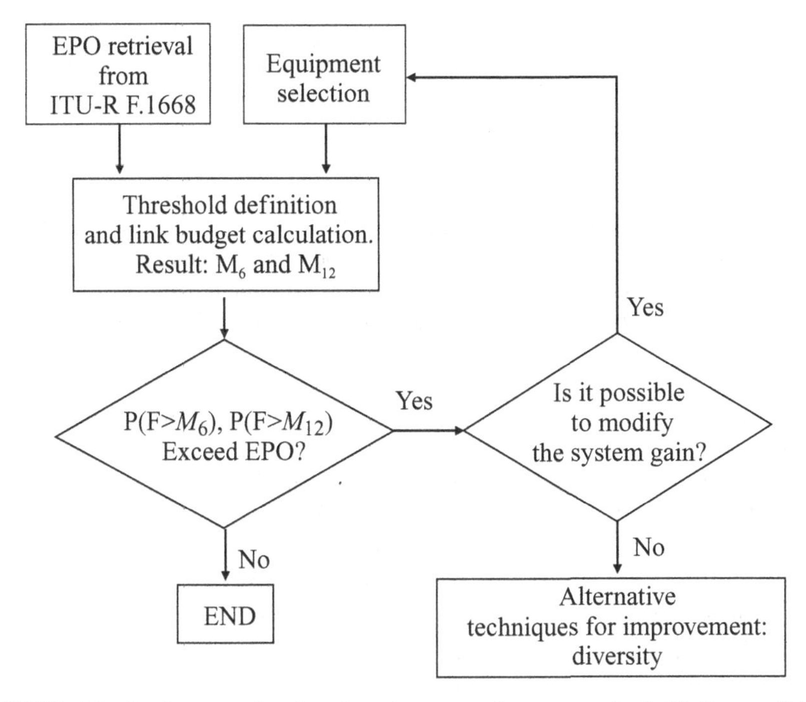

FIGURE 8.2 Design procedure for adequate error performance under flat fading conditions.

variations can be compensated by AGC circuits. In this regard, the design of the AGC systems is a compromise between the circuit inertia (input integration time), gain and internal noise put at the output port. These criteria are specific for each manufacturer and will not solve any flat fading situation.

In consequence, flat fading has to be considered in those geographical areas where anomalous refraction will be probable, in order to ensure that flat fading do not degrade the radio link below EPOs.

Two main causes of flat fading have been described in previous sections: anomalous refraction and rain. Traditionally, rain has not been considered in radio-relay link error performance calculations and flat fading has been only related to anomalous refraction conditions in clear sky conditions. The procedure is summarized in Figure 8.2.

8.4.2.1 Error Performance and Flat Fading Caused by Refraction

Flat fading caused by anomalous refraction conditions is taken into account in the radio link design by means of a power margin above the error performance threshold. This calculation will be carried out using the statistical distribution of fade depths. If P(F) is the probability of exceeding a fading depth F, then:

$$P(F) = P_0 \cdot 10^{-F/10} \tag{8.13}$$

where P_0 is a parameter called *multipath occurrence factor* and depends on radio link length, frequency, terrain roughness and weather conditions. The probability of fading exceeding a M value, flat fading margin, is usually written:

$$P(F) = P_0 \cdot 10^{-M/10} \tag{8.14}$$

This margin has to be decided for each design in order to comply with EPOs. As previously explained, we should expect that the margin calculated according to availability objectives to protect the system against rain attenuation should be more restrictive (wider margin). Anyway, it is advisable to follow the next calculation steps:

1. Obtain from ITU-R F.1668 the performance objectives that should not be exceeded during a certain time (SESR and BBER). Identify the associated BER values as explained in previous sections. The threshold limits will usually be $1.7 \cdot 10^{-5}$ or 10^{-6} (SESR) and 10^{-12} (BBER).
2. Find receiver power thresholds associated with the objective BER objectives. If the data are not available from the manufacturer, they might be obtained from modulation error correction techniques. Let us assume that:
   - T_{h6} is the threshold associated with a BER of 10^{-6}
   - T_{h5} is the threshold associated with a BER of $1.7 \cdot 10^{-5}$
   - T_{h12} is the threshold associated with a BER of 10^{-12}
3. Calculate equivalent margins associated with T_{h6} (or T_{h5}) and Th_{12}:

$$\begin{aligned} M_5 &= C - T_{h5} \\ M_6 &= C - T_{h6} \\ M_{12} &= C - T_{h12} \end{aligned} \tag{8.15}$$

   where C is the received power calculated according to availability criteria.
4. Calculate equivalent error performance margins associated with T_{h6} (or T_{h5}) and T_{h12}:

$$\begin{aligned} \mathrm{EPO_{SESR}} &= P(M_6) = P_0 \cdot 10^{-M_6/10} \\ \mathrm{EPO_{BBER}} &= P(M_{12}) = P_0 \cdot 10^{-M_{12}/10} \end{aligned} \tag{8.16}$$

5. Check that M_6 (or M_5) and M_{12} are less restrictive than the M_3 obtained for complying with availability objectives related to rain. If they would be more restrictive than the one required for rain, these ones will be the reference.

In Chapter 7 different methods for obtaining flat fading statistics were described as a function of the features of the geographical area where the radio-relay link that is being designed. These methods offer data for average worst month anywhere in the world and use expressions derived from equation (8.13). The methods differ in the procedure for calculating P_0, the multipath occurrence factor.

Among the methods available from different bibliographic sources, we recommend using the two methods in described in ITU-R Rec. 530-13. The first method is suitable for calculating flat fading during small percentages of time. The method is also adequate in systems where very deep fades are expected. The algorithm has two versions of different accuracy. The first version is an approximation that usually it will be applied to preliminary studies, that is, presenting studies associated with frequency use license requests, and the second one, more accurate, would be used for final design purposes.

The second method of the ITU-R Rec. P.530-13 is applicable to any fading depth statistics as it combines the procedure for very deep fading presented in the previous point with an empirical interpolation procedure for small amplitude fading occurrences of decreasing value close to 0 dB.

8.4.2.2 *Error Performance and Rain* Flat fading caused by rain will degrade error performance of any radio-relay link. It is nevertheless usual to disregard its effect in the design steps that comply to EPOs. The reason is that rain is supposed to be related to events during more than 10 s and thus, they are included in the design through availability.

In fact, the assumption behind this simplification is that the gross margin for availability will be at the same time protecting the link against the degradation on the error performance. Figure 8.3 illustrates this situation.

The diagram assumes that the gross margin of the system (margin that will cope with intense rain events longer than 10s that create unavailable periods) is set in 25 dB (M_3). Analyzing typical modulations and error correction schemes, the difference

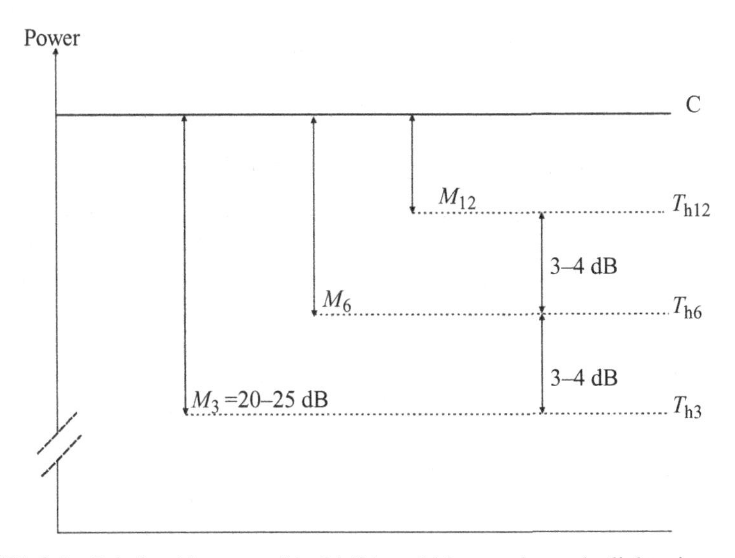

FIGURE 8.3 Relationship among M_3, M_5/M_6 and M_{12} margins under light rain conditions.

between different thresholds (10^{-3}, 10^{-4}, and 10^{-12}) will be in the range of 3–4 dB. In consequence, there will be a margin of 12 dB ($C - M_3 - (M_6 + M_{12})$) that would remain for absorbing lighter rain intensities and keeping the BER below 10^{-6}.

In the case of having data that suggests that rain calculations should also be considered for EPOs, the following procedure could be applied:

1. Assume that a previous study on availability exists with a criterion of a 10^{-3} being the BER not exceeded during a time percentage (According to ITU-R F.1703).
2. The EPO of the radio-relay link will be obtained from ITU-R F.1668 (equivalent BER values of $1.7{\cdot}10^{-5}/10^{-6}$ and 10^{-12}).
3. Assume that flat fading due to anomalous refraction and that caused by rain are independent events and, therefore, each one could be assigned a portion of the objective:

$$\text{EPO}_{\text{Flat\_fading}} = \text{EPO}_{\text{refraction}} + \text{EPO}_{\text{rain}} \tag{8.17}$$

4. The probability of having refraction caused fading depths associated with M_5/M_6 and M_{13} are calculated with equation (8.16). Under normal conditions, the obtained probability values should be below the EPOs. The difference to the maximum EPO is then assigned to rain events.
5. Once the portion of EPO assigned to rain is obtained, we will check that the probability of having a rain attenuation amplitude equivalent to M_5/M_6 and M_{13} margins is under the EPO allocated to rain.

It is important to mention that the 10^{-12} EPO is considered as the threshold under fading free conditions by manufacturers, so, in practice, the previous steps would only include the BER objective of $1.7{\cdot}10^{-5}$ or 10^{-6} in the EPO.

8.4.3 Error Performance and Selective Fading Due to Multipath

Digital radio-relay links with medium and high capacities (higher than 10 Mbps) where either the number of radio channels is relevant or the width of each spectra is remarkable will suffer (sometimes in addition) distortion or multipath selective fading. In these systems, the analysis of flat fading and its impact on the EPOs, it is not enough to ensure that these EPOs are fulfilled. In this case, even in the case of not observing relevant variations on the received signal (flat fade) the radio channels might be suffering relevant distortion and the SESR could be higher than the accepted values over significant time percentages.

In consequence, if multipath fading has already reduced the eye diagram amplitude to zero, an increase in the gross margin will not improve error performance in high capacity digital systems. This means that increasing either the transmitter power increase or the antenna gain is not an effective countermeasure in those cases, and other techniques are necessary. These techniques can be base band equalizers or frequency diversity systems.

The characterization of the distortion caused by multipath fading is done in terms of outage probability. The outage probability of the transmission is here defined as the probability of BER exceeding a specific threshold value.

Three approaches have traditionally been considered in order to predict outages affecting the radio-relay link error performance:

- Linear amplitude dispersion methods.
- Net-fade margin method.
- Methods based on signature curves.

According to ITU-R, there is not enough data to recommend any of the three options. The authors nevertheless recommend the use of the signature method because the information found about its application and related parameters is provided in detail by ITU-R Rec.P.530 and several other reports and handbooks (see Chapter 7). This method takes as input the equipment signature curve that will be used in the radio link and combined with the statistic parameters related to multipath (such as multipath appearance factor and the average delay between rays). The signature curve provides a graphical description of the equipment robustness against different selective fading occurrences within the spectrum of the radio channel.

Let us remember that the output of the signature method (equation (8.18)) is the probability exceeding a BER value that will be associated with the specific signature curve of the equipment involved:

$$P_s = 2.15\eta\left(W_M \times 10^{-B_M/20}\frac{\tau_m^2}{|\tau_{r,M}|} + W_{NM} \times 10^{-B_{NM}/20}\frac{\tau_m^2}{|\tau_{r,NM}|}\right) \tag{8.18}$$

where

η = multipath activity factor
$W_{M/NM}$ = signature width (GHz)
$B_{M/NM}$ = signature depth (dB)
$\tau_{r,M/NM}$ = reference delay (ns) associated with the signature
= M and NM subscripts refer to the minimum phase and non-minimum phase situations respectively

Once outage probability is calculated (exceeding of a specific BER threshold), we will be able to evaluate whether that degradation complies with the EPO objectives established in radio link specifications or not. Otherwise, corrective measures will be needed. There are two types of measures: adaptive equalizers and diversity techniques.

8.4.3.1 Net-Fade Margin Method The use of a fixed margin model to calculate the impact of multipath selective fading is based on the already described model of equation (8.19):

$$P(F) = P_0 10^{-\frac{F}{10}} \tag{8.19}$$

According to this equation, the concept of *net* or *effective* margin for digital systems is defined, which consider the combined effect of flat and selective fading occurrences caused by multipath within the same parameter:

$$P_T(F = Me) = P_0 10^{-\frac{Me}{10}} \tag{8.20}$$

P_T (in%) in the equation (8.20) is the overall probability, which includes flat and selective fading. P_0 is the multipath occurrence factor already described in previous sections (in this case converting the factor to the associated percentage).

The method requires a new parameter, the net margin (in opposition to gross margin that copes with flat fading) represented by M_e. The net margin can be calculated using equation (8.21):

$$M_e = M_3 - 10\log\left(1 + \frac{P_s}{P_0} 10^{-\frac{M_3}{10}}\right) \tag{8.21}$$

M_e is obtained from the net margin M_3, calculated previously following availability criteria. In the case of small values of M_3, the value of M_e remains constant and almost equal to M_3. For higher values, the net margin tends asymptotically to a constant value described by equation (8.22):

$$\lim_{M_e \to \infty} M_e = 10\log\left(\frac{P_0}{P_s}\right) \tag{8.22}$$

8.4.4 Occurrence of Simultaneous Fading on Multihop Links

Under clear sky conditions, it will be assumed that the multipath occurrences on consecutive tandem links are uncorrelated. This supposition is applied not only to flat but also to selective fading. Under this supposition, the total outage probability can be obtained as the sum of outage probabilities of each hop.

The ITU-R P.530 recommendation offers an empirical equation for radio-relay links in the band of 4–6 GHz that accounts for some correlation degree between hops. Equation (8.23) represents the probability of exceeding a fading of A (dB) in a radio-relay link of n hops:

$$P_T = \sum_{i=1}^{n} P_i - \sum_{i=1}^{n-1} (P_i P_{i+1})^C \tag{8.23}$$

where

P_i = outage probability of the i-th hop
n = number of hops of the radio-relay link

d_i = path length of the i-th hop
C = empirical factor between 0.8 and 1 as a function of fading amplitudes
$C = 0.5 + 0.0052A + 0.0025(d_i + d_{i+1})$ if $A < 40$ dB and $(d_i + d_{i+1}) <$ 120 km
$C = 1$ otherwise

8.4.5 Cross-Polar Discrimination (XPD) Reduction Outages

So far, different attenuation and fading phenomena and associated statistics have been described in relation to the system design process.

In addition, systems working in the frequency bands allocated to fixed wireless system (FWS) require an analysis of the XPD reduction that happens under clear sky conditions as well as under rain fading periods. Following the usual assumption of uncorrelated propagation phenomena, the analysis associated with XPD reduction under clear air and XPD reduction under rain conditions will be carried out independently.

Finally, the impact on the system outage (error performance) will depend on whether the receivers include cross-polarization interference cancellation (XPIC) circuits or not. XPIC modules are polarization correction circuits that reduce depolarization effects (see Chapter 4).

Chapter 7 described the calculation steps for depolarization in both cases: XPD_{clear_air} and XPD_{rain}. According to the frequency range in which hydrometeors are relevant for link design, if the system operates below 10 GHz, the relevant XPD will be associated with clear air (refraction effects) and at higher frequencies, the important XPD source is rain.

Finally, we must remember that the design parameter used in the radio-relay link design will be the outage probability associated with depolarization (P_{XP}) obtained by applying to the methods described in Chapter 7.

8.4.6 Overall Error Performance

The overall outage probability under clear sky conditions can be calculated by a combination of the outage probabilities related to the following propagation factors: flat fading, selective fading and XPD reduction. Equation (8.24) illustrates the calculation procedure:

$$P_t = \begin{cases} P_{ns} + P_s + P_{XP} & \\ P_d + P_{XP} & \text{if diversity schemes are used} \end{cases} \tag{8.24}$$

where

P_{ns} = non-selective (flat fading) outage probability under clear sky
P_s = selective outage probability (distortion, selective fading)
P_{XP} = outage probability due to XPD
P_d = outage probability due to multipath in diversity systems

Under rain conditions, the outage probability associated with the rain phenomena might be considered as be the maximum of two outage probabilities: the one associated with rain attenuation (P_{rain}) and the outage probability due to XPD decrease (P_{XP}).

Nevertheless, despite the absence of a recommendation from ITU-R, the practical procedures are usually to apportion the outages associated with rain attenuation to availability, dismissing its impact on error performance, and consider the XPD reduction for error performance calculation purposes.

8.5 DESIGN IN PROBLEMATIC PROPAGATION ENVIRONMENTS

Attenuations and distortions produced as a consequence of multipath propagation and other anomalous propagation conditions can be mitigated by using two strategies that are not mutually exclusive. Both tools, applied on the receiver side, are diversity techniques and adaptive equalizers.

In both cases the gain associated with these techniques is called *improvement factor* and it is defined as ratio between the outage time associated with the standard system and the one associated with the enhancement design. Improvement factors will depend on the system specifications for maximum outage probability assumed during the design stages.

8.5.1 Diversity Techniques

Diversity techniques were introduced in Chapter 4 along with protection schemes that increase the system availability. There is a wide range of diversity techniques, among which the most usual ones are frequency and space diversity. The improvement of the diversity technique will depend upon different parameters that include the influence of in-band amplitude dispersion and inter-system interference on channel performance, and the algorithms that are used in the implementation of diversity.

Current diversity systems use combiners at radiofrequency, intermediate, or baseband stages, or by using switches at intermediate or baseband frequencies. In the case of signal combining techniques equal gain, maximum power, or minimum dispersion control algorithms are used.

Current microwave LOS links use either of the minimum dispersion combining devices (MID) or dual reception receivers. In the latter, the diversity signals are combined by maximum power combining algorithms (MAP) and the aggregate signal will be enhanced by adaptive equalizing (EQ). The simultaneous use of maximum power combining and adaptive equalizing algorithms provides an enhancement that is higher than the addition of the individual gains associated with each technique.

This section describes the calculation procedure to obtain the improvement factor associated with each technique. The result of this process will be a new outage probability function that will depend on key diversity parameters: spacing between antennas in the case of spatial diversity and spacing between frequencies in the case of frequency diversity.

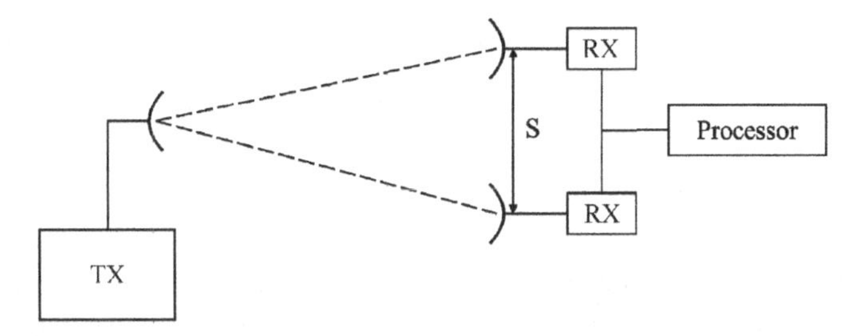

FIGURE 8.4 Simplified diagram of the equipment in a hop with spatial diversity.

8.5.1.1 Space Diversity Space diversity systems assume that selective (or flat) fading occurrences at each one of the two (or more) antennas are not correlated in time.

The improvement factor of a spatial diversity system depends on the spatial separation of the antennas, on propagation factors and also on the robustness of the diversity scheme. This robustness depends also strongly on the techniques used for the combination of the different diversity components at the receiver. Figure 8.4 shows a simplified diagram of a hop with spatial diversity where S (spacing between antennas) parameter is outlined, very important in the obtained gain due to diversity.

According to ITU-R Rec. F.752, there are three families of methods used for combining space diversity signals on the basis of the maximum power principle. These systems use receivers with MAP + EQ. The choices are:

- *Maximum power equal gain combining*: This method, which uses linear addition.
- *Maximum power equal gain/switch combining*: This method uses linear addition and there is no signal/noise ratio loss at the combiner output when the input signal from the first receiver is switched off.
- *Maximum power optimized combining*: The diversity branches that input the combiner contain electronic attenuators, the attenuation of which varies automatically in proportion to the ratio of the powers of the space diversity signals received.

As mentioned previously, MIDs are also used in radio-relay links. In this case, the combining algorithm is the minimum BER or weighted strategy that is based on processing base band or IF signals for minimizing BER.

In order to calculate the improvement factor associated with spatial diversity, we must differentiate between narrow band radio links (capacities of 2 Mbps or lower) or broadband radio links.

8.5.1.1.1 Improvement Factor Due to Spatial Diversity in Narrow Band Links The method shown below is based on empirical data provided by the ITU-R SG 3 for radio-relay links that meet the following conditions:

Hop distance d	$43 \leq d \leq 240$ km
Frequency range	$2 \leq f \leq 11$ GHz
Spacing between antennas	$3 \leq S \leq 23$ m

Despite the above specifications, the ITU-R considers that data obtained with this method will be valid for path lengths up to 25 km. The improvement factor is calculated by the equation (8.25) in this case:

$$I = \left[1 - \exp\left(-0.04 \times S^{0.87} f^{-0.12} d^{0.48} p_0^{-10.4}\right)\right] 10^{(A-V)/10} \quad (8.25)$$

where

$V = V = |G1 - G2|$ and G_1, G_2: gains of both antennas (dBi).
$A =$ fading depth (dB) of the unprotected path
$p_0 =$ multipath occurrence factor (%)
$S =$ vertical spacing of the receiver antennas (m), from center to center
$f =$ frequency (GHz)
$d =$ path length (km)

8.5.1.1.2 Improvement Factor Due to Spatial Diversity in Broadband Links

In the case of protection for broadband radio links, it is usual to use maximum power ratio combinations. The ITU-R P.530 Recommendation suggests a calculation procedure for obtaining the improvement factor for this type of combiner. The recommendation itself mentions the fact that other more complex combiners could achieve a better improvement factor. The step-by-step calculation procedure is written below and it is an enhanced explanation of the calculation steps provided in ITU-R P.530:

1. The multipath activity parameter η, will be calculated by means of equation (8.26), where $P_0 = p_w/100$ is the multipath occurrence factor associated with the percentage of the time $p_w(\%)$ of exceeding $A = 0$ dB in the average worst month (more details in Chapter 7):

$$\eta = 1 - e^{-0.2(P_0)^{0.75}} \quad (8.26)$$

2. The square of the correlation nonselective coefficient k_{ns}, is calculated following equation (8.27). I_{ns} is the improvement factor for narrowband systems from the equation (8.25) for a fading depth, A(dB). A(B) corresponds to the flat fading margin, F(dB). P_{uns} is the transmission outage probability by the nonselective component of multipath (flat fading):

$$k_{ns}^2 = 1 - \frac{I_{ns} \cdot P_{uns}}{\eta} \quad (8.27)$$

3. The correlation selective coefficient, k_s is calculated by means of equation (8.28) and r_w is the correlation coefficient of the relative amplitudes in this equation, which in turn will be calculated using equation (8.29):

$$\begin{cases} k_s^2 = 0.8238 & r_w \leq 0.5 \\ k_s^2 = 1 - 0.195 \cdot (1 - r_w)^{0.109 - 0.13 \cdot Log_{10}(1 - r_w)} & 0.5 < r_w \leq 0.9628 \\ k_s^2 = 1 - 0.3957 \cdot (1 - r_w)^{0.5136} & r_w > 0.9628 \end{cases} \quad (8.28)$$

$$\begin{cases} r_w = 1 - 0.9746 \left(1 - k_{ns}^2\right)^{2.170} & \text{para} \quad k_{ns}^2 \leq 0.26 \\ r_w = 1 - 0.6921 \left(1 - k_{ns}^2\right)^{1.034} & \text{para} \quad k_{ns}^2 > 0.26 \end{cases} \quad (8.29)$$

4. The nonselective outage probability of the transmission, P_{dns}, is calculated using the improvement factor related to flat fading and P_{uns}, the outage probability due to the nonselective component of fading without space diversity:

$$P_{dns} = \frac{P_{uns}}{I_{ns}} \quad (8.30)$$

5. Selective outage probability of the transmission, P_{ds}, is calculated from multipath propagation activity factor η, the selective correlation coefficient k_s and P_{us}, the outage probability of the selective component of fading without spatial diversity (calculated by the signature method for example as described in Chapter 7):

$$P_{ds} = \frac{P_{us}^2}{\eta \cdot \left(1 - k_s^2\right)} \quad (8.31)$$

6. The total outage probability is calculated from equation (8.32):

$$P_d = \left(P_{ds}^{\frac{\alpha}{2}} + P_{dns}^{\frac{\alpha}{2}}\right)^{\frac{2}{\alpha}} \quad (8.32)$$

8.5.1.2 Angle Diversity The angle diversity in a hop consists of having two beams pointing towards the receiver station with two slightly different elevation directions, either through two antennas or through a single parabolic antenna with double feeder. In the case of two antennas, this diversity technique is differentiated from spatial diversity because in this case the distance between the two antennas is not important for obtaining a significant improvement factor.

A various experimental tests have been carried out to determine the gain of the angle diversity technique in comparison to the frequency diversity. There is not a regular pattern result in all tests although we can conclude that when the perturbation factor is associated with multipath fading distortion, both types of diversity (spatial and angular) provide similar results. In the case of radio links that are also affected by thermal noise, spatial diversity gives better results.

Similarly to the case of spatial diversity, ITU-R P.530 recommendation presents an empirical method for calculating the improvement factor from the most relevant parameters of the angular diversity system, which are the radiation patterns features and the difference of pointing angles. The calculation procedure is detailed as follows.

1. The average angle-of-arrival μ_θ (degrees) is calculated from equation (8.33) were G_m is the average value of refractivity gradient (units N/km). If there is a significant reflection component on the ground, μ_θ will be the angle-of-arrival of the reflected signal.

$$\mu_\theta = 2.89 \times 10^{-5} G_\mathrm{m} d \tag{8.33}$$

2. The nonselective reduction parameter r, is calculated from equation (8.34):

$$r = \begin{cases} 0.113 \sin[150(\delta/\Omega) + 30] + 0.963 & \text{for} \quad q > 1 \\ q & \text{for} \quad q \leq 1 \end{cases} \tag{8.34}$$

$$q = 2505 \times 0.0437^{(\delta/\Omega)} \times 0.593^{(\varepsilon/\delta)}$$

where

δ = angular separation between the two radiation patterns
ε = elevation angle of the upper antenna (positive towards ground)
Ω = half-power beam width of the antenna patterns

3. The nonselective correlation parameter Q_0, is calculated from equation (8.35):

$$Q_0 = r\left(0.9399^{\mu_\theta} \times 10^{-24.58\,\mu_\theta^2}\right)\left[2.469^{1.879^{(\delta/\Omega)}} \times 3.615^{\left[(\delta/\Omega)^{1.978}(\varepsilon/\delta)\right]} \times 4.601^{\left[(\delta/\Omega)^{2.152}(\varepsilon/\delta)^2\right]}\right] \tag{8.35}$$

4. Multipath activity factor η, is calculated following the already well known equation:

$$\eta = 1 - \mathrm{e}^{-0.2(P_0)^{0.75}} \tag{8.36}$$

5. The probability nonselective fading of depth F (dB) is calculated from equation (8.37):

$$P_\mathrm{dns} = \eta Q_0 \times 10^{-F/6.6} \tag{8.37}$$

6. The selective correlation coefficient, k_s is calculated from equation (8.38):

$$k_\mathrm{s}^2 = 1 - \left(0.0763 \times 0.694^{\mu_\theta} \times 10^{23.3\mu_\theta^2}\right)\delta\left(0.211 - 0.188\mu_\theta - 0.638\mu_\theta^2\right)^{\Omega} \tag{8.38}$$

7. The probability of selective transmission outage, P_{ds}, is calculated using the multipath propagation activity factor η, the selective correlation coefficient k_s and P_{us}, the outage probability associated with the selective component of fading without spatial diversity (calculated using the signature method for example):

$$P_{ds} = \frac{P_{us}^2}{\eta \cdot \left(1 - k_s^2\right)} \tag{8.39}$$

8. The total outage probability is calculated from equation (8.40):

$$P_d = \left(P_{ds}^{\frac{\alpha}{2}} + P_{dns}^{\frac{\alpha}{2}}\right)^{\frac{2}{\alpha}} \tag{8.40}$$

8.5.1.3 Frequency Diversity

The frequency diversity enhancement depends on the ratio of the number of active over reserve radio channels. There are two commonly used alternatives in this regard: 1+1 systems and N+1 systems. Figure 8.5 shows a block diagram of both types of diversity.

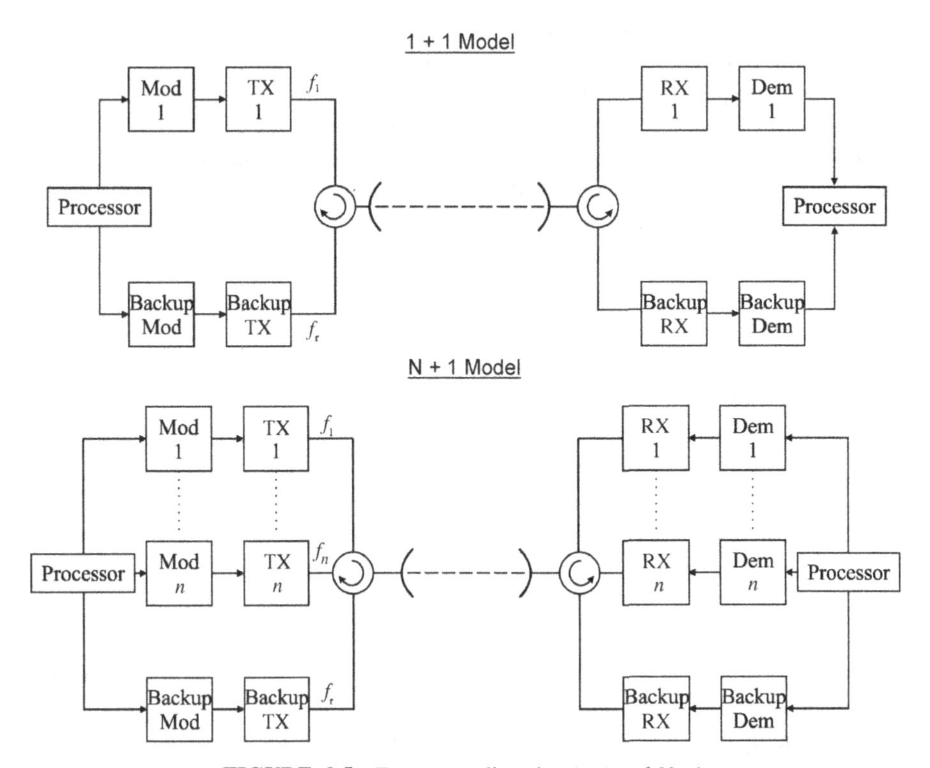

FIGURE 8.5 Frequency diversity. 1+1 and N+1.

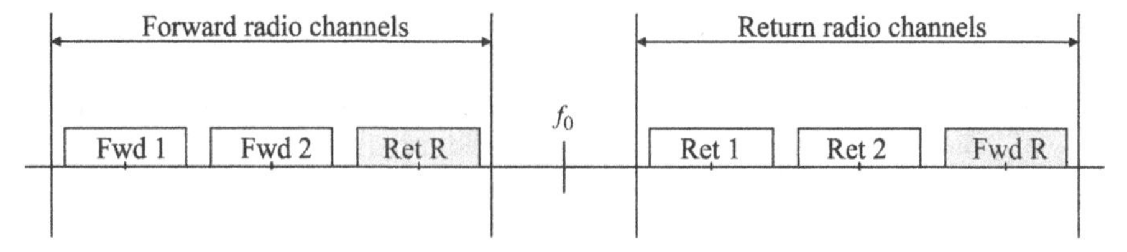

FIGURE 8.6 Cross-band 2+1 frequency diversity.

The improvement factor is increased significantly if the diversity scheme is based on cross-band frequency diversity, where signals of the reserve radio channel are located in the bands reserved for the return path on each hop, as illustrated in Figure 8.6.

The system performance improvement obtained with 1+1 diversity configurations generally depends on the correlation of degradations (fading depth, amplitude dispersion and group delay, for example) between both radio channels. In N+1 systems, the improvement factor due to frequency diversity that can be applied to an in-service channel decreases as the number of channels increases. The switching section in a radio-relay link is usual to include more than one hop. Consequently, in addition to correlation among degradation factors, it is necessary to consider correlation of degradations in different hops of the same switching section.

ITU-R P.530 recommendation suggests a calculation method of the improvement obtained with frequency diversity techniques in radio-relay links with 1+1 configurations. These are based on empirical studies carried out during the decades of 1970 and 1980. The complete calculation process is very similar to the one recommended for spatial diversity, exception made for the calculation of the non-selective improvement factor I_{ns}, which is obtained from the equation (8.41):

$$I_{\mathrm{ns}} = \frac{80}{fd}\left(\frac{\Delta f}{f}\right) 10^{F/10} \tag{8.41}$$

where

Δf = separation between active and reserve radio channels (GHz). If $\Delta f > 0.5$ GHz, we assume $\Delta f = 0.5$

f = center frequency of the band (GHz)

F = flat fading margin (dB)

The formula applies to radio-relay links that meet the following conditions:

Hop distance d	$30 \leq d \leq 70$ km
Frequency margin	$2 \leq f \leq 11$ GHz
Relative spacing between active/reserve radio channels	$\Delta f / f \leq 5\%$

8.5.1.4 Combined Diversity: Space and Frequency

As far as especially difficult radio-relay links is concerned, (i.e., hops with the entire path over the sea), they may need combined space and frequency techniques. These techniques are called hybrid diversity. There are two options for these cases: hybrid diversity with two and with four receivers that are illustrated in Figure 8.7.

In both cases, the improvement factor is calculated through the non-selective correlation coefficients that correspond to the space and frequency diversity arrangements. In the case of combined diversity with two receivers, spatial calculation is used as a model, by simply replacing the calculation of $k_{ns,s}$ *(non-selective correlation coefficient of spatial diversity)* with a new k_{ns} that is included in the process by equation (8.42):

$$k_{ns} = k_{ns,s} \times k_{ns,f} \tag{8.42}$$

where

$k_{ns,s}$ = non-selective correlation coefficient for spatial diversity
$k_{ns,f}$ = non-selective correlation coefficient for frequency diversity

In the case of hybrid diversity arrangements with tour receivers, calculation requires some additional steps:

1. Calculate multipath activity factor η as explained in previous sections (see also Chapter 7).
2. Calculate the diversity parameter, m_{ns} in function of non-selective correlation coefficients of space and frequency $k_{ns,s}$ and $k_{ns,f}$:

$$m_{ns} = \eta^3 \left(1 - k_{ns,s}^2\right)\left(1 - k_{ns,f}^2\right) \tag{8.43}$$

3. Calculate probability of non-selective outage P_{dns} with diversity, from the probability of non-selective outage without diversity P_{ns}:

$$P_{dns} = \frac{P_{ns}^4}{m_{ns}} \tag{8.44}$$

4. Calculate the non-selective correlation coefficient k_{ns}:

$$k_{ns}^2 = 1 - \sqrt{\eta}\left(1 - k_{ns,s}^2\right)\left(1 - k_{ns,f}^2\right) \tag{8.45}$$

5. Calculate the selective correlation coefficient, k_s from equation (8.46). The parameter r_w is the correlation coefficient of the relative amplitudes in this

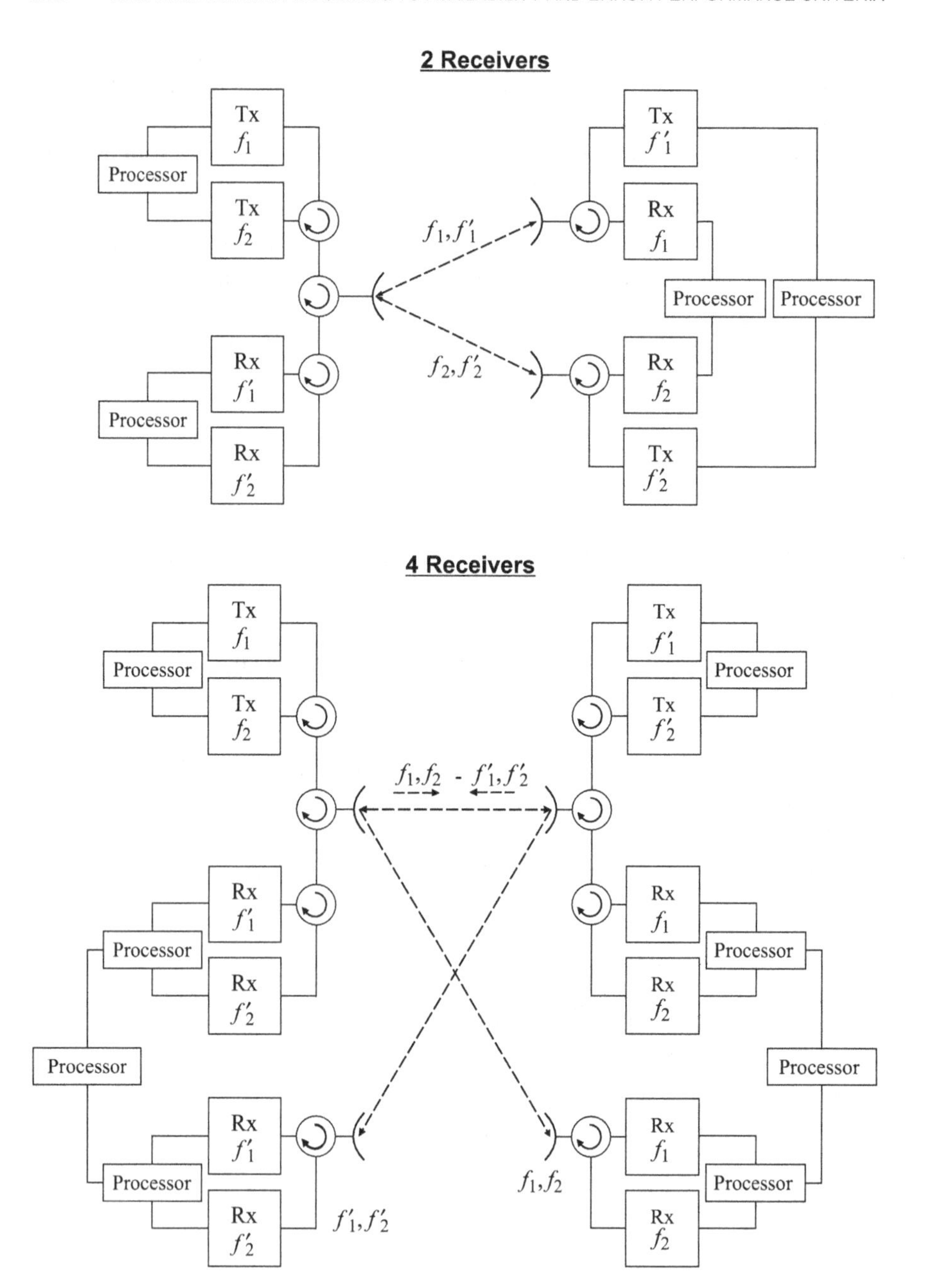

FIGURE 8.7 Combined space and frequency diversity with two and four receivers.

equation, which is calculated with the equation (8.47).

$$\begin{cases} k_s^2 = 0.8238 & r_w \leq 0.5 \\ k_s^2 = 1 - 0.195 \cdot (1 - r_w)^{0.109 - 0.13 \cdot \mathrm{Log}_{10}(1 - r_w)} & 0.5 < r_w \leq 0.9628 \\ k_s^2 = 1 - 0.3957 \cdot (1 - r_w)^{0.5136} & r_w > 0.9628 \end{cases} \tag{8.46}$$

$$\begin{cases} r_w = 1 - 0.9746\left(1 - k_{ns}^2\right)^{2.170} & \text{para} \quad k_{ns}^2 \leq 0{,}26 \\ r_w = 1 - 0.6921\left(1 - k_{ns}^2\right)^{1.034} & \text{para} \quad k_{ns}^2 > 0{,}26 \end{cases} \tag{8.47}$$

6. Calculate the selective outage probability of transmission, P_{ds}, by using the multipath activity factor η, the selective correlation coefficient k_s and P_{us}, the outage probability due to selective component of fading of the system without spatial diversity (for example, calculated with the signatures method).

$$P_{ds} = \frac{P_{us}^2}{\eta \cdot \left(1 - k_s^2\right)} \tag{8.48}$$

7. Calculate the total outage probability:

$$P_d = \left(P_{ds}^{\frac{\alpha}{2}} + P_{dns}^{\frac{\alpha}{2}}\right)^{\frac{2}{\alpha}} \tag{8.49}$$

8.5.2 Adaptive Channel Equalization

Having studied the improvement factors achieved with different diversity this section describes the use of adaptive equalizers in the receiver and the associated benefits. An effective equalizer that compensates channel distortion should follow the propagation channel variations that usually characterize multipath. It is necessary to remark that except from certain cases of fading associated with reflection components on the Earth's surface, the multipath will be always a highly variable phenomenon. Equalization techniques can be categorized into two families depending upon the domain where calculations are carried out: frequency domain equalization and time domain equalization (see Chapter 4).

8.5.3 Adaptive Equalization Combined with Spatial Diversity

Combining adaptive equalization and spatial diversity provides significant reductions of the impact of outages caused by multipath. The measured improvement usually exceeds the product of the corresponding individual improvements. If a quantification of this effect would be required for a certain design, the joint gain of frequency diversity and equalization is approximately equal to the product of the spatial diversity improvement achieved with by the square of the improvement value obtained with the equalizer.

8.5.4 Considerations on System Design in the Presence of Ducts

It is well known that there are certain geographical areas were elevated ducts are a more or less common phenomenon. If a link should be designed in those areas, where the probability of extremely low refractivity gradient values (super-refractive troposphere) is high, the following factors should be carefully considered within the design:

- Position and antenna pointing.
- Maximum beamwidth of the antenna in order to minimize the amount of radiated energy towards upper reflection layers as well as the amount of energy received from those layers.
- Type of modulation used for enabling as long as possible symbol durations.
- Path geometry for minimizing the probability of destructive reflection components.

8.6 QUALITY AND AVAILABILITY CALCULATION GUIDELINES IN REAL LINKS

This section summarizes the design methodology in real radio-relay links according to fidelity and availability objectives. It is usual to carry out independent calculations associated with each type of objective, and furthermore, it is also common to first follow the design steps associated with availability calculations and then, in the second step, refine the design according to the EPOs.

8.6.1 Availability

Unavailability due to equipment and propagation must be taken into account in a radio-relay links. Since they are supposed to be independent statistical processes, they are separately calculated and added for the calculation of total unavailability according to equation (8.50):

$$U_{\text{link}} = U_{\text{P}} + U_{\text{E}} \tag{8.50}$$

This equation is evaluated along the total radio-relay link. Unavailability due to equipment will be then separately calculated for each switching section. Afterwards, being sections independent from each other, the total unavailability will be the sum of unavailability values of each section.

In the case of unavailability associated with propagation, it will be calculated for each hop, instead of each switching section. As in the previous case, we assume that the propagation phenomena between hops is statistically independent and therefore, the overall unavailability is the sum of unavailability values of each hop. It is worth reminding here that the predominant factor for unavailability above 10 GHz is rain. For lower frequencies, (flat or selective) multipath will the dominant perturbation. Usually, for practical design purposes, it is assumed that flat fading events will be shorter than 10 consecutive seconds and thus only regarded as error performance events. In consequence, equipment will be the main cause of unavailability below 10 GHz. In the case of frequencies above 10 GHz, we proceed as follows:

1. The process is started using Recommendation ITU-R F.1703 for defining the unavailability objectives that apply to the specific radio link. It will be necessary to determine where the link will be installed (which part of the hypothetical reference digital path (HRDP) model) for objective allocation purposes. Additionally the BER value associated with SESR will be specified. In most cases, and for availability calculation purposes, the BER associated with a SESR is supposed $10^{-3.}$
2. Obtain the value of the reception threshold Th_3 associated with BER of 10^{-3}. This value will be provided by the manufacturer as the sensitivity for 10^{-3}. If this value would not be available, it might be calculated from theoretical BER-E_b/N_0 curves of the modulation and coding scheme used in the link. The theoretical value will be then increased in 3 dB in order to account for implementation losses.
3. Calculate the gross margin on each hop. In the calculation, the basic propagation loss will be the free space loss, any losses due to vegetation, as well as oxygen and water vapor absorption:

$$T_{h3} - M_3 = P_{tx} - L_{tt} + G_t - L_b + G_r - L_{tr} \tag{8.51}$$

4. Calculate the rain attenuation exceeded during 0.01% of the time $A_{0.01}$. This value is used then to obtain the probability $p\%$ of having a equivalent rain attenuation that equals the gross margin M_3:

$$M_3 = A_p = 0.12 A_{0.01} \quad p^{-(0.546+0.043 \log p)} \tag{8.52}$$

5. Solve the equation by successive approximations or any other numerical method. The obtained value is the unavailability due to rain associated with the hop. If there were several hops, we should calculate all the values hop by hop, adding the unavailability percentages of each hop for obtaining the propagation unavailability of the radio-relay link. The result of the process is $U_P(\%)$.
6. The total unavailability of the radio-relay link is the sum of the unavailability due to equipment and the unavailability due to propagation. This total unavailability should be finally compared with the specified unavailability objectives for the total length of the radio-relay link.

8.6.2 Error Performance

The inclusion of EPO EPOs in a radio-relay link design procedure starts with EPO objectives from ITU-R F.1668 recommendation. The general guidelines for studying the error performance are the following:

1. The objectives of ITU-R Recommendation F.1668 have to be translated into BER values not exceeded for different time percentages. This question has been discussed in previous sections of this chapter.
2. Error performance will only be affected by degradations caused by propagation mechanisms.
3. For each propagation perturbation that creates signal fading, the associated outage probability will be calculated using the methods described in this chapter and also in Chapter 7.
4. It will be assumed that each fading mechanisms will not be simultaneously happing in different hops, so, the probability of exceeding a BER in a radio link due to that mechanism will be equal to the sum of the probabilities of individual hops.
5. Within each hop, the signal degradation mechanisms will be statistically independent, so specific outages will be the sum of percentages associated with each mechanism.
6. Flat fading as well as selective fading have to be included in the calculation process. We will assume that, for each hop, the total outage probability associated with multipath P_{TT}, will be equal to the sum of individual flat fading and selective fading (P_{TP} and P_{TS} respectively). If it was the case, the outage probability due to XPD degradation would be also considered (see Chapter 7).

$$P_{\mathrm{TT}} = P_{\mathrm{TP}} + P_{\mathrm{TS}} \tag{8.53}$$

The evaluation of each propagation mechanism is made on a hop-by-hop basis and probabilities will be later added. The final total outage will be then compared with the EPOs at the specified BER. If they would not be fulfilled, additional enhancement techniques would be then analyzed (i.e., diversity).

8.7 INTERFERENCES

8.7.1 Interferences in the Link Design Process

The degradation produced by interferences in a microwave LOS link consists of a modification of the curves that relate error performance parameters and the required C/N ratios.

The presence of interfering signals increases the minimum threshold level required for a given BER. The interfering signal will reduce the eye diagram aperture in the receiver demodulating circuits.

The interfering signal alone does not usually have enough level to create errors, as the eye diagram will not be completely closed by the interference. Nevertheless, if the eye aperture is closed, the impact of other perturbation sources on the decision process (i.e., noise) will be more critical. As a consequence, the probability of incorrect detection will be increased, or if the problem is analyzed from the point of view of sensitivity: interferences will increase the required level to maintain the same error performance as in an interference free situation.

Interference calculations are usually limited, in most cases, to intra-system interference calculations. These will be usually related to the design or/and validation of the radio-relay link channel plan feasibility reality will provide a variety of different cases where the exact calculation of interference will be different. Some links will duplicate capacity using cross-polarized co-channel arrangements, some others will use cross-polarized co-channel arrangements in alternate hops, some other configurations will based on four and six frequencies, etc. Furthermore, the problem will get more complex in applications that make an intensive use of fixed service frequencies in a reduced area. This is the case of radio-relay links that support the access network in mobile communications. In these cases, different links within the same geographic area will be regarded as parts of the same system and the interference calculations will be considered "intra-system" interference study cases.

In any case, the evaluation of interferences will be based on carrier to interference ratio (C/I) calculations. The C/I ratio relates the desired power C with the total interfering power I. Radio-relay link specifications usually include a minimum value for the C/I ratio that must be respected in every hops of the link, in order not to degrade the receiver threshold over a maximum limit. The common specification is to provide minimum C/I values (maximum acceptable interference values) that do not degrade the system threshold in a value higher than 1 or 3 dB. The exact amount of acceptable degradation will depend on the frequency band and equipment manufacturer.

The impact of the interference will be evaluated then by calculating the C/I ratios in every hop of the link (or at least at stations where interference is expected). The resulting values will be compared with the minimum C/I specified by the manufacturer, and if the maximum interference values are not exceeded, the frequency plan will be validated. In some cases, if the ratios are not fulfilled, it will be necessary to change either frequency channels or, change antenna positions or use higher performance equipment (antennas discrimination, filters, etc.). After checking that the C/I values are above the threshold, it will not be necessary to include any interference calculations when dimensioning the link budget, except from correcting the sensitivity with the degradation value associated with the minimum C/I value, which, as mentioned above, is 1 or 3 dB.

In specific situations, when radio-relay link planning is based on frequency bands not intensively used in nearby areas, a different approach can be taken. First, the interference level at the receiver input I (dBm) will be calculated, and estimate the degradation produced in the gross thresholds of the system associated with the specific value of I. This task is not trivial since degradation depends on each specific unit. ITU-R handbook *Digital Radio-Relay Systems* suggests a method for considering the interference effect in the gross margin as a degradation of the system C/N. This is done through theoretical equations where the effect of different interference sources (co-channel, adjacent channel, etc.) is considered similar to thermal noise.

Some other references provide empirical equations similar to the equation (8.54) that relate reception threshold degradation to the C/I ratio:

$$T_{\text{hI}} = T_{\text{h}} + 10\log\left(1 + 10^{\frac{-T_{\text{h}}+C+I}{10}}\right) \tag{8.54}$$

where

T_{hI} = degraded threshold due to interference
T_{h} = threshold on interference free conditions
C = figure of merit under interference conditions (it can be regarded a similar parameter as the minimum C/I ratio)
I = interference level at the receiver input

In special cases, when inter-system interference is a relevant factor, either due to other radio-relay links or satellite communication systems, correction of the EPOs of the system might be needed, with the aim of including the effect of external interference. ITU-R F.1565 Recommendation contains reference values for this calculation.

8.7.2 Reference C/I Values and Threshold Degradation

The usual C/I values are within the range of 15–25 dB depending on the modulation scheme and coding. As a general criterion for radio-relay link design, ITU-R recommendations usually consider maximum allowable sensitivity degradation values associated with interference in the range of 0.5–1 dB, which provides reference values. Tables 8.7 and 8.8 provide an example of interference related specifications taken from the catalog of a manufacturer under co-channel and adjacent channel interference.

8.7.3 Interference Propagation

The mechanisms that affect the propagation an interfering signal are obviously the same as those of any radio signal operating in the same frequency, regardless of its desired or unwanted character.

Nevertheless, specific calculation methods are used for estimating levels and statistics associated with the interference. The reason relies on the fact that interference

TABLE 8.7 Specifications of C/I Degradation Due to Co-Channel Interference

	C/I (dB)			
	BER 10–3 Threshold Degradation		BER 10–6 Threshold Degradation	
Capacity	1 dB	3 dB	1 dB	3 dB
All capacities	18	15	23	19

TABLE 8.8 Specifications of C/I Degradation Due to Adjacent Channel Interference

		C/I (dB)			
		BER 10^{-3} Threshold Degradation		BER 10^{-6} Threshold Degradation	
Capacity	Channel Spacing (MHz)	1 dB	3 dB	1 dB	3 dB
2 × 2 Mbit/s	3.5	−5	−8	0	−4
	5.0	−24	−27	−19	−23
4 × 2 Mbit/s	7.0	−5	−8	−1	−5
	7.5	−9	−12	−5	−9

will be created in many cases by marginal effects, which may not be relevant on the average case for analyzing the desired signal, but they may be the main reason for interference occurrence. Interference propagation usually happens through several simultaneous mechanisms. Table 8.9 summarizes propagation mechanisms and conditions related to interference.

8.7.4 Carrier to Interference (C/I) Calculation Model

In the simplest scenario, calculation of C/I ratio consists of a crossed link budget between two systems, one corresponding to the desired signals and the other one corresponding to the interfering signal or signals.

It will be supposed that desired and interfering signals will be uncorrelated, even for signals that travel following very similar path directions if they have of slightly different elevation. The diagram shown in Figure 8.8 is the reference scheme for interference calculations.

Building the link budge to calculate the interfering signal power originated by the transmitter of station C that arrives to the received in B the resulting equation is:

$$P_{\mathrm{IB}} = P_{\mathrm{TC}} - L_{\mathrm{TTC}} - L_{\mathrm{FC}} - G_{\mathrm{TC}} - A_{\mathrm{C}}(\beta) - L_{\mathrm{b}} + G_{\mathrm{RB}} - A_{\mathrm{B}}(\alpha) - L_{\mathrm{TRB}} - L_{\mathrm{FB}} \quad (8.55)$$

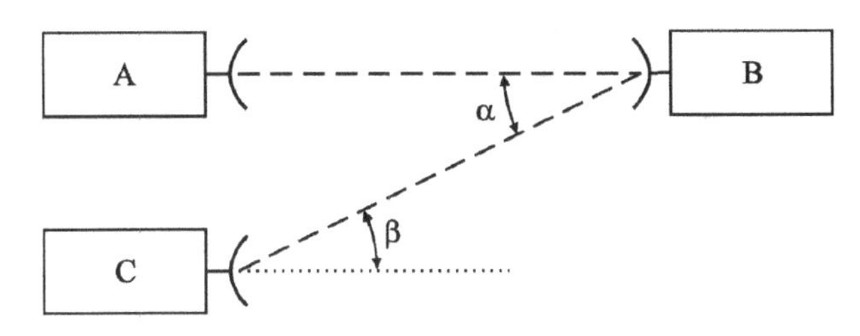

FIGURE 8.8 General diagram for calculating signal to interference ratio.

TABLE 8.9 Interference Propagation Mechanisms

Duration	Mechanism	Description	Systems	Calculation
Permanent	Line of sight propagation	Line of sight propagation under standard conditions. The received level depends on simultaneous diffraction and multipath in addition to the LOS path	Intra-system FWS→Satellite, FWS→FWS, RADAR→FWS	Generic Models (Chapter 7) Interference models for line of sight (focusing and multipath corrections) ITU-R P.452
	Diffraction	Terrain diffraction interference in non LOS situations	Intra-system FWS→Satellite, FWS→FWS, RADAR→FWS	Interference Models for clear sky diffraction. ITU-R P.452
	Tropospheric scattering	Energy scattering due to nonhomogeneous troposphere. It causes long distance interferences (100–150 km.). Important in high power systems (RADAR)	RADAR→FWS	Formulae for tropospheric dispersion under interference conditions. ITU-R P.452
Intermittent	Surface Ducts	Intense sub-refraction. Most important short-term mechanism over bodies of water and flat coastal land areas. Long distance severe interference (500 km above the sea).	Intra-system FWS→Satellite, FWS→FWS, RADAR→FWS	Propagation model for ducts and tropospheric layers. ITU-R P.452
	Scattering due to hydrometeors	Scattering by hydrometeors. Reduced interference signal levels	**FWS→Satellite** RADAR→FWS	Specific Interference model for hydrometeors ITU-R P.452
	Refraction and reflection elevated layers and ducts	Caused by anomalous refraction gradient values, usually super-refractive conditions. Long distance interference (250–300 km.)	**Intra-system** FWS→Satellite, FWS→FWS, RADAR→FWS	Propagation model through tubes and tropospheric layers. ITU-R P.452

where

P_{IB} = interfering power at the input of B receiver
P_{TC} = available power of transmitter C
L_{TTC} = losses of the antenna branching circuits in C
L_{FC} = additional losses in IF filters in C (negligible in co-channel interference)
G_{TC} = antenna gain of the interfering station C
$A_C(\beta)$ = additional attenuation in antenna C on the output direction β
L_b = basic propagation loss
G_{RB} = antenna gain of interfered station B
$A_B(\alpha)$ = additional attenuation in B receiver antenna in the α direction
L_{TRB} = losses in antenna branching in B
L_{FB} = additional losses in IF filters of B (negligible in co-channel interference)

Basic propagation losses are usually calculated assuming free space conditions for the interfering path. For practical purposes, and exception made for high capacity links in the lower bands (6, 8 GHz) the specific propagation methods for the interfering component provided by ITU-R Rec. 452 are usually not applied, at least in the first calculation stage.

The ideal result would be that the obtained C/I is far from the values that create degradation in the receiver sensitivity. If that was not the case, it would be required to consider including other attenuation mechanisms in calculations in order to avoid overestimation of the interference problem. These fixed losses in the interfering path will be usually losses due to diffraction in obstacles, irregular terrain, etc.

Equation (8.55) assumes that all signals have the same polarization and that polar discrimination in transmission is infinite (energy is only transmitted in the co-polar component). As shown in Chapter 3, frequency plans usually include radio channels with both polarizations as a tool either for minimizing the impact of intra-system interference within different hops of the link or to increase the system capacity with co-channel and alternate channel arrangements.

We have also to consider in parallel that in a real antenna, a certain level of cross-polar signal will be always transmitted, which will also contribute to the total interference. As far as receiver antennas are concerned, the case is similar; they do not have perfect (infinite) cross-polar signal discrimination. If we extend equation (8.55) to a situation with co-polar and cross-polar transmitted components, levels in transmission are given by equations (8.56) and (8.57):

$$\mathrm{EIRP}_{Cp} = P_{TC} - L_{TTC} - L_{FC} - G_{TC} - A_{C-Cp}(\beta) \tag{8.56}$$

$$\mathrm{EIRP}_{Xp} = P_{TC} - L_{TTC} - L_{FC} - G_{TC} - A_{C-Xp}(\beta) \tag{8.57}$$

where EIRP_{Cp} and EIRP_{Xp} are EIRP values in polar and cross polar polarizations, and $A_{C\text{-}Cp}(\beta)$ and $A_{C\text{-}Xp}(\beta)$ are transmitter antenna discrimination value in polar and cross-polar polarizations for the angle β. If we now evaluate the link budget from the point of view of reception, gain for both polarizations can be calculated from equations (8.58) and (8.59):

$$\text{Reception Gain}_{\text{Cp}} = G_{\text{RB}} - A_{\text{B-Cp}}(\alpha) - L_{\text{TRB}} - L_{\text{FB}} \tag{8.58}$$

$$\text{Reception Gain}_{\text{Xp}} = G_{\text{RB}} - A_{\text{B-Xp}}(\alpha) - L_{\text{TRB}} - L_{\text{FB}} \tag{8.59}$$

At the receiver, the interference level is calculated as the sum of polar and cross-polar components. If interfering and interfered systems have crossed polarizations:

$$P_{\text{IB-Xp}} = P_{\text{TC}} - L_{\text{TTC}} - L_{\text{FC}} - G_{\text{TC}} - A_{\text{C-Cp}}(\beta) - L_{\text{b}} + G_{\text{RB}} - A_{\text{B-Xp}}(\alpha) - L_{\text{TRB}} - L_{\text{FB}} \tag{8.60}$$

$$P_{\text{IB-Cp}} = P_{\text{TC}} - L_{\text{TTC}} - L_{\text{FC}} - G_{\text{TC}} - A_{\text{C-Xp}}(\beta) - L_{\text{b}} + G_{\text{RB}} - A_{\text{B-Cp}}(\alpha) - L_{\text{TRB}} - L_{\text{FB}} \tag{8.61}$$

$$P_{\text{IB-Total}} = 10\log\left(10^{P_{\text{IB-Cp}}/10} + 10^{P_{\text{IB-Xp}}/10}\right) \tag{8.62}$$

Polar and cross-polar radiation patterns will be necessary to obtain the antennas discrimination values.

8.7.4.1 Multiple Interferences There are usually different sources that contribute to interference on dense networks. In these cases, it is necessary to calculate an aggregated C/I that takes into account all interfering sources. The combination of different interfering sources on a single receiver is a complex topic. The degree of correlation and the specific statistics of all the components have usually led to a simplification in practice. So far, the usual practice is a simple uncorrelated power addition of all the interfering sources. According to this approach, the global C/I ratio is:

$$\frac{\text{C}}{\text{I}}(\text{dB}) = P_{\text{rD}}(\text{dBm}) - 10\log\left[\sum_{\text{i}=1}^{\text{n}} p_{\text{rIi}}\right] \tag{8.63}$$

where p_{rIi} is the received power in the interfered receiver of the i-th source in linear units. This power will be calculated hop-by-hop using the equations for cross-polar or polar interference situations, depending on each specific case.

8.7.4.2 Common Interference Situations Figure 8.9 illustrates the most usual interference situations. They are called triangular interference (Figure 8.9 (b)) and overshoot interference (Figure 8.9 (a))

8.7.4.2.1 Nodal/Triangular Interference Figure 8.10 illustrates the interferences associated with in this situation. On the one hand, there is a potential interference from systems transmitting from the nodal station (A1 and A2) towards the receivers at end stations B and C. On the other hand, there is an interference from B and C on receivers of systems in the node station (A1 and A2).

In the case of interference produced on the systems of the nodal station (on receiver circuits corresponding to the B-A1 hop) by the system in C the interfering signal level

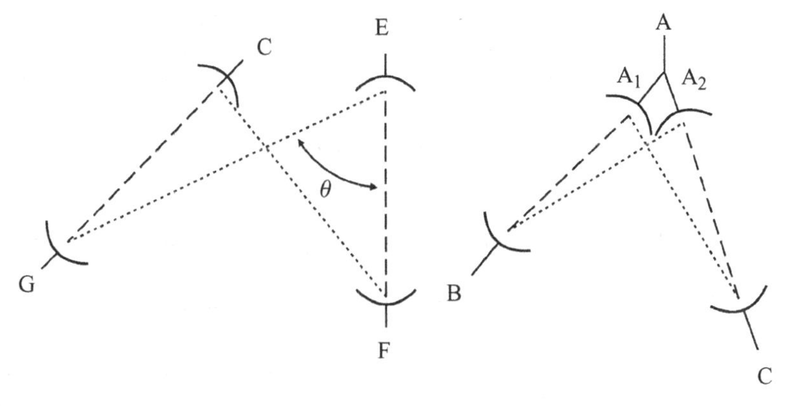

FIGURE 8.9 Reference models for intra-system interference calculation.

(on A1) can be written as a function of the desired signal level L_{RX}:

$$P_I = L_{RX} + \Delta R_X + \Delta G - A_G \tag{8.64}$$

where

ΔR_X = difference between the desired signal level in reception in (A2) interfering and (A1) interfered systems

ΔG = difference between antennas gains of the A1-B and A2-C systems in the nodal station

A_G = antenna discrimination in the nodal station A, on the direction θ (angle formed by both hops in the nodal station)

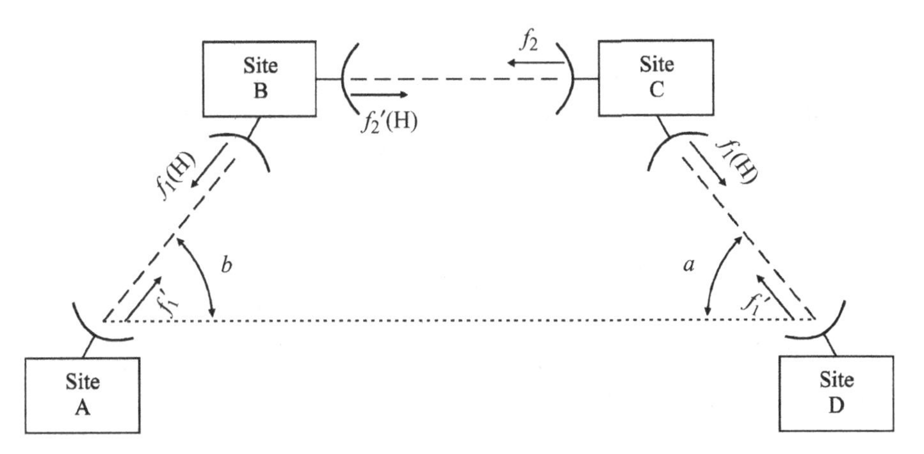

FIGURE 8.10 Overshoot interference in nonconsecutive hops.

Considering the other interference case, where a transmitter in the nodal station (A1 for example) creates an interference to a receiver in one of the end stations (C for example), the resulting equation is:

$$P_{\mathrm{I}} = L_{\mathrm{RX}} + \Delta T_{\mathrm{X}} + \Delta G - A_{\mathrm{G}} \tag{8.65}$$

where

ΔT_{X} = difference on transmitting power between interfering (A1) and interfered (A2) systems

ΔG = difference between maximum gains of the antennas associated with A1-B and A2-C systems in the nodal station

A_{G} = difference on antenna gains of the nodal station associated with the departure angle θ (angle formed by both hops in the nodal station)

The desired signal levels received in the nodal station are usually similar, either because radio links have similar lengths or because the dimensioning of both radio links will be in any case similar in a fading free environment. The same can be stated for the transmission features (similar transmitted power levels). In addition, antennas may also be the same. Under these conditions, in a fading free environment, $\Delta G = 0$, $\Delta T_{\mathrm{X}} = 0$ and $\Delta R_{\mathrm{X}} \approx 0$. So, the interfering signal is:

$$P_{\mathrm{I}} = L_{\mathrm{RX}} - A_{\mathrm{G}} \tag{8.66}$$

Equation (8.66) is equivalent to state that the ratio between the desired to interfering signal level in a fading free environment will be the antenna discrimination $\mathrm{A_G}$. This conclusion is important for the antennas choice and their installation on the tower. Discrimination A_{G} must account for the minimum C/I associated with the maximum allowable degradation of the reception threshold, plus a margin for possible fading of the desired signal (assuming the worst case, when the interfering signal and the desired signal are not correlated).

8.7.4.2.2 Overshoot Interference Overshoot interference is calculated in radio links with several hops, where frequencies are reused in nonconsecutive hops. Two frequency plans are a typical case of this problem (see Chapter 3). Figure 8.10 illustrates the problem.

We can assume that polarization in the first and third hop is the same in the simplest case, and polarization discrimination in transmission is infinite (energy is only transmitted in the polar component). In this case, equation (8.55) is directly applied, which assumes there is no cross-polar component in transmission and there is no propagation effect that causes depolarization (rain or multipath). Discrimination is due to departure and arrival angles of the interference path in transmission and reception, under these conditions. We can obtain C/I ratio from equation (8.55) in a fading free environment. If fading should occur, the worst case will be an uncorrelated behavior of desired and interfering signal. Uncorrelated fading will usually create C/I ratios less than the minimum requirement, and it will be it will be necessary to take countermeasures.

The most obvious one consists of changing polarization in the third hop. This involves recalculating the C/I ratio, applying equations (8.60) and (8.61). In most cases, this change will be enough to solve the problem. In dense reuse scenarios, a different choice of antenna models, and/or antenna installation characteristics and/or channel arrangements and/or a different frequency band, might be required.

8.7.5 Interference Calculations in Dense Frequency Reuse Scenarios

The interference calculation process is today the most relevant and complex task for designing microwave LOS links in dense frequency reuse scenarios. The paradigmatic example of this scenario is found in access networks of mobile broadband systems. In these cases, the interference calculation is the tool for selecting the appropriate radio channel(s) that create the minimum interference in other links of the same network. In this case, interferences between different links are considered intra-system interference.

In those cases, the number of radio links that share the same frequency band in a reduced area is remarkable and therefore, the number of possible interference situations. This choice is made through an iterative calculation process that takes into account the following factors:

1. Coordination range around the station that is being designed.
2. Maximum C/I allowed by manufacturer to ensure a degradation of the reception threshold of 1 or 3 dB for a BER of 10^{-6}.
3. Margin above maximum C/I.
4. Adaptive modulations and automatic transmit power control (ATPC) systems.

Coordination distance is the radius of the circle that limits the geographic area whose radio links have to be taken into account for interference calculation. Table 8.10 shows an example of coordination distances in the design of radio-relay links of the access section of a cellular communications network.

TABLE 8.10 Coordination Distances as a Function of the Frequency Band

Frequency (GHz)	Distance (km)
7–6	120
13	80
18–15	60
38–26	40

TABLE 8.11 Link Engineering Summary Procedure

Phase I. Preliminary theoretical and field studies

Step 1 Analysis of availability and EPOs

- (a) Establish the applicable network model (HRDP) and its relationship with the radio link.
- (b) Identify availability objectives either according to ITU-R F.1703 or defined by radio link operator.
- (c) Identify EPOs either according to ITU-R F.1668 or defined by radio link operator.

Step 2 Study of the appropriate frequency band, equipment selection, and specification studies (This step will be an specification rather than a calculation in many cases)

- (a) Analyze the capacity specifications.
- (b) Identify potential channel arrangements from ITU-R recommendation.
- (c) Analysis of the capacities provided by different manufacturers and models, baseband and multiplexation options, system upgrade and extension possibilities, diversity and redundancy schemes allowed, etc.
- (d) Preliminary equipment choice.

Step 3 Preliminary choice of the radio-electric route and site selection

- (a) Identify physical path alternatives.
- (b) Investigate the feasibility of using existing infrastructures (sites, towers).
- (c) Decide the preferred alternative routes.
- (d) Decide the k value statistical model and values that apply in the desired area. Draft path profiles drafting and evaluate clearance criteria.
- (e) Evaluate availability applying the preferred prediction model.
- (f) Evaluate error performance applying the preferred prediction model.
- (g) Compare evaluated quality with radio link objectives (availability and error performance).
- (h) Calculate preliminary installation costs of the alternative routes.
- (i) Decision on the optimum alternative.
- (j) If satisfactory results are not obtained, repeat the corresponding steps.

Step 4 Field-based inspections

- (a) Confirm sites, access, and potential obstacles.
- (b) Evaluate characteristics and geographical conditions of transmitting sites: stations, access roads, power delivery, etc. (photographs are useful at this stage).
- (c) Complete a report of the field-based review and provide design recommendations.

Phase II Radio link design procedure

Step 1 Distribution and allocation of availability and EPOs

- (a) Allocate quality objectives according to ITU-R F.1668 and F.1703 Recommendations.
- (b) Deduce design objectives for each one of the hops of the link. Translate ESR, SESR, and BBER into BER values.
- (c) Include a design security margin above the objectives.

TABLE 8.11 (*Continued*)

Step 2 Choice of frequency bands and channel arrangements

Choose the frequency band according to the existing national frequency plan and to the commercially available equipment

Step 3 Draw route profiles

Draw exact profiles, including the existing towers, if it is the case

Step 4 Calculate the optimum antenna heights

(a) Find and calculate reflection points.
(b) Define final clearance criteria according to expected refraction statistics.
(c) Calculate clearance for predicted minimum and maximum k factors, for single and diversity antenna systems (if that was the case).
(d) Use hills/mountains of terrain to block reflected rays, if possible.
(e) Try to achieve path inclinations as acute as possible.

Step 5 Link budget

(a) Define minimum reception thresholds associated with the threshold BER (equipment data).
(b) Calculate predicted interference level coming from co-sited and neighboring radio links.
(c) Calculate the worst case of attenuation due to rain for the specific weather.
(d) Choose the antennas to obtain the appropriate gains, XPD and C/I ratio, calculating the radio link power balance related to the threshold BER, with the aim of reaching a flat fading margin.
(e) Repeat Step d (if necessary) to achieve an optimum gain, XPD and C/I. (in dense reuse scenarios, i.e., links in cellular networks, this task is a relevant step in the calculation process and will be done at the same stage as the radio channel selection stage)

Step 6 Error performance calculations

(a) Calculate the predicted quality (outages), according to the preferred prediction model, refining the previous calculations done on preliminary route analysis and power balances of Phase I
(b) Compare with the design objectives, deciding on the need to include reception diversity, equalizers, or both.
(c) Modify initial performance calculations with enhancement factors associated with diversity and equalization.
(d) Compare with design objectives, adopting a decision on diversity and equalization.
(e) If quality is worse than design objectives, adjust design parameters and repeat Steps 4 to 6.
(f) Check designs comply with design objectives.

TABLE 8.11 *(Continued)*

Phase III Installation and System Tests

Step 1 Equipment set up and installation

Step 2 System tests

(a) System tests to evaluate background bit error rates and system threshold checks.

(b) Identification and measurement of unexpected interference problems. Antenna adjustments.

(c) If abnormal refractivity is probable in the area, perform different tests over a period of time, taking measurements on nonconsecutive days.

Step 3 Operation and maintenance

This area obviously depends on the operation frequency, since it is directly related to propagation losses associated with the interference path.

Planning tools usually have a data base with all radio links of the network, and are able to identify all potentially interfering and interfered stations within the coordination area. The database has all equipment information, antennas, and used radio channel frequencies. This information will be then used to evaluate the radio channel has the lowest impact on network. Accuracy, updating and management of the database is crucial for planning radio-relay links in very dense networks.

Previous sections have described the effect of C/I on the system threshold levels (sensitivity). From a practical perspective, the minimum C/I will be a combination of the parameter provided by the equipment manufacturer and a safety margin over this reference. This margin is usually in the range of 3–5 dB. In this case, translated to the radio channel search problem, a specific channel will be usable if the associated C/I ratios created on other links (and from the combination of those links on the receiver of the new candidate) are at least 3 or 5 dB above the minimum C/I threshold provided by the manufacturer.

The impact of using adaptive modulations on the interference estimation is an additional factor to consider. The changing nature of modulation involves variability in the required threshold and therefore, a variation on the impact that a specific interference causes on the system. Currently, there is no recommended criterion to apply in adaptive modulation environments. From a conservative point of view, the highest order modulation is considered the reference design that represents the worst case. Nevertheless, a certain degree of correlation between interfered an interfering paths could be included in order to enable realistic planning in dense reuse networks.

The second factor to consider is the use of ATPC. There is also a lack of widely accepted recommendations in this case. The worst case would be to assume that transmitted power is the maximum of the adjustment range of transmitters. Currently, network designs usually ignore the presence of the ATPC subsystems with the aim of simplifying calculations. Unfortunately, this practice, even used for coordination, leads to pessimistic network designs.

8.8 LINK ENGINEERING SUMMARY PROCEDURE

In order to summarize the design process of a microwave LOS link and for aiming at giving step-by-step guidelines for the radio-planning engineer, an abbreviated design guide (Table 8.11) is provided in this section. The guideline makes reference to appropriate sections of this and previous chapters.

BIBLIOGRAPHY

140 Mbit/s digital radio field experiment–Further results. J.C. Campbell, A.L. Martin and R.P. Coutts. (ICC'84). IEEE International Conference on Communications. Amsterdam. 1984.

ITU-R Handbook: Radiowave propagation information for designing terrestrial point-to-point links. International Telecommunication Union. Radiocommunication Sector. ITU-R. Geneva. 2008.

ITU-R Rec.F.752: Diversity techniques for point-to-point fixed wireless systems. International Telecommunication Union. Radiocommunication Sector. ITU-R. Geneva. 2006.

ITU-R Rec. F.758: Considerations in the development of criteria for sharing between the terrestrial fixed service and other services. International Telecommunication Union. Radiocommunication Sector. ITU-R. Geneva. 2005.

ITU-R Rec. F.1668: Error performance objectives for real digital fixed wireless links used in 27 500 km hypothetical reference paths and connections. International Telecommunication Union. Radiocommunication Sector. ITU-R. Geneva. 2007.

ITU-R Rec. F.1703: Availability objectives for real digital fixed wireless links used in 27 500 km hypothetical reference paths and connections. International Telecommunication Union. Radiocommunication Sector. ITU-R. Geneva. 2005.

ITU-R Rec. P. 530-8: Propagation data and prediction methods required for the design of terrestrial line-of-sight systems. International Telecommunication Union. Radiocommunication Sector. ITU-R. Geneva. 1999.

ITU-R Rec. P. 530-13: Propagation data and prediction methods required for the design of terrestrial line-of-sight systems. International Telecommunication Union. Radiocommunication Sector. ITU-R. Geneva. 2009.

ITU-T Rec. G.821: Error performance of an international digital connection operating at a bit rate below the primary rate and forming part of an Integrated Services Digital Network. International Telecommunication Union. Telecommunication Standardization Sector. ITU-T. Geneva. 2002.

ITU-T Rec. G.826: End-to-end error performance parameters and objectives for international, constant bit-rate digital paths and connections. International Telecommunication Union. Telecommunication Standardization Sector. ITU-T. Geneva. 2002.

ITU-T Rec. G.827: Availability performance parameters and objectives for end-to-end international constant bit-rate digital paths. International Telecommunication Union. Telecommunication Standardization Sector. ITU-T. Geneva. 2003.

ITU-T Rec. G.828: Error performance parameters and objectives for international, constant bit-rate synchronous digital paths. International Telecommunication Union. Telecommunication Standardization Sector. ITU-T. Geneva. 2000.

ITU-T Rec. G.829: Error performance events for SDH multiplex and regenerator sections. International Telecommunication Union. Radiocommunication Sector. ITU-T. Geneva. 2002.

Manual ITU-R: Digital Radio-Relay Systems. International Telecommunication Union. Radiocommunication Sector. ITU-R. Geneva. 1996.

Microwave Radio Links: From Theory to Design. C. Salema. John Wiley & Sons, New Jersey. 2003.

Radio Regulations. International Telecommunication Union. Radiocommunication Sector. ITU-R. Geneva. 2008.

The Impact of G.826. M. Shafi and P. Smith. IEEE Communications Magazine. IEEE Piscataway, NJ. September. 1993.

ITU-R Rec. P.452: Prediction procedure for the evaluation of interference between stations on the surface of the Earth at frequencies above about 0.1 GHz. International Telecommunication Union. Radiocommunication Sector. ITU-R. Geneva. 2010.

CHAPTER 9

LINK OPERATION AND MONITORING

9.1 INTRODUCTION

Having finished the detailed planning and design phases of a microwave LOS link, the third and final phase of the total implementation process consists of equipment provisioning, installation, and commissioning.

After installation, the systems must undergo a test period (TP) in order to check if they fulfill the necessary requirements for bringing into service (BIS). The validation and acceptance tests can be accomplished in two steps: a first validation stage carried out at the manufacturer premises and a second stage after equipment has been installed on the field.

Having tested that the parameters that describe equipment performance are within the required quality range required for BIS, the link equipment and installation will be accepted and the system will be considered on the exploitation and service operation status. During the exploitation and operation period, and over the entire life of the system, a preventive and corrective maintenance will be required.

The objective of maintenance is to avoid or minimize any system performance degradation that might appear due to very diverse causes that might be originated by external factors, such adverse propagation conditions, or by failures (faults) and equipment aging. The tasks associated with maintenance are identification, localization, and correction of any fault, or degradation source that might cause deviations in the system performance parameters. In this case, we are considering performance in its wide general meaning and not tied to "error performance."

Microwave Line of Sight Link Engineering, First Edition. Pablo Angueira and Juan Antonio Romo.

These parameters will have a range of acceptable values that will be monitored and they will be the reference to trigger appropriate correction measures when the values are out of range. In this case, when the correction measures provoke a maintenance action, they are called corrective prevention actions. They will consist of a list of standardized procedures and faults and anomaly solving protocols that are provided as part of the system description and operation documents.

The test and validation procedures applied to radioelectric sections of the point-to-point systems are described by the International Telecommunication Union, Radiocommunication Sector (ITU-R) Recommendations F.1566 and F.1668, where in addition to the validation procedures, the system performance limits are also described. The specifications are provided according to both BIS and maintenance period criteria.

This chapter provides an analysis and summary of the content of the recommendations mentioned above as well as other the International Telecommunication Union, Telecommunication Standardization Sector (ITU-T) reference criteria that can be considered complementary to those of the ITU-R. The operation and maintenance tasks will be also described in relation to supervision and continuous monitoring resources and tools of current commercial systems for network management.

9.2 REFERENCE PERFORMANCE OBJECTIVES (RPO)

ITU-T Recommendation M.2100 has defined the system performance limits for BIS and maintenance that should be applied to plesiochronous digital hierarchy (PDH) international paths and connections. In turn, ITU-T Recommendation M.2101 contains limits for international synchronous digital hierarchy (SDH) paths international SDH multiplex sections based on ITU-T Rec. G.826 and G.828 respectively. The limits that are applied for BIS and maintenance phases are based on the end-to-end reference performance objectives (RPO) and are applicable to any real digital microwave LOS link of an arbitrary length d. The RPO values are a function of the error performance parameters already described in Chapter 5 and 8: errored second (ES), severely errored second (SES), background block error (BBE). Table 9.1 shows the objectives for digital paths and Table 9.2 the corresponding ones to digital sections.

TABLE 9.1 RPO for End-to-End Paths

PDH	Primary	Secondary	Tertiary	Quaternary	
SDH (Mbit/s)	1.5–5	> 5–15	> 15–55	> 55–160	> 160–3500
ESR for paths (G.826)	0.02	0.025	0.0375	0.08	N.A.
ESR for paths (G.828)	0.005	0.005	0.01	0.02	N.A.
SESR	0.001	0.001	0.001	0.001	0.001
BBER for SDH paths (G.826)	N.A.	N.A.	N.A.	N.A.	N.A.
BBER for SDH paths (G.828)	2.5×10^{-5}	2.5×10^{-5}	2.5×10^{-5}	5×10^{-5}	5×10^{-5}

TABLE 9.2 RPO for End-to-End SDH International Multiplex Sections

Rate	STM-0	STM-1	STM-4
Blocks	64 000	1 92 000	7 68 000
ESR (according to G.826)	0.0375	0.08	N.A.
ESR (according to G.828)	0.01	0.02	N.A.
SESR	0.001	0.001	0.001
BBER (according to G.826)	N.A.	N.A.	N.A.
BBER (according to G.828)	2.5×10^{-5}	5×10^{-5}	5×10^{-5}

In order to allocate the RPO in international paths, the objectives have been distributed in geographical sections called path core elements (PCE) that in turn have been classified according to the following categories:

- International path core elements (IPCE), between an international gateway and a frontier station in a terminating country, or between frontier stations (FS) of an intermediate country.
- Inter-country PCE (ICPCE) between two adjacent frontier stations of the two countries involved. The ICPCE corresponds to the highest order digital path carried on a digital transmission system linking the two countries.

The percentage of objectives allocated in an end-to-end digital sections is 0.2. In the case of paths, the percentage depends upon the number of PCEs along the path, as well as the type and distances between the elements. The total A% is calculated using equation (9.1) The reference values for practical calculations can be obtained from Table 9.3.

$$A = \sum_{1}^{N} a_{\mathrm{n}}\% \tag{9.1}$$

The distances indicated in Table 9.3 are real route lengths or air-route distances multiplied by an appropriate routing factor (R*f*), whichever is less; for multiplex sections the length d refers to the actual distance only.

The distribution of RPO to paths and sections originate allocated performance objectives (APO). The APO values are different for each error performance event and they are a function of A% and the measurement TP (the period where the system will be evaluated), either for BIS or maintenance. The TP is expressed in seconds: 86400 s for a TP of 24 h and 604800 s for a TP of 7 days. Table 9.4 summarizes the APO calculation procedure for different cases.

The error performance events (ES, SES, BBE) in the expressions of Table 9.4 are the same error events as defined for error performance objectives (EPO) i.e., are defined errored block second, severely errored block second, and BBE.

TABLE 9.3 RPO Allocation in End-to-End SDH International Multiplex Sections

A% Value (End-to-End Allocation)		PCE Classification (km)	Allocation to Each IPCE and ICPCE on the path (% of End-to-End RPO) a_n
International paths	Equation (9.1) $\sum_{1}^{N} a_n\%$	IPCE Terminating/ transit national network:	
		$d \leq 100$ km	1.2
		100 km $< d \leq 200$ km	1.4
		200 km $< d \leq 300$ km	1.6
		300 km $< d \leq 400$ km	1.8
		400 km $< d \leq 500$ km	2.0
		500 km $< d \leq 1000$ km	3.0
		1000 km $< d \leq 2500$ km	4.0
		2500 km $< d \leq 5000$ km	6.0
		5000 km $< d \leq 7500$ km	8.0
		$d > 7500$ km	10.0
		ICPCE $d \leq 300$ km	0.3
International multiplex section	0.2		

TABLE 9.4 APO Calculation Procedures

	Path	Multiplex Section
APO_{ES}	(A% × RPO_{ES} × TP)/100	
APO_{SES}	(A% × $RPOS_{SES}$ × TP)/100	
APO_{BBE} (VC-1 and VC-2)	(A% × RPO_{BBE} × TP × 2000)/100	–
APO_{BBE} (VC-3, VC-4, VC-4-Xc)	(A% × RPO_{BBE} × TP × 8000)/100	–
APO_{BBE} (STM-0)	–	(A% × RPO_{BBE} × TP × 64000)/100
APO_{BBE} (STM-1)	–	(A% × RPO_{BBE} × TP × 192000)/100
APO_{BBE} (STM-4)	–	(A% × RPO_{BBE} × TP × 768000)/100

9.3 BRINGING INTO SERVICE (BIS)

The methodology for BIS a microwave LOS link is described in Recommendation ITU-R F.1330. This reference proposes a series of steps to perform tests and evaluate the system performance specifically designed for the BIS phase. Associated with the method, specific system performance thresholds have been defined for BIS. The objectives are called bringing into service performance objectives (BISPO). These limit values are defined as in the previous section for paths and multiplex international sections, and for both PDH and SDH systems.

9.3.1 BIS Performance Objective Values Calculation Procedure

The BISPO values for both paths and multiplex sections can be calculated according to the expression from equation (9.2):

$$\text{BISPO} = \text{APO}/F_{\text{m}} \tag{9.2}$$

where

APO = allocated performance objective
F_{m} = maintenance margin

The maintenance margin F_{m} is an additional margin that is added in order to reduce as much as possible further maintenance actions. The values of this parameter depend on the type of route and the propagation conditions. Table 9.5 provides an estimation of the values for F_{m}.

The APO values are different for each error performance event and will depend on the RPO values, the allocation percentage A% and the TP duration. They will be obtained by following the procedures provided in Table 9.4.

9.3.2 BIS Test and Evaluation Procedure

The methodology proposed for testing and validating a microwave LOS system, in order to give acceptance in the bringing into serviced phase is summarized by Figure 9.1.

TABLE 9.5 Approximate Maintenance Margin Values (Fm)

	Standard Propagation Conditions	Adverse Propagation Conditions
PDH Paths and Sections	2	0.5
SDH Paths	2	0.5
PDH Transmission Paths	10	0.5
SDH Multiplex Sections	10	0.5

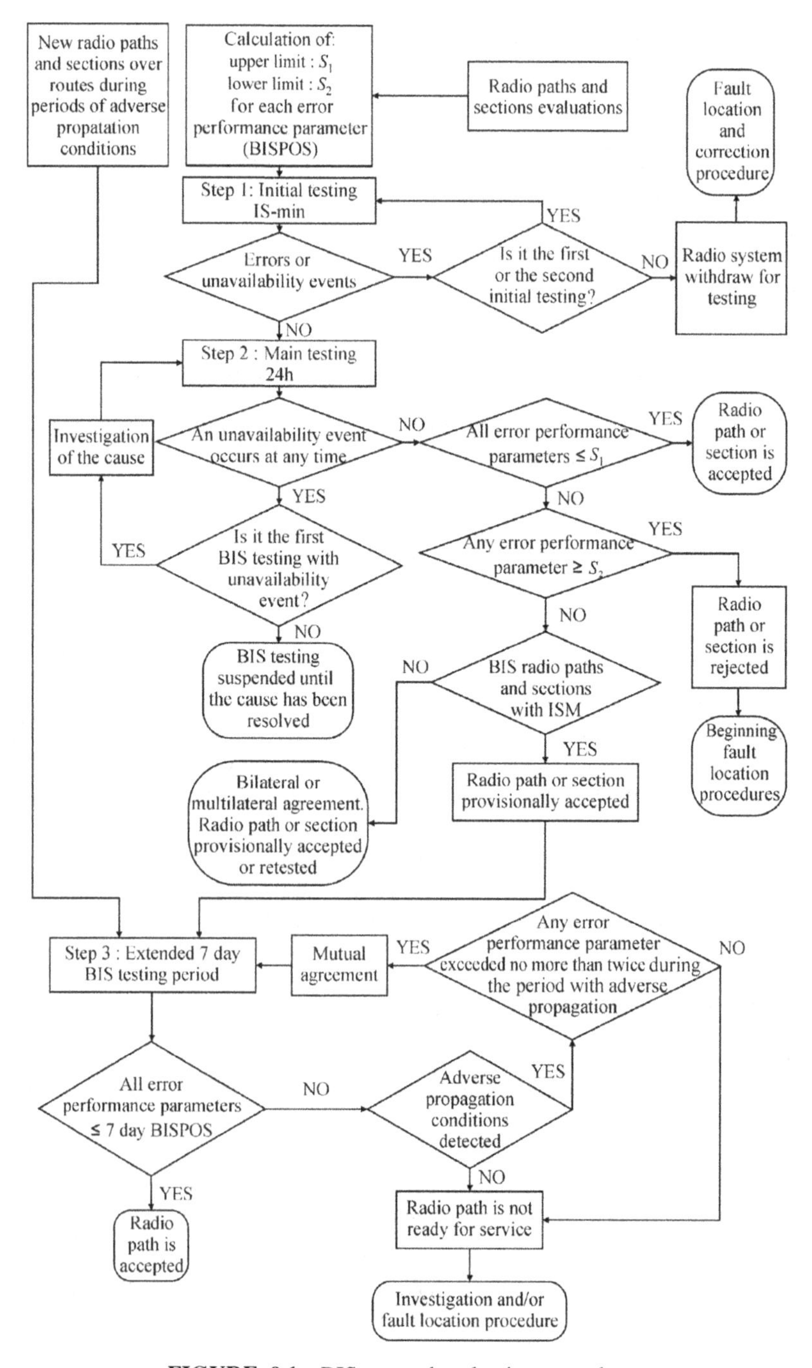

FIGURE 9.1 BIS test and evaluation procedure.

The procedure starts identifying the error performance limit values. Two threshold values have been defined: upper limit S_1 and S_2. These values expressed by equations (9.3) and (9.4) are a function of the r BIS system performance objectives:

$$S_1 = \text{BISPO} - 2\sqrt{\text{BISPO}} \tag{9.3}$$

$$S_2 = \text{BISPO} + 2\sqrt{\text{BISPO}} \tag{9.4}$$

The BISPO values will be obtained for each error performance values following the methodology described in previous sections of this chapter. The test procedures can be organized into three different stages:

- An initial TP of 15 min.
- A main TP of 24 h.
- A possible extension of 7 days.

During the initial TP there should not be and availability errors or events. The measurements must be carried out using a bit error rate (BER) tester that will use pseudo-random bit sequence (PRBS) encapsulated on a frame structure. It is recommended that the tests are carried out during clear air conditions and trying to avoid periods and day hours where intense multipath activity is expected. With respect to multipath, studies in West Europe suggest that the most appropriate season is winter and the two preceding months. If the tests should be carried out during summer, the minimum probability of multipath appears to be between 10:00 and 14:00 summer local time in West Europe.

After success in Phase 1 of the tests, the main tests will take place. At the end of the 24-hour period, the results will be compared to the BIS limits S_1 and S_2. The main tests can be performed using real traffic if the in-service monitoring (ISM) capability is available. If this tool was not available, the tests would be carried out following the same procedure and equipment as in the initial 15-min phase. If during the main TP, an unavailability event occurs, the cause should be investigated and a new BIS test rescheduled. In the case that a further unavailability event occurs in the second BIS test, then BIS testing should be suspended until the cause is identified and, if possible, solved.

The extended tests over 7 days will be carried out for those paths and sections with ISM functions available whose performance during the 24 h tests is between the limits S_1 and S_2. The 7-day tests are also recommended over new routes without previous radio paths, where performance results were not available.

The outcome of the tests over 7 days should comply with the BISPO for each one of the parameters. It should be reminded here that the BISPO will have different values depending upon the TP. (see Table 9.4).

The test procedures can be organized into three different stages:

- An initial TP of 15 min.
- A main TP of 24 h.
- A possible extension of 7 days.

During the initial TP there should not be and availability errors or events. The measurements must be carried out using a BER tester that will use PRBS encapsulated on a frame structure. It is recommended that the tests are carried out during clear air conditions and trying to avoid periods and day hours where intense multipath activity is expected. With respect to multipath, studies in West Europe suggest that the most appropriate season is winter and the two preceding months. If the tests should be carried out during summer, the minimum probability of multipath appears to be between 10:00 and 14:00 summer local time in West Europe.

Having succeeded phase 1 of the tests, the main tests will take place. At the end of the 24 h period, the results will be compared to the BIS limits S_1 and S_2. The main tests can be performed using real traffic if the ISM capability is available. If this tool were not available, the tests would be carried out following the same procedure and equipment as in the initial 15-min phase. If during the main TP, an unavailability event occurs, the cause should be investigated and a new BIS test rescheduled. In the case that a further unavailability event occurs in the second BIS test, then BIS testing should be suspended until the cause is identified and, if possible, solved.

The extended tests over 7 days will be carried out for those paths and sections with ISM functions available whose performance during the 24-hour tests is between the limits S_1 and S_2. The 7-day tests are also recommended over new routes without previous radio paths, where performance results were not available.

The outcome of the tests over 7 days should comply with the BISPO for each one of the parameters. It should be reminded here that the BISPO will have different values depending upon the TP. (see Table 9.4).

9.4 MAINTENANCE

The maintenance of a microwave LOS link aims at identifying, locating and correcting faults and errors that degrade the system performance. Most radio-relay systems today are furnished with integrated management and supervision systems, already introduced in Chapter 4.

These management and supervision systems provide a continuous ISM of the overall link performance that is used in the maintenance strategy and procedures. ISM is provided by existing dedicated modules (monitors) that continuously check the performance of each path or transmission path. This architecture enables system operation quality data gathering and storage, threshold setting and programmed periodic reports of instantaneous, historical and exception events.

Some scenarios present conditions not strictly associated with the definition of ISM. These non-ISM systems are, for example, those with shared supervision and thus not continuous in time, or simply those systems without supervision. These cases are referred to as "pre-ISM" situations.

The system performance limits and the measurement procedures for radio-relay link maintenance in international PDH and SDH sections or paths are described in ITU-R Rec. F.1566. The method is coherent with the procedures described in ITU-T

M2100 and ITU-T M2101 that have been adapted to the specifics of a microwave LOS link.

Actually, the procedures to detect and locate failures, with and without ISM are based on ITU-T M.2120 that deals with those procedures applied to paths, sections and digital transmission systems associated with multiple operator network portions.

An aspect that has not been developed by ITU-R F.1566 is the influence of propagation conditions on the procedures to detect and locate failures. This question is an open study issue within the ITU-R Study Groups.

9.4.1 Calculation Method for Maintenance Performance Limits (MPL)

Any system or subsystem can operate in a limited number of predefined conditions that will depend on the system performance. These conditions are referred to as system performance levels. Three different levels are defined:

- Unacceptable.
- Degraded.
- Acceptable.

In addition to the limits that differentiate between these performance conditions, Recommendation ITU-T M.35 defines a special limit known as system performance after "fault clearance," i.e., failure reparation. The specification of the levels and associated limits are described in recommendations ITU-T M.2100 and ITU-T M.2101.

According to the general maintenance limit classification three different maintenance performance limits (MPLs) are defined in radio-relay systems:

- Unacceptable maintenance performance limit (MPL_{UP}).
- Degraded maintenance performance limit (MPL_{DP}).
- Performance after reparation (MPL_{PAR}).

Each limit is calculated for the path or multiplex section and for each one of the appropriate error performance events (ES, SES and BBE) following the general equation (9.5):

$$MPL = APO \times PLF \tag{9.5}$$

where PLF is the performance level factor. The values of PLF depend on the (APO), the MPLs, and the specific portion of the network where it is being applied. Table 9.6 shows the values recommended by ITU-R Rec. F.1566.

The values corresponding to APOs, will be obtained from the general expressions following the procedure summarized in Table 9.4., for each type of path or section, error performance event, RPO, and TP.

TABLE 9.6 MPL Values and Performance Level Ranges

		SDH Multiplex Sections PDH Transmission Systems			SDH Paths PDH Sections and Paths		
		ESR	BBER	SESR	ESR	BBER	SESR
Limits (MPL)	*Performance after Repair MPL_{PAR}*	0.1 APO		0.5 APO		0.5 APO	
	Performance Objective		APO			APO	
Performance Level Range	*Acceptable*		< 0.5 APO			< 0.75 APO	
	Degraded		> 0.5–< 10 APO			> 0.75–< 10 APO	
	Unacceptable		> 10 APO			> 10 APO	

9.4.2 Maintenance Procedures

From a maintenance perspective, when a specific performance limit is given a specific value (in terms of ES, BBE, and/or SES) the value becomes a threshold. Each performance limit or threshold will have an associated measurement time.

A threshold report (TR) is a nonrequested error performance report produced by a maintenance entity (ME) with information corresponding to a period of either 15 min or 24 h.

Recommendations ITU-T M.20 and M.34 describe the general strategy to handle and manage the system performance monitoring data and the thresholds. Additionally, ITU-T M.2100 (PDH) and M.2101 (SDH) describe in detail the concept of system performance thresholds and measurement periods.

In digital radio-relay systems, the MPLs are calculated by combining the ES, SES, and BBE values over different threshold monitoring periods. These data are used to detect transitions to and from the unacceptable performance level under standard propagation conditions. When the levels with degraded or unacceptable system performance characteristic are reached, a maintenance action should be taken independently of the system performance measurement, while other thresholds might be used for long-term system performance studies. Operation and Maintenance systems use real time processing to determine the priorities of different occurrences of threshold crossings.

Three types of threshold monitoring periods are considered for microwave LOS link system performance: T1, T2 and T3. Each one is defined in the following subsections.

9.4.2.1 T1 Threshold Supervision The supervision period T1 is set in 15 min. A TR is considered when the number of ES, SES, BBE is exceeded during this supervision period. A reset TR is produced when the number of ES, BBE and SES is equal or lower than the reset threshold. The reset threshold is an optional feature of maintenance equipment.

9.4.2.2 T2 Threshold Supervision The duration of the T2 monitoring period is 24 h. The TR is produced when the ES, SES, or BBE count exceed the threshold limit. The period T2 is also the reference when the system (path or section) is restored after a maintenance option even under standard propagation conditions. The period T2 should be used also as the initial TP in order to confirm that the section or path satisfies the MPLs after repair actions (MPL_{PAR}).

After a T2 period where the system performance is above the thresholds the system should be monitored during 7 more days (T3 period) in order to confirm that the maintenance action has been successful.

9.4.2.3 T3 Threshold Supervision The period T3 extends over 7 days. In the same way as the previous T1 and T2, the TR is delivered if the ES, BBE or SES count exceeds the performance limits.

9.4.2.4 Long-Term Measurement and Monitoring Long-term measurement and monitoring implies keeping report data over a sufficiently long period of time. The usual practice is to keep historic reports during a period of 1 year (at least).

9.4.3 Availability

The error performance parameters and limits are only relevant if the system is "on service." In consequence, they should only be evaluated over the periods where the link is available.

The criteria to describe the transition to/from an unavailability period are defined for maintenance purposes in ITU-T Rec. M.2100 and M.2101. Even in the case of bidirectional links, the availability criteria only apply to one direction of the communication. It will be supposed that correction of faults and availability periods affecting one of the system directions supposed to not being affected by the behavior of the other direction.

Once this specific condition associated with maintenance, the definition of available and unavailable periods remains the same as the standard performance objective. An available period starts with ten consecutive non-SES events. These 10 s are considered part of the available period.

In order to define the transition to and from the unavailability period, it is necessary to compile SES data. ITU T Recommendation M.2120 describes the criteria for evaluating unavailability events when undertaking performance limits for maintenance measurements.

Currently, the availability thresholds are under study within the ITU-T and if required, they will be agreed between involved operators.

9.4.4 Evaluation of Error Performance Parameters

The calculation of these parameters is described in detail in ITU-T Recommendations M.2100 and M.2101. The evaluation of error performance parameters for maintenance purposes during adverse propagation conditions should take due account of the effect of propagation.

9.4.5 Fault Location Procedures

The test procedures to detect and locate system fault in systems with or without ISM are defined by ITU-T Recommendation M. 2120. This recommendation can be used for section, paths and PDH/SDH digital radio-relay systems. There will be two study cases: pre-ISM and ISM systems. In both cases the general statements apply:

- The procedure applies to periods where fading is inexistent or at least limited. The test procedures for situations with relevant fading occurrences have not been studied yet.
- The efficiency of the fault location procedure will depend to a great extent on the type of information associated with each stream (cyclic redundancy check, parity bits, frame header information, etc).
- The fault location will be strongly dependent on the resources available to the MEs.
- Once an ME has been repaired, it will be necessary to ensure that the performance is adequate. Depending upon the type, cause and repairing process, this validation could consist on transmitting a test signal. In some cases a new BIS sequence might be required. As soon as the path is again in-service, it should be monitored at least for 24 h.

9.4.5.1 Fault Location in pre-ISM Environments In a pre-ISM environment there is not necessarily any mechanism that could help in detecting a fault in a transmission system or a multiplex section. In addition, most probably, a historic record and storage maintenance database will not exist either.

In these environments, the fault location process is usually trigger by the complaint or report issued by a user and thus the monitoring is always after the fault has been occurred. A relevant consequence of this approach is that it will not be possible to guarantee the original problem that has caused the fault, particularly if the cause is a transient effect.

If there would be more than one section involved in the fault, the fault location will usually start from the most degraded section. If additional resources are available, a parallel approach could speed up the process. In any case, it will be necessary a complex control that coordinates the actions, so activities, tests, etc, carried out by one team do not mask the problem being analyzed by other teams.

If the traffic is completely interrupted and if ISM equipment is not available the fault location process it will be based on PRBS equipment (if possible with a frame structure). The feeding points and the measurement locations should be chosen so the fault location·process is efficient.

9.4.5.2 Fault Location in ISM Environments ISM environments enable structured procedures for fault detection and maintenance action triggering. When a degraded or unacceptable performance limit is reached, the following steps are usually taken:

- Send a message to the control stations of the paths carried by the transmission system or the multiplex section.
- Store the message for access by those control stations that do not receive the message immediately. The storage will normally be at the fault report point.
- Initiate the fault localization capability of the MEs to find the faulty maintenance sub-entity. This should be done in a time frame appropriate to the prompt or deferred maintenance alarm.

The station that controls the path, once the problem is known should:

- Initiate corrective measures in a time lapse adequate to the relevance of the alarm.
- Confirm the characteristics of the performance level: unacceptable or degraded. This will be carried out based on historical maintenance data associated with the affected path.

BIBLIOGRAPHY

ITU-R Rec. F.1330: Performance limits for bringing into service the parts of international plesiochronous digital hierarchy and synchronous digital hierarchy paths and sections implemented by digital fixed wireless systems. International Telecommunication Union. Radiocommunication Sector. ITU-R. Geneva. 2006.

ITU-R Rec. F.1566: Performance limits for maintenance of digital fixed wireless systems operating in plesiochronous and synchronous digital hierarchy-based international paths and sections. International Telecommunication Union. Radiocommunication Sector. ITU-R. Geneva. 2007.

ITU-R Rec. F.1705: Analysis and optimization of the error performance of digital fixed wireless systems for the purpose of bringing into service and maintenance. International Telecommunication Union. Radiocommunication Sector. ITU-R. Geneva. 2005.

ITU-T Rec.G.821: Error performance of an international digital connection operating at a bit rate below the primary rate and forming part of an Integrated Services Digital Network. International Telecommunication Union. Telecommunication Standardization Sector. ITU-T. Geneva. 2002.

ITU-T Rec.G.826: End-to-end error performance parameters and objectives for international, constant bit-rate digital paths and connections. International Telecommunication Union. Telecommunication Standardization Sector. ITU-T. Geneva. 2002.

ITU-T Rec. G.828: Error performance parameters and objectives for international, constant bit-rate synchronous digital paths. International Telecommunication Union. Telecommunication Standardization Sector. ITU-T. Geneva. 2000.

ITU-T Rec. M.2100: Performance limits for bringing-into-service and maintenance of international multi-operator PDH paths and connections. International Telecommunication Union. Telecommunication Standardization Sector. ITU-T. Geneva. 2004.

ITU-T Rec. M.2101: Performance limits for bringing-into-service and maintenance of international multi-operator SDH paths and multiplex sections. International Telecommunication Union. Telecommunication Standardization Sector. ITU-T. Geneva. 2006.

ITU-T Rec. M.2110: Bringing into service international multi-operator paths, sections and transmission systems. International Telecommunication Union. Telecommunication Standardization Sector. ITU-T. Geneva. 2007.

APPENDIX: PRACTICAL EXAMPLES

1 INTRODUCTION

This section contains two examples that illustrate the application of the procedures described through all the chapters of this book.

The first example illustrates the application of Microwave LOS links in cellular access networks. Specifically, the example is a group of three links in an urban environment that form part of the access section of a 3G universal mobile telecommunications system (UMTS) transport network. This use case shows the factors that apply to short distance medium capacity links in high-frequency bands (between 18 and 30 GHz). This first example provides detailed information about the availability and error performance objective (EPO) calculations. It also illustrates the fact that link path engineering in short urban links reduces to Line of Sight analysis in reality.

The second example is a high capacity link in one of the lowest bands where general-purpose microwave LOS links operate today: the 6 GHz band. This example provides detailed interference calculations and shows how different techniques can be applied to mitigate interference problems, evaluating the gain obtained using each technique.

Microwave Line of Sight Link Engineering, First Edition. Pablo Angueira and Juan Antonio Romo.

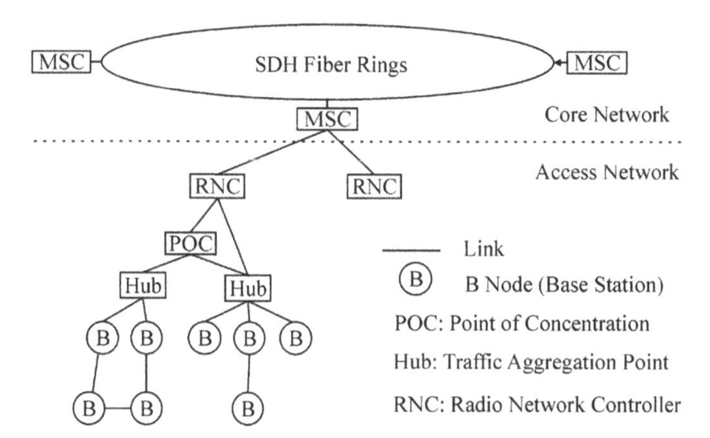

FIGURE A.1 Links as part of the access network in UMTS systems.

2 MICROWAVE LOS LINKS IN CELLULAR ACCESS NETWORKS

2.1 Case Description

This section contains an example that illustrates the application of Microwave LOS links in cellular access networks. Specifically, the example is a group of three links in an urban environment that form part of the access section of a 3G UMTS transport network. A brief description of the problem is presented in the following paragraphs.

The 3G network, in line with the structure of current cellular standards, is built by two network sections: the access network and core network. The core network is composed of all the switching centers, control offices and backbone transport infrastructure, where as the access network is composed by base stations (called B Nodes), the base station controllers (called radio network controllers (RNCs)) and the links and traffic concentration infrastructure. Figure A.1 illustrates the network architecture. The B Nodes are the destination of an origin of the network traffic, that will be aggregated before it is delivered to the core network through the RNC. The traffic can be aggregated either in hubs or points of concentration (POC). Even though the specifics of those nodes are out of the scope of this book, they can be seen for the purposes of this example as two hierarchical levels of aggregation, where different traffic sources from hubs are aggregated at the POC. Finally, several POC sources will concentrate at the RNC that in turn will forward the traffic to the Core Network.

Microwave LOS links play an important role in the design and deployment of the described network as the one of most widespread technical solutions to connect B Nodes, Hubs, POCs, and RNCs. This example consists of three B nodes connected to a hub using microwave links in a real environment of the North of Spain.

2.2 Initial Requirements and Specifications

2.2.1 Network Structure The city of Bilbao is located on a hilly area of the Bizkaia province. The city itself lies along the Nervion River with irregular terrain

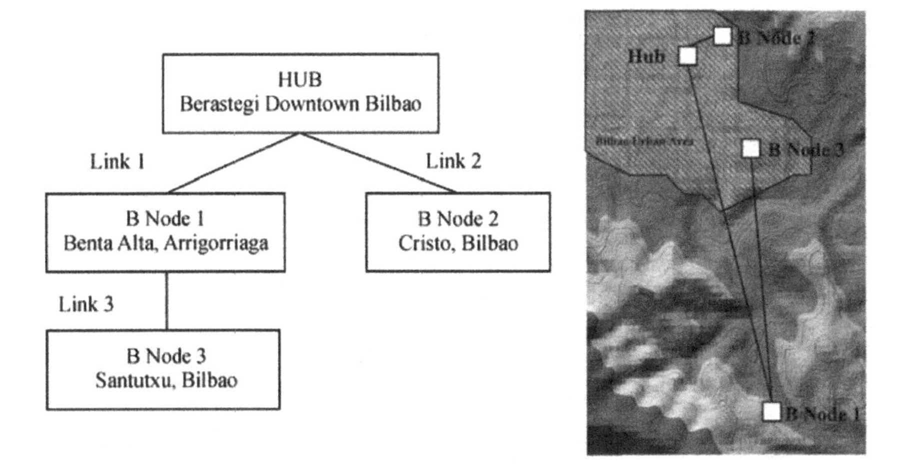

FIGURE A.2 Functional diagram and geography showing the links between the B Nodes and the hub.

heights in different areas of the city. This irregularity will be analyzed in the section related to link path engineering. The example is a network composed of three links between three B nodes connected to a hub. The network connections and sites are described in Figure A.2.

The hub is located in downtown Bilbao, on the top of a building on one of the center streets of the city. The B nodes 3 and 2 are also installed on the top of two buildings in two additional districts of the city, and the third one B Node 1, is located outside the city, close to a communications tower on a rural area, at an elevated hill 5 km away from the city.

2.2.2 Traffic and Network Capacity

The traffic matrix is shown in Table A.1. The values are number of E1 (2048 Kbps) units between each pair of stations. In this case, the manufacturers refer to the capacity using the notation #E1 × 2. For example, the capacity of a system with 4 E1, with a throughput of 4 × 2048 Kbps will be referred to as a "4 × 2".

TABLE A.1 Traffic Matrix (Number of E1 Circuits)

	Berastegi (Hub)	Benta Alta (B Node 1)	Cristo (B Node 2)	Santutxu (B Node 3)
Berastegi (Hub)	–	17 × 2	8 × 2	–
Benta Alta (B Node 1)	17 × 2	–	–	15 × 2
Cristo (B Node 2)	8 × 2	–	–	–
Santutxu (B Node 3)	–	15 × 2	–	–

The usual format for expressing the number of E1 in manufacturers' documentation is #E1 × 2, indicating that each E1 system is equivalent to a bit rate of 2048 Kbps (roughly 2 Mbps).

TABLE A.2 Equipment Technical Specifications. Transmitter

Protection	1+1 Hot Stand-by	
ATPC option	Yes	
Adaptive modulation	Yes	
Radio channel bandwidth options	3.5, 7, 14, 28, 40, and 56 MHz	
	Max. RF Output Power (dBm)	
	Frequency Band (GHz)	
Modulation	26	38
256 QAM	19	16
128 QAM	20	17
64 QAM	20	17
16 QAM	21	18
C-QPSK	24	21
Minimum RF output power (dBm)	−10	−10

This traffic structure represents the traffic originated at each B node or Hub with a specific destination on Table A.1. It should be taken into account that according to the link structure in Figure A.2, the traffic in Link 3 (B Node 3–B Node 1) will aggregate to the traffic originated in B Node 1 with destination Berastegi Hub.

2.2.3 Equipment The equipment is part of the specifications of the mobile operator and the planning engineer will not be able to make a selection of the radio units. It should be noted that the example is part of a dense network of microwave LOS links that are only in the area of Bizkaia, within 30 km around the city of Bilbao will be a few dozens. Table A.2 summarizes the equipment characteristics.

This radio model is a hybrid radio (See Chapter 4) that enables aggregation of time division multiplex (TDM) native traffic and Ethernet. A typical network evolution use case would start with TDM traffic and increasingly add Ethernet sources when the traffic increases, moving to a full Ethernet traffic at the final network stage.

As shown in Table A.2, the units that will be used in the links have the adaptive modulation option activated. Usually, the planner will be able to select which modulation options will be used on each link. Having several modulations implies different thresholds and gross margins for each case. In consequence, it will be necessary to define an engineering modulation reference that will determine the maximum transmitted power and the system thresholds associated with 10^{-3} and 10^{-6} bit error rate (BER) values. The engineering reference modulation will be one that exceeds the minimum guaranteed capacity on the specifications (see Table A.3).

2.2.4 Antenna The antenna will be chosen by the planning engineer among a set of defined diameters for both single and double polarization in the 26 and 28 GHz bands (see Table A.4). The antennas used in both bands are high performance parabolic antennas. The recommendation of the mobile operator is to use antennas with a diameter or 0.3 m. If required, in order to reduce potential interference

TABLE A.3 Equipment Technical Specifications. Capacity, Modulation Options, and Thresholds

Capacity	Modulation	Threshold 10^{-6} (dBm) Bands (GHz)		Capacity	Modulation	Threshold 10^{-6} (dBm) Bands (GHz)	
		26	38			*26*	*38*
2E1	C-QPSK/3.5 MHz	−90	−88	35E1	128-QAM/14 MHz	−71	−69
4E1	C-QPSK/7 MHz	−87	−85	39E1	256-QAM/14 MHz	−65	−63
4E1	4-QAM/7 MHz	−90	−88	16E1	C-QPSK/28 MHz	−81	−79

Capacity	Modulation	Threshold 10^{-6} (dBm) Bands (GHz)		Capacity	Modulation	Threshold 10^{-6} (dBm) Bands (GHz)	
		26	38			26	38
8E1	16-QAM/7 MHz	−85	−83	32E1	16-QAM/28 MHz	−79	−77
10E1	16-QAM/7 MHz	−83	−81	46E1	16-QAM/28 MHz	−77	−75
15E1	64-QAM/7 MHz	−76	−74	65E1	64-QAM/28 MHz	−71	−69
14E1	64-QAM/7 MHz	−76	−74	75E1	128-QAM/28 MHz	−68	−66
17E1	128-QAM/7 MHz	−72	−70	STM-1	128-QAM/28 MHz	−68	−66
8E1	C-QPSK/14 MHz	−84	−82	80E1	256-QAM/28 MHz	−64	−62
17E1	16-QAM/14 MHz	−82	−80	80E1	16-QAM/56 MHz	−77	−75
20E1	16-QAM/14 MHz	−80	−78	STM-1	16-QAM/56 MHz	−77	−75
22E1	16-QAM/14 MHz	−80	−78	80E1	128-QAM/56 MHz	−65	−63
30E1	64-QAM/14 MHz	−74	−72	80E1	256-QAM/56 MHz	−61	−59

Manufacturers will also provide the thresholds associated with a BER of 10^{-3}

TABLE A.4 Antennas

0.3 m High Performance	Band (GHz)	
	26	38
Gain	37.3	40.4
Front-to-back	62	61
XPD	27	27
3 dB beamwidth	3	2

0.6 m high performance	Band (GHz)	
	26	38
Gain	41.5	44.3
Front-to-back	67	63
XPD	27	27
3 dB beamwidth	1.5	2

Radio Channels available for the mobile operator: 38 GHz Band (Forward/Return).																	
Start (MHz)	RADIO CHANNEL NUMBER																End (MHz)
37758	E1								E2								37870
37758	D26				D27				D28				D29				37870
37758	B51		B52		B53		B54		B55		B56		B57		B58		37870
37758	C101	C102	C103	C104	C105	C106	C107	C108	C109	C110	C111	C112	C113	C114	C115	C116	37870
Radio Channels available for the mobile operator: 38 GHz Band (Return/Forward).																	
Start (MHz)	RADIO CHANNEL NUMBER																End (MHz)
39018	E1'								E2'								39130
39018	D26'				D27'				D28'				D29'				39130
39018	B51'		B52'		B53'		B54'		B55'		B56'		B57'		B58'		39130
39018	C101'	C102'	C103'	C104'	C105'	C106'	C107'	C108'	C109'	C110'	C111'	C112'	C113'	C114'	C115'	C116'	39130

E Radio Channels are 56 MHz wide
D Radio Channels are 28 MHz wide
B Radio Channels are 14 MHz wide
C Radio Channels are 7 MHz wide

FIGURE A.3 Radio Channels available for the mobile operator: 38 GHz Band.

problems, 0.6 m antennas will be also possible, with reduced beamwidth and increased gain. The branching losses will be 1.7 dB and 3.7 dB on the 1+1 Hot Stand-by main and backup paths respectively.

2.2.5 Frequency Band Selection The mobile operator has gained access to two frequency bands to roll out the access network (see Figure A.3 and A.4). The possible radio channels are part of the bands for fixed services in 26 and 38 GHz. The preferred configuration according to the internal procedures of this operator is the use of the 38 GHz band for links in urban environments and keep the 26 GHz band for suburban use. In line with these criteria, the links 1 and 3 can be considered suburban links and Link 2 is an urban link, thus Links 1 and 3 will operate on the 26 GHz band and Link 2 will operate in the 38 GHz band.

The channel arrangement reference schemes are based on the International Telecommunication Union. Radiocommunication Sector (ITU-R) F.748 and F.749 recommendations, as well as the Spanish Regulatory framework compiled by the Cuadro Nacional de Asignación de Frecuencias–National Frequency Allocation Chart. The mobile operator has granted access to the frequency ranges as shown by Figures A.3 and A.4. Tables do not include 3.5 MHz arrangements (the minimum capacity requirement is 8E1 as described in previous subsections).

An initial radio channel plan will be proposed according to the capacity requirements. Usually, the planning engineer makes the channel choice based on his experience over the terrain, considering nearby links and past interference issues in the same area. The design will be accomplished with this initial channel assignments, and as a consequence of the interference evaluation process, the channels will be

Radio Channels available for the mobile operator: 26 GHz Band (Forward/Return).																	
Start (MHz)	RADIO CHANNEL NUMBER																End (MHz)
25179	E1								E2								25291
25179	D1				D2				D3				D4				25291
25179	B46		B47		B48		B49		B50		B51		B52		B53		25291
25179	C91	C92	C93	C94	C95	C96	C97	C98	C99	C100	C101	C102	C103	C104	C105	C106	25291
Radio Channels available for the mobile operator: 26 GHz Band (Return/Forward).																	
Start (MHz)	RADIO CHANNEL NUMBER																End (MHz)
26187	E1'								E2'								26299
26187	D1'				D2'				D3'				D4'				26299
26187	B46'		B47'		B48'		B49'		B50'		B51'		B52'		B53'		26299
26187	C91'	C92'	C93'	C94'	C95'	C96'	C97'	C98'	C99'	C100'	C101'	C102'	C103'	C104'	C105'	C106'	26299

E Radio Channels are 56 MHz wide
D Radio Channels are 28 MHz wide
B Radio Channels are 14 MHz wide
C Radio Channels are 7 MHz wide

FIGURE A.4 Radio Channels available for the mobile operator: 26 GHz Band.

either validated or canceled. If the interference validation is not passed, a new channel proposal will be proposed and the process will be started iteratively. If after the interference calculations the interference is still remarkable, other frequency bands can be explored: 23 GHz, 18 GHz, and 13 GHz.

In this example, the initial radio channel bandwidth proposal would be 28 MHz for Link 1, 14 MHz for Link 2 and 28 MHz for Link 3. If those choices were possible, the modulation that fulfills the capacity requirements would be in all cases the Constant Envelope Offset Quadrature Phase Shift Keying (C-QPSK) option, which is the most robust alternative.

In reality, the 28 MHz radio channel bandwidth is not possible in this case because the interference levels created on already existing links exceed the maximum C/I requirements.

The solution requires first to reduce the channel bandwidth and find out any 14 MHz or 7 MHz channel that does not suffer or produce C/I levels lower than specification. At the same time, the bandwidth reduction will imply an increase in the modulation order that enables transport of the 15×2 Mbps capacity. This fact will reduce the gross margin M_3 and rain fading can become a problem. If that was the case, the analysis will be repeated with radio channels in lower bands as mentioned before.

After the interference analysis, the 14 MHz bandwidth choice is not feasible either. The 7 MHz option provides enough C/I margin and thus that will be the final selection. In order to fulfill the capacity requirements, the engineering reference modulation has to be raised to 64-QAM. The final choice is summarized in Tables A.5, A.6 and Figure A.5.

TABLE A.5 Capacity and Radio Channel Bandwidths

Link	Traffic	Required Capacity (Mbps)	Modulation	Channel Bandwidth (MHz)	System Capacity (Ethernet) Mbps
Link 1 (Hub–B Node 1)	17 × 2	34	C-QPSK	28	33.4–40.8
Link 2 (Hub–B Node 2)	8 × 2	16	C-QPSK	14	16.2–19.8
Link 3 (B Node 1–B Node 3)	15 × 2	31	64-QAM	7	31.3–38.3

Required capacity values are approximate TDM capacity requirements according to the number of E1. System capacity values are specified by the manufacturer for each modulation/channel bandwidth combination (see Table A.3)

TABLE A.6 Radio Channel Selection (26 and 38 GHz bands)

Link	Band (GHz)	Channel (MHz)	Radio Channel (Forward)/Polarization	Radio Channel (Return)/Polarization
Link 1 (Hub–B Node 1)	26	28	D3′ H (26257 MHz)	D3 H (25249 MHz)
Link 2 (Hub–B Node 2)	38	14	B56 H (37835 MHz)	B56′ H (39095 MHz)
Link 3 (B Node 1–B Node 3)	26	7	C96 H (25217.5 MHz)	C96′ H (26225.5 MHz)

<table>
<tr><td colspan="16">RADIO CHANNEL NUMBER</td></tr>
<tr><td colspan="8">E1</td><td colspan="8">E2</td></tr>
<tr><td colspan="4">D1</td><td colspan="4">D2</td><td colspan="4">D3</td><td colspan="4">D4</td></tr>
<tr><td colspan="2">B46</td><td colspan="2">B47</td><td colspan="2">B48</td><td colspan="2">B49</td><td colspan="2">B50</td><td colspan="2">B51</td><td colspan="2">B52</td><td colspan="2">B53</td></tr>
<tr><td>C91</td><td>C92</td><td>C93</td><td>C94</td><td>C95</td><td>C96</td><td>C97</td><td>C98</td><td>C99</td><td>C100</td><td>C101</td><td>C102</td><td>C103</td><td>C104</td><td>C105</td><td>C106</td></tr>
</table>

FIGURE A.5 Radio Channel Selection Example (only those at 26 GHz Band shown).

2.3 Link Path Engineering

The link path engineering phase will consist of validating the clearance conditions and identifying any potential obstacle along the path of any of the three links of the network under study.

In this case, the antenna heights and site selection are not a choice of the radio planner. The data are summarized in Table A.7.

TABLE A.7 Antenna Height Values

Site	Height (m)
Hub	25
B Node 1	20
B Node 2	26
B Node 3	30

TABLE A.8 Refractivity and *k* Factor Statistics (ITU-R P.453)

Coordinates	Berastegi Hub
ΔN (50%)	−35.92
AN (1%)	<−250
k (50%)	1.29
k (99%)	0.7–0.8

This situation is quite common when designing radio links in mobile access networks. The sites correspond to the location of the B Nodes, and the antenna height will be mostly conditioned by the building heights, with little for elevating these antennas over the roof where the equipment is being installed. This is the case of the sites associated with the Hub, B Node 1 and B Node 3, installed on rooftops. B Node 2 is installed on a communications tower and the available slot on the tower for microwave LOS link antennas is at 20 m a.g.l.

In order to check the clearance conditions and according to recommendation ITU-R P.453, the statistics of the refractivity in the region are summarized in Table A.8.

The value k (99%) cannot be obtained from the classical approximation given by equation (A.1), because the values not exceeded during 1% of the time are out of the range where the approximation applies (−157 N Units). In consequence and from a practical perspective, the effective $k_e = k$ (99%) shown in Table A.9, will be obtained from Recommendation ITU-R P.530:

$$k = \frac{157}{157 + \Delta N} \tag{A.1}$$

The clearance conditions in this example will be fulfilled in all cases. The only case where there is a possible obstacle along the path is Link 1. In this link, the maximum value of the first Fresnel ellipsoid radius is 0.11 m (link distance 5.09 km, $f = 26$ GHz). In addition, and due to the short length of the link the maximum equivalent Earth correction associated with subrefractive conditions is almost negligible ($6\ 10^{-7}$ m). In consequence, the link path analysis turns into an optical line of sight evaluation.

TABLE A.9 Refractivity and *k* Factor Statistics for Practical Calculations

Coordinates	Berastegi Hub
k (50%)	1.33
k (99%)	0.7–0.8 (depending on the link range: 3–5 km)

TABLE A.10 SESR and BER for Error Performance and Availability Calculations

Case	B.E.R.
SESR associated with availability	10^{-3}
SESR associated with EPO	10^{-6}

Following the same argument, the other two links, shorter and without remarkable irregularities, will also be evaluated by a simple optical line of sight clearance validation.

2.4 System Dimensioning

This section will describe the availability and error performance analysis of each one of the links. The first step will consist of identifying the reference availability and EPOs applicable to these links. The reference values will be provided by the ITU-R recommendations. Obviously the equipment were designed after 2005, and thus the applicable references are ITU-R F.1668 and F.1703, for error performance and availability objectives respectively.

The association of severely errored second rate (SESR) and BER thresholds, as discussed in Chapter 8, will be as described in Table A.10.

2.4.1 Error Performance Objectives (EPO) The links are part of the access section of the network and the EPOs will be independent of the distance. The SESR objective will be provided by Table 5.16 in Chapter 5 (EPOs for synchronous digital hierarchy (SDH) fixed wireless links forming all of the access network section of the national portion of the hypothetical reference path (HRP) according to ITU-R Rec.G.828).

The value is 0.00015 (0.002 × C, with C = 0.075), 0.015 in percentage. That value is equivalent to 389 severely errored second (SES)/worst month. The ITU-R recommendations do not specify the distribution of this objective. Nevertheless, this operator states a maximum of 9 hops in tandem that suggests a maximum objective of [0.015/9] = 0.0016% (389/9 SES/worst month). This value applies to any of the three links under study: Link 1, Link 2, and Link 3.

2.4.2 Availability Objectives The availability objectives will be provided by ITU-R F.1703. Based on the same conditions one can apply equation (A.2) with B_5 and C_5 the coefficients for national portion and access section links, 0 and 5 10^{-4} (see Table 5.19 on Chapter 2). In that case the maximum allowable availability ratio will be:

$$\mathrm{AR} = 1 - \left(B_5 \frac{L_{\mathrm{link}}}{L_{\mathrm{R}}} + C_5\right) = 1 - \left(0 \frac{50}{2500} + 5 10^{-4}\right) = 0.9995\ (99.95\%) \quad \text{(A.2)}$$

TABLE A.11 Radio Specifications for Reference Engineering Modulations (Adaptative Modulation)

	Link 1	Link 2	Link 3	Units
Band	26	38	26	GHz
Radio Channel (Ref. Frequency)	25753	39095	25721.5	MHz
Capacity/BW	17 × 2/28 MHz	8 × 2/14 MHz	15 × 2/14 MHz	–
Reference Modulation	C-QPSK	C-QPSK	64-QAM	–
Maximum Transmitted Output Power	24	21	20	dBm
Minimum Transmitted Output Power	−10	−10	−10	dBm
Branching Losses (Main Path of 1+1 hsby)	1.5	1.5	1.5	dB
Antenna Gain (0.3 m)	37.3	40.4	37.3	dB
Antenna Gain (0.6 m)	41.5	44.3	41.5	dB
Threshold 10^{-3} Th_3	−85	−88	−77	dBm
Threshold 10^{-6} Th_6	−81	−84	−76	dBm
MTBF	50 000	50 000	50 000	h
MTTR	5	5	5	h
Equipment Unavailability $U_E = \frac{MTTR}{MTBF + MTTR}$	1.0E–08	1.0E–08	1.0E–08	–

MTTR, mean time to repair

Considering also the possibility of a maximum concatenation of 9 hops, the availability objective will be $[100 - (5\ 10^{-2}/9)] = 99.995\%$ if availability events are considered uncorrelated on different hops.

In reality, this operator differentiates between traffic sources carried by any link and the most restrictive availability value is associated with voice services as 99.999%.

2.4.3 Thresholds, Link Budget, and System Margin This section describes the link budget calculation process. First, the reference engineering modulation has to be identified. Each modulation will have an associated maximum transmitted output Power and sensibility thresholds. Data are summarized on Tables A.11, A.12 and A.13.

TABLE A.12 Link Path Data

	Link 1	Link 2	Link 3	Units
Link distance	5.09	0.53	3.68	km
Site height –A	17	17	50	m
Antenna height A	25	25	30	m
Site height –B	194	9	194	m
Antenna height B	20	26	20	m
Overall height A	42	42	80	m
Overall height B	214	35	214	m
Elevation	0.042	0.066	0.058	(h/d ratio)
Elevation	0.042	0.066	0.058	rad (arctg h/d)
Minimum antenna height h_l	42	35	80	m

TABLE A.13 Basic Propagation Loss, Absorption, and Field Margin

	Link 1	Link 2	Link 3	Units
FSL	134.8	118.8	132.0	dB
γ_0	0.015	0.04	0.015	dB/km
γ_w	0.7	0.081	0.7	dB/km
Gas absorption	3.6	0.1	2.6	dB
Field margin	2	2	2	dB
Total fixed loss	136.4	116.8	132.6	dB

Applying the equation from Chapter 8, the link budget expression follows:

$$P_{rx} = P_{tx} - L_{tt} + G_t - L_b + G_r - L_{tr} \tag{A.3}$$

The losses on the branching network at the receiver are negligible. The propagation loss will be composed of the free space loss, the atmospheric absorption caused by gases, and the field margin. This margin is an implementation loss that in this case is established in 2 dB by the operator. The expressions will be the free space loss formula (A.4) and the gas absorption expression (A.5) described in Chapter 7:

$$L_b = 20 \log \left(\frac{4\pi d}{\lambda} \right) \tag{A.4}$$

where

d = link distance (m)
λ = wavelength (m)

$$A = \gamma_a d = (\gamma_a + \gamma_w)\ d \tag{A.5}$$

where

γ_a = total specific attenuation due to atmospheric gases (dB/km)
γ_o = specific attenuation due to oxygen (dB/km)
γ_w = specific attenuation due to water vapor (dB/km)

And from these fixed losses and the equipment thresholds, the system margins and gross margins are immediate. Table A.14 summarizes the results.

2.4.4 Availability Validation: Rain and Equipment Failures The availability objective of any individual hop of the network was 99.995% considering a maximum of 9 hops in tandem.

In general, the overall unavailability ratio of the link will be associated with rain and equipment failures. In our case, the equipment failures and associated unavailability

TABLE A.14 Link Budget

	Link 1	Link 2	Link 3	Units
Receiver input power 0.3 m antenna	−40.8	−18.0	−41.0	dBm
Receiver Input Power 0.6 m antenna	−32.4	−10.2	−32.6	dBm
System gain (Th_3)	109	109	97	dB
System gain (Th_6)	105	105	96	dB
M_3 0.3 m antenna	44.2	70.0	36.0	dB
M_6 0.3 m antenna	40.2	66.0	35.0	dB
M_3 0.6 m antenna	52.6	77.8	44.4	dB
M_6 0.6 m antenna	48.6	73.8	43.4	dB

were shown in Table A.11. The equipment unavailability U_E is close to 10^{-8} %, well below the objective, and thus it will be considered negligible. In consequence, the only cause of unavailability periods will be the rain.

The rain calculations will be based on Section 8 in Chapter 7. The calculation process will consist of obtaining the $A_{0.01}$, i.e., the attenuation associated with a probability of 0.01% and from this value, obtain the attenuation associated with the unavailability objective percentage p. If the resulting A_p is higher than the margin M_3 the link will satisfy the objectives, otherwise, measures would be necessary. The results are shown in Table A.15.

The rain rate is an empirical data obtained by the operator from the experience in installing links in Spain. If this value was not available, reference data are provided by ITU-R P.837. The value provided by the ITU is slightly more pessimist, between 45 and 50 mm/h for the area under study.

The last two rows of the table are the gross margin values obtained previously and have been pasted here for comparison purposes. It can be observed that the

TABLE A.15 Rain Calculations

	Link 1	Link 2	Link 3	Units
Rain rate $R_{0.01}$	43	43	43	mm/h
Polarization	Horizontal	Horizontal	Horizontal	
Reference distance d_0	18.4	18.4	18.4	km
Distance factor r	0.8	1.0	0.8	–
Effective distance d_{eff}	4.0	0.5	3.1	km
Constant k	0.1685	0.4235	0.1681	
Constant α	0.9910	0.8736	0.9913	
γ_R	7.0	11.3	7.0	dB/km
$A_{0.01}$	27.9	5.8	21.4	dB
Availability objective (p %)	0.005	0.005	0.005	%
Ap	35.8	7.5	27.5	dB
M_3 0.3 m antenna	44.2	70.0	36.0	dB
M_3 0.6 m antenna	52.6	77.8	44.4	dB

rain fading is well above the margin in all cases. Nevertheless, and if a 99.999% availability objective was applied the associated A_p is 59.7 dB. In that case, the rain fades in Link 1 will be deeper than the gross margin, and as shown in previous table, increasing the antenna diameter will not be enough for fulfilling the objective. Taking into account that the transmitted output power and the receiver thresholds cannot be modified; the best solution will be probably to change the operating frequency band. The possible target would be the 18 GHz band that would reduce the free space losses in a few dBs (around 3–4 dB), but at the same time would reduce the rain fading A ($p = 0.001$) to 34.7 dB, much lower than the previous 59.7 dB value.

2.4.5 Error Performance: Flat and Selective Fading Outages Having validated the link design according to availability objectives the EPO limits will be evaluated now. The objective for each individual hop is 0.0016% (389/9 SES/worst month). This value applies to any of the three links under study: Link 1, Link 2 and Link 3. The total outage will be caused by flat fading and selective fading (distortion). In this case, the third potential outage cause associated with cross-polar discrimination (XPD) reduction is neglected.

2.4.5.1 Geoclimatic Factor First the geoclimatic factor, K, will be calculated. This factor will be a function of the refractivity statistics, link situation (coastal or inland) and the path profile characteristics. This parameter calculation was made by the equations described in Chapter 7. The calculations are summarized in Table A.16:

$$K = 10^{-4.6-0.0027\,\mathrm{dN_1}} \tag{A.6}$$

$$K = 10^{-4.4-0.0027\mathrm{dN_1}}\,(10 + s_a)^{-0.46} \tag{A.7}$$

where

$\mathrm{dN_1}$ = refractivity gradient not exceeded for 1% of the average year in the lower 65 m of the troposphere (data available in ITU-R P.453).

s_a = roughness of terrain in the area, defined as the standard deviation of terrain heights (m) in a 110 km × 110 km area with resolution of 30 s

In this case, the operator has its own empirical data that take into account the terrain path profiles. The dN1 parameter is a bilinear interpolation of the values provided by ITU-R and calculated for the coordinates of the hub. The distances are short enough to consider this parameter constant for the three links.

TABLE A.16 Geoclimatic Factor Data

Geoclimatic and Refraction Data	Link 1	Link 2	Link 3	Units
dN1	−345.66	−345.66	−345.66	N units
Geoclimatic factor K (Equation A.6)	2.15E–04	2.15E–04	2.15E–04	–
Geoclimatic factor (Operator data)	8.43E–05	1.00E+00	2.00E+00	–

TABLE A.17 Multipath Occurrence Factor and A_t

Multipath Occurrence Factor p_0(%)	Link 1	Link 2	Link 3	Units
Initial planning	0.39113	0.00048	0.12965	%
Detailed planning	0.65229	0.00041	0.19921	%
A_t	24.5	21.0	23.9	dB
M_6	40.2	66.0	35.0	dB

2.4.5.2 Outage Due to Flat Fading The calculation of the flat fading outage depends on the time percentage under study. In Chapter 7, two methods were proposed, one for small percentages of time and a second one for all percentages.

The ITU-R P.530 suggests a parameter A_t that will provide a reference to make the method choice. This parameter is calculated as:

$$A_t = 25 + 1.2 \log p_0 \tag{A.8}$$

where p_0 is the multipath occurrence factor (%) that can be calculated according to any of the following equations (recommended for initial and accurate planning respectively). The calculation is summarized in Table A.17.

$$p_0 = K\ d^{3.1} \left(1 + |\varepsilon_p|\right)^{-1.29}\ f^{0.8}\ 10^{-0.00089h_L} \tag{A.9}$$

$$p_0 = K\ d^{3.4} \left(1 + |\varepsilon_p|\right)^{-1.03}\ f^{0.8}\ 10^{-0.00076h_L} \tag{A.10}$$

Considering the difference between A_t and M_6, it is clear that the fade depth values of interest, quite higher in magnitude than the A_t value will be associated with small percentages of time. In consequence, the flat fading outage will be calculated using equations (A.11) and (A.12) for detailed and initial planning respectively (Table A.18):

$$p_w = p_0\, 10^{-\frac{A}{10}} = K\, d^{3.4} \left(1 + |\varepsilon_p|\right)^{-1.03}\ f^{0.8}\ 10^{-0.00076h_L - A/10} \tag{A.11}$$

$$p_w = p_0\, 10^{-\frac{A}{10}} = K\, d^{3.1} \left(1 + |\varepsilon_p|\right)^{-1.29}\ f^{0.8}\ 10^{-0.00089\, h_L - A/10} \tag{A.12}$$

where

K = geoclimatic factor
d = hop distance (km)
f = frequency (GHz)
$|\varepsilon_p|$ = path inclination (mrad)
p_0 = multipath propagation occurrence factor (%)

TABLE A.18 Flat Fading Outage

Flat Fading	Link 1	Link 2	Link 3	Units
Initial planning	0.000038	0.000000	0.000041	%
Detailed planning	0.000063	0.000000	0.000063	%

TABLE A.19 Signature Data (Minimum and Non-minimum Phase Behavior Identical in This Case)

Modulation	Capacity	Reference Delay	Th3 Bn (dB)	Th3 W (GHz)	Th6 Bn (dB)	Th6 W (GHz)
C-QPSK	8 × 2	6.3 ns	23	0.0100	20	0.0105
	17 × 2	6.3 ns	17	0.0200	14	0.0210
QAM	8 × 2	6.3 ns	28	0.0070	27	0.0075
	17 × 2	6.3 ns	24	0.0140	23	0.0150

2.4.5.3 Outage Due to Selective Fading The selective fading outage will be calculated using the Signature Method described in Chapter 7. This method requires the signature data of the receiver measured by the manufacturer for each modulation and radio channel bandwidth scheme. Data are summarized in Table A.19.

The selective outage probability associated with multipath distortion will be:

$$P_s = 2.15\eta\left(W_M \times 10^{-B_M/20}\frac{\tau_m^2}{|\tau_{r,M}|} + W_{NM} \times 10^{-B_{NM}/20}\frac{\tau_m^2}{|\tau_{r,NM}|}\right) \tag{A.13}$$

where

$W_{M/NM}$ = signature width (GHz)
$B_{M/NM}$ = signature depth (dB)
$\tau_{r,M/NM}$ = reference delay (ns) used to obtain the signature

M and NM subscripts refer to minimum and nonminimum phase situations respectively. (In our case are equal)

The average delay τ_m (ns) is obtained from equation (A.14):

$$\tau_m = 0.7\left(\frac{d}{50}\right)^{1.3} \tag{A.14}$$

whereas the multipath activity factor η is calculated from P_0 from equation (A.15):

$$\eta = 1 - e^{-0.2(P_0)^{0.75}} \tag{A.15}$$

with P_0 (%) is the multipath occurrence factor (obtained from equations A.11 or A.12). The results are shown in Table A.20.

TABLE A.20 Selective Fading Outage

Selective Fading	Link 1	Link 2	Link 3	Units
Initial planning	1.152E–06	5.324E–12	5.491E–08	%
Detailed planning	1.350E–07	0.000E+00	0.000E+00	%

TABLE A.21 Composite Outage

	Link 1	Link 2	Link 3	Units
Flat fading				
Initial planning	0.000038	0.000000	0.000041	%
Detailed planning	0.000063	0.000000	0.000063	%
Selective Fading				
Initial planning	1.152E–06	5.324E–12	5.491E–08	%
Detailed planning	1.350E–07	0.000E+00	0.000E+00	%
Composite Outage				
Initial planning	0.000039	0.000000	0.000041	%
Detailed planning	0.000063	0.000000	0.000063	%
EPO	0.0016	0.0016	0.0016	%

2.4.5.4 Composite Outage and EPO Validation Once the flat and selective fading outage probabilities have been obtained, those should be combined to obtain the composite outage probability. The ITU-R P. 530 recommends simple addition of probabilities:

$$P_t = P_{non\_selective} + P_{selective} \tag{A.16}$$

The results are shown in Table A.21. The composite outage percentages are well above the objectives. The result is in line with the expected values. The low outage probability is associated with a series of factors: the short length of all three links, the high-frequency bands where multipath outage should not be a significant propagation effect and the receiver robustness against multipath provided by current equalizing techniques.

2.5 Comments on Interference Calculations

The interference calculations will provide information to validate the frequency band, the radio channel bandwidth and the specific radio channel or channels on each link and hop. The access section of a cellular network is a good representative of an intense frequency reuse. In Chapter 8 guidelines for evaluating interference in this scenario were provided.

First of all, the maximum allowed interference has to be specified. There are two ways of specifying this limit:

1. Define a minimum Interference ratio (C/I ratio) usually associated with a maximum degradation value (1 or 3 dB depending on the frequency band. Usually lower bands have higher allowance.)
2. Define a maximum degradation value (threshold degradation (TD)), that will be a function of the exact C/I ratio, the radio equipment, the band and the

modulation scheme. An example illustrating the type of empirical expressions used for degradation calculations was provided in Chapter 8:

$$\mathrm{TD} = T_{\mathrm{hI}} - T_{\mathrm{h}} = 10\log\left(1 + 10^{\frac{-T_{\mathrm{h}}+C+I}{10}}\right) \tag{A.17}$$

where

$T_{\mathrm{h\,I}}$ = degraded threshold due to interference
T_{h} = threshold on interference free conditions
C = figure of merit under interference conditions (it can be regarded a similar parameter as the minimum C/I ratio)
I = interference level at the receiver input

From a theoretical perspective, the calculations in both cases are the same: interference level (I) and desired signal level (C). The difference will rely on the aggregate interference calculations. In the first case the calculation will be an aggregate C/I and in the second case the result will be an aggregate TD. In both cases and simplifying the reality, uncorrelated interference and desired paths will be assumed in most cases. Under this assumption, the combination will be simple addition in the linear domain.

In our case, the C/I objective is 30 dB, and the equivalent threshold degradation associated with the minimum required C/I is 1 dB. It is usual to differentiate between maximum allowed degradation from an individual interference source and the overall aggregate degradation from multiple interferers. Usually, both criteria have to be fulfilled. We will suppose that the maximum composite degradation will be 3 dB in this example.

After the requirements have been identified, one should decide on the modulation that will be analyzed, considering that the transmitted power and the receiver threshold (will be different for each one of the modulation choices available from the adaptive modulation technique). The usual calculation procedure is usually done for the highest order modulation at both sides of the interference problem. Other criteria can be also considered, such as lowest order modulation, engineering reference and interferences from higher order modulation signals into lowest order modulation receivers.

If automatic transmit power control (ATPC) is used there will be also different possibilities, depending on the power value or range associated with the interfering and interfered station for C/I calculation. The exact number of choices will depend on the software tool that the radio engineer will use. In the specific case of the links presented in this example, the operator usually makes the calculations with the maximum transmitted power on both interfered and interfering stations.

Once the C/I target, the maximum degradation and the modulation and ATPC options have been identified, the calculation process starts identifying the number of potential interfering and interfered stations to/from the link under design. The operators will usually have different sizes of the coordination area as a function of the frequency band. In the case of the bands used in this example the coordination area will be 40 km around the link under design.

TABLE A.22 Interference Calculation Result Example (Individual Interference Between Two Links)

Interference Calculation	Modulation	Interference Power (dBm)	Single TD (dB)	Composite TD (dB)
D→A	(16-QAM)→(128-QAM)	−95.85	1.39	25.38
B→C	(128-QAM)→(16-QAM)	−89.20	6.14	6.14
C→B	(16-QAM)→(128-QAM)	−83.29	8.85	23.93

The C/I will be evaluated for each one of the interference cases within the 40 km circle. The C/I is obtained using interference calculation equations described in Chapter 8 (equations 8.40, 8.41 and 8.42). The number of potential analysis can be up to a few dozens inside a 40 km wide area of an access network.

As the overall data of the network in Bilbao was not available, the following paragraphs show an interference problem in another area of the network of the same operator.

Table A.22 shows an example of the data that is obtained from this calculation procedure. The table assumes a link under design composed of stations A and B, and an interfering (and interfered) link composed of stations C and D. Figure A.6 shows the geographical configuration. (A and C are co-sited).

The last column in Table A.22 represents the composite degradation, obtained from simulation of the propagation

Data in Table A.22 show that interferences occurring from D and C stations into A and B are created by multiple links, not only by D–A link. The composite degradation in both cases area 25.38 and 23.93 dB respectively. In fact, data suggest that either D is not the main interferer into station A. On the contrary, the individual and composite interference in the case of station B into C are the same, which means that this is the only interferer into the C station.

Similar tables will be obtained for each one of the links using channels that might create interferences on the link under design and those that could be interfered by this new link within the 40 km area. Additionally, radio planning software usually

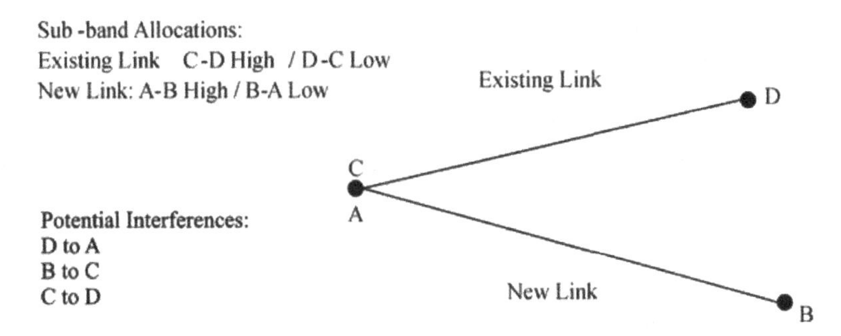

FIGURE A.6 Individual interference example.

TABLE A.23 Composite Interference Information (New Link Under Design A-B)

Interfering Link	Channel/ Polarization	Interference Power A (dBm)	TD (dB) Site A	Interference Power B (dBm)	TD (dB) Site B
TR1163(H)-TR0025P(L)	D5V/D5V	−68.3	23.03	−71.83	19.85
TR0075(L)-TR0117(H)	D6V/D6V	−95.85	1.39	−83.39	8.85
TR9874(L)-TR8525(H)	D6V/D6V	−99.71	0.63	−98.24	0.83
TR1161(H)-TR7305(L)	D5V/D5V	−101.7	0.47	−109.64	0.07

provides summarize information on the overall interference problems. Table A.23 shows an example of these reports.

Table A.23 shows only the first four rows of a table that in reality has more than 35 entries (representing the 35 potential interference situations within the study area). In fact the second row corresponds to the A–B/C–D example shown previously by Table A.22.

Recovering now the example in Bilbao, a similar calculation procedure was performed at the time of selecting the radio channel for each one of the links (Link 1, Link 2, and Link 3). According to the final radio channel structure, Link 1 and 2 with radio channels of 28 and 14 MHz wide (see Table A.6) provided composite TD values below 3 dB, whereas the initial search of a 28 MHz channel to be used in Link 3 provided TD values higher than 3 dB for all radio channel choices. The process was repeated for 14 MHz without success and finally, a 7 MHz radio channel was found with TD values below the threshold.

3 MICROWAVE LOS LINKS IN TRANSPORT NETWORKS

3.1 Initial Basic Data and Specifications

This example describes a small section of the transport network of a telecommunications operator that connects a transit switching center in Brasilia (Brazil) with three local access and switching centers in three surrounding cities: Celandia, Planaltina de Goiás and Luziania.

The topographical features of the area suggest that the microwave LOS link solution has important advantages over other possible solutions as fiber links. In any case, if fiber infrastructure would be already deployed, this link could be regarded as a backup system in a critical part of the network.

Figure A.7 shows a node interconnection diagram on the terrain height map of the area under study.

Table A.24 shows the initial capacity requirements, expressed in number of E1 circuits for each route.

3.1.1 Specifications The operator specifies that the link should be designed according to availability, error performance and clearance objectives provided by

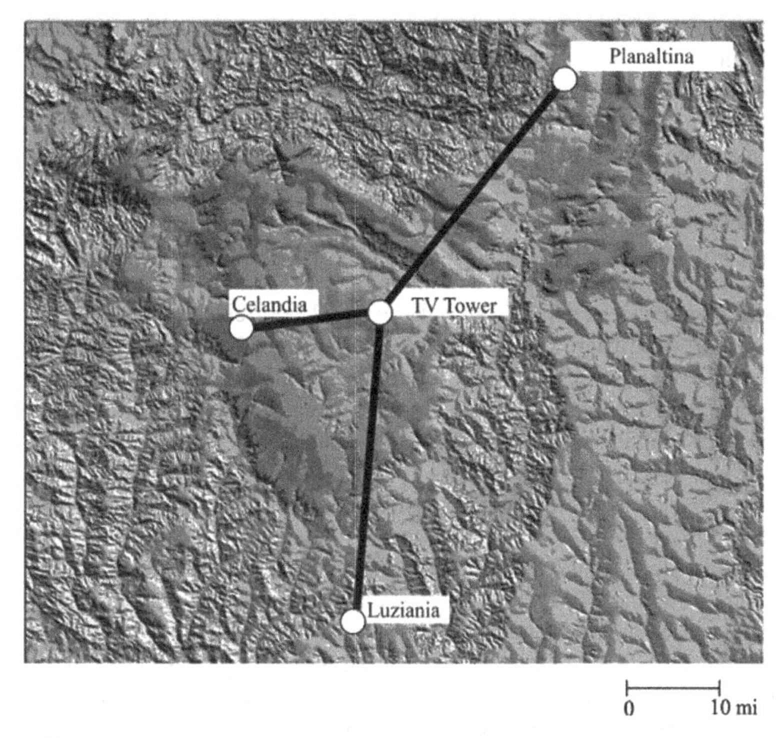

FIGURE A.7 Location map of the centers that will be connected by the link.

the ITU-R and ITU-T. Additionally, the capacity, the frequency band choices, and channel arrangements will also be designed according to ITU-R specifications.

The overall capacity will be dimensioned with an allowance of an additional 25% over the initial traffic requirement. In addition, each link (hop) will have a backup radio channel. Finally, the system will consume as less frequencies as possible.

3.1.2 Initial Dimensioning The link requires a 25% of excess capacity over the initial traffic requirements. This condition implies that each route should carry $1.25 \times (32 + 4 + 4)$ E1 frames on the service channels. The link will carry a baseband

TABLE A.24 Traffic Matrix (Number of E1)

Route	# E1
Celandia–Brasilia	32×2
Planaltina de Goiás–Brasilia	32×2
Luziania–Brasilia	32×2
Celandia–Planaltina de Goiás	4×2
Celandia–Luziania	4×2
Planaltina de Goiás–Luziania	4×2

STM-1 frame and the backup configuration will be 1 + 1 (an active plus a backup radio channel).

The frequency band selection will be done based on two fundamental criteria. First, the area has intense, short convective rain occurrences during the rain season, and selecting a frequency band over 10 GHz would have associated very high values of gross margin. Second, the hop distances should be as long as possible. In consequence, the band should be the lowest among those made available for this operator in the local frequency regulations. The band U6 (higher part of the 6 GHz band) is a good candidate that fulfills the previous conditions.

In this example, it is assumed that the external interference sources within this area are negligible. There is not any link operating in these frequencies within a distance that could create a potential interference. Additionally, there are not earth stations operating satellite services in close frequencies within the influence area of the link under design. The usual reference distance, for analyzing interferences (both terrestrial and satellite) in this band is 120 km. In consequence, the interferences studied in this example are created by different stations of the links that form the radio routes of the project.

Finally, aiming at reducing the number of required frequencies, and as an initial design phase, the link will be planned to use the same radio channels in every hop. The antennas used will be standard parabolic antennas, with single polarization.

If the interference levels present in different hops of the system would exceed the specifications, alternative solutions might be possible: use antennas with better performance or apply different polarizations. The last option would be to design a radio channel arrangement using more frequencies. In this case, additional calculations would be required to provide the minimum channel separation for each antenna type.

3.2 Equipment Selection

Table A.25 contains the most relevant parameters of the equipment that will be used in this link design. The operating band ranges from 6.425 GHz to 7.11 GHz. The Table includes the different choices for parameters where more than one value is possible as well as the selected value at the initial design phase.

In order to ensure that the minimum thresholds are not degraded more than 1 dB, the combined interference level should be equivalent to a C/I of 18 dB for a BER of 10^{-3} and 23 dB for a BER of 10^{-6}. Additional catalog data will be described, as they are required during the design process developed in further sections.

3.3 Link Path Engineering

The link is designed to interconnect four nodes. The station in Brasilia is sited at a transmitting site known as TV Tower, located in the center of the city. This site is not located close to the switching node origin and destination of the traffic of the link. The connection between the microwave link site and the switching node is a fiber link.

TABLE A.25 Equipment Data as Described by the Manufacturer and Initial Choice Values

Parameter	Possible Values	Design Choice
Modulation	128-QAM 64-QAM	64-QAM
Transmission capacity (per RF channel)	$1 \times$ STM-1 $2 \times$ STM-1	$1 \times$ STM-1
Gross bit rate of the aggregated radio signal	$1.08 \times$ Transmission capacity	
RF channel arrangements (ITU-R)	F.384	
RF channel spacing (MHz) STM-1	40	
Adaptive equalizer	19 TAPS	
Coding type	MLC	
Spectrum shaping	Raised cosine Roll-off factor $\alpha = 0.20$	
Transmitted power		
ATPC (maximum) (dBm)	32	
ATPC range (dB)	17	
Receiver threshold BER $= 1 \times 10^{-3}$ (dBm)	−73	
Receiver threshold BER $= 1 \times 10^{-6}$ (dBm)	−71	
Receiver threshold BER $= 1 \times 10^{-12}$ (dBm)	−68	
Switching configuration	1+1	
Branching losses T+R (dB) 1+1 single polar (STM-1)	4	

The sites in Celandia, Planaltina de Goiás and Luziania are colocated with the local switching centers that will be connected by the link. Table A.26 contains the geographical characteristics of each site. The antenna heights in Table A.26 correspond to available installation heights for this link.

The Digital Elevation Model data available from SRTM (Shuttle Radar Topography Mission) database has been used as the reference for profile analysis in this example. The link path profiles are shown in Figure A.8. Each profile has been calculated for standard, subrefractive and super-refractive atmosphere conditions. The k factor values used for each refraction condition are 1.34, 0.83, and 2.75

TABLE A.26 Initial Site Data

	Coordinates (Lat/Lon)				
Site	Latitude	Longitude	Height asl (m)	Antenna height agl (m)	Location
TV Tower	15° 47′ 26″ S	47° 53′ 35″ W	1132.5	40	Brasilia
Celandia	15° 48′ 53″ S	48° 06′ 19″ W	1289.5	40	Celandia
Planaltina	15° 25′ 17″ S	47° 35′ 39″ W	1133.5	30	Planaltina de Goiás
Luziania	16° 15′ 19″ S	47° 55′ 48″ W	991	35	Luziania

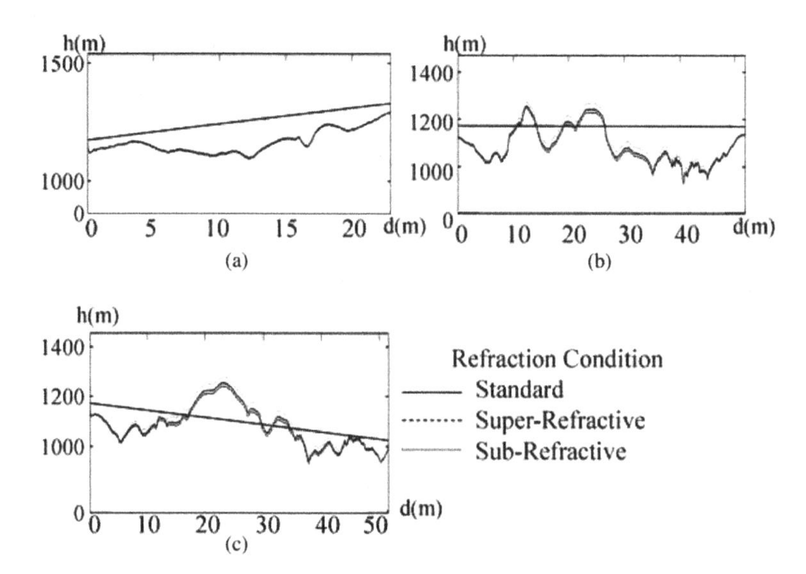

FIGURE A.8 Terrain path profiles (a) Route TV Tower–Celandia (b) Route TV Tower–Planaltina (c) Route TV Tower–Luziania.

respectively. Those data have been obtained from local refractivity data as describe by ITU-R Recommendation P.453.

The Line of Sight is obstructed for hops TV Tower–Planaltina and TV Tower–Luziania. In both cases, additional intermediate repeater sites will be required to ensure that clearance conditions are fulfilled. After an analysis of the local geography and evaluating the existing infrastructure and both antenna and equipment installation feasibility, the intermediate repeaters will be installed in Sobradinho and Gama. Table A.27 summarizes the geographical features of both stations and Figure A.9 shows a new link diagram that includes the new repeater sites.

The antenna heights must fulfill the clearance criteria described in ITU-R P.530 for non-diversity installations in tropical climate areas. The clearance should be equal to the first Fresnel ellipsoid radius, R, for $k = 4/3$ and 0.6 R for the effective k (k_e). Figure A.10 shows the new terrain profiles for standard, subrefractive and super-refractive conditions.

TABLE A.27 Repeater Sites

Site	Latitude	Longitude	Height[a] (m)	Antenna height[b] (m)	Location
Sobradinho	15° 38′ 33″ S	47° 53′ 50″ W	1303.6	30	Sobradinho
Gama	15° 58′ 48″ S	48° 01′ 15″ W	1273.9	35	Gama

[a] Above median sea level
[b] Above ground level

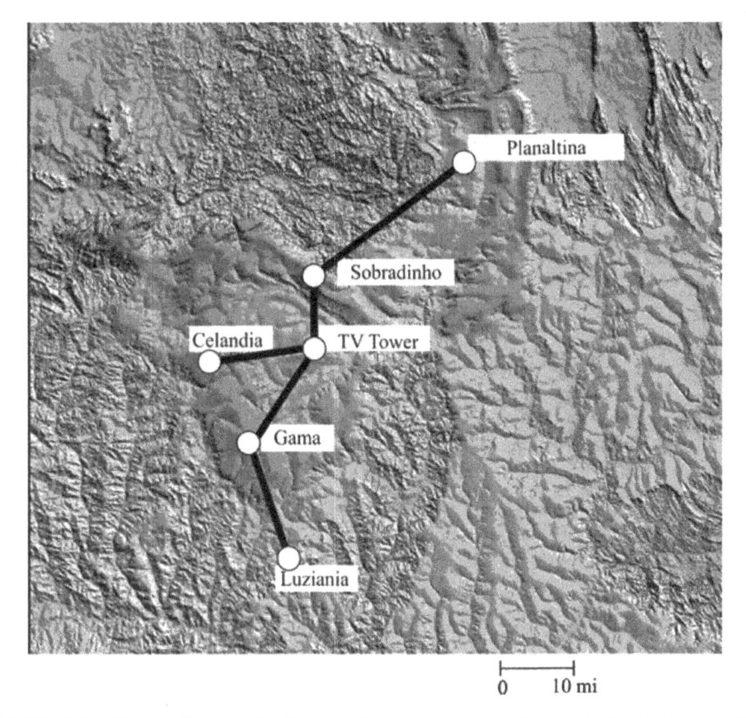

FIGURE A.9 Route diagram including the new repeater sites in Sobradinho and Gama.

3.4 Modulation, Capacity and Channel Arrangements

The number of states and symbol rate of the modulation scheme should enable a capacity equal to the minimum requirements of the system. This choice will be directly related to the system threshold that in turn will influence the maximum hop distance. Additionally, the symbol rate and modulation scheme should be also selected according to the maximum radio channel bandwidth of the radio channel arrangement plan.

3.4.1 Bandwidth Calculation The necessary bandwidth of any radio channel is given by equation (A.18)

$$B_{\mathrm{RF}} = (1+\alpha)\, V_{\mathrm{b}} \frac{1}{\log_2 M} \tag{A.18}$$

where

α = roll-off factor $\alpha = 0.20$

V_{b} = gross bit rate. Usually, this value will be equal to the aggregate radio frame bit rate (8% higher than the net bit rate). In our case $V_{\mathrm{b}} = 1.08 \times 155.52$ Mbps = 167.9 Mbps

M = number of different modulation symbols

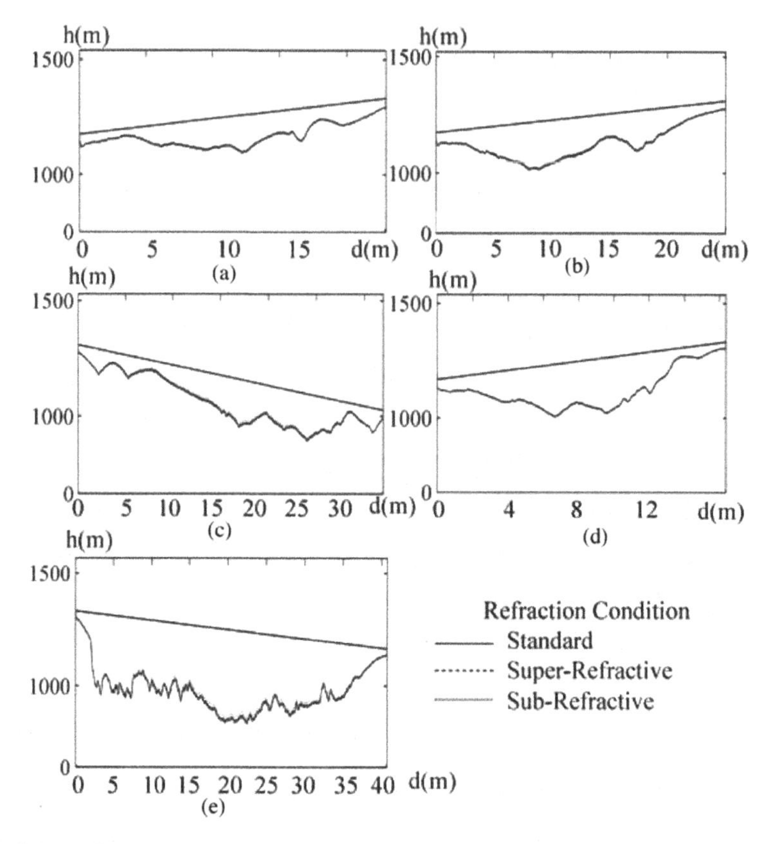

FIGURE A.10 Path profiles (a) TV Tower–Celandia (b) TV Tower–Gama (c) Gama–Luziania (d) TV Tower–Sobradinho (e) Sobradinho–Planaltina.

3.4.2 ITU- R Radio Channel Arrangement Plan The radio channel arrangement plan of the 6U band described by ITU-R F.384 will be used as the reference design. This plan is appropriate for SDH bitrates with a maximum of 8 radio channels. The center frequencies of the spectra at the high and low subbands in this arrangement are

$$\text{Low subband: } f_n = f_0 - 350 + 40\,n \text{ MHz}$$
$$\text{High subband: } f'_n = f_0 - 10 + 40\,n \text{ MHz}$$

where

$$n = 1, 2, 3, 4, 5, 6, 7 \text{ or } 8 \text{ with } f_0 = 6.770 \text{ GHz}.$$

The separation between adjacent channels is $\Delta = 40$ MHz and thus, in order to work with as low adjacent channel as possible, the radio channel bandwidth should be lower than 40 dB. According to Table A.28, this condition will be fulfilled if the

TABLE A.28 Different Radio Channel Bandwidth as a Function of M (M-QAM)

M	B_{RF} (MHz)	M	B_{RF} (MHz)
8	67.18	64	33.59
32	40.31	128	28.80

modulation choice is 64-QAM or higher. A single polarization scheme is proposed in the initial phase.

The Recommendation ITU-R F.384 also suggests that in the case of using Standard Antennas, the usual channel selection will be four frequencies in the low band and four in the high part: $n = 1, 3, 5$ and 7 or $n = 2, 4, 6$ and 8 in both sub bands. This condition implies that two radio channels should be separated more than two times the adjacent channel distance provided by the recommendation. The recommendation also suggests that adjacent channels of the same subband would require alternate polarization.

3.4.3 Frequency Assignments The assignments of frequencies to the link stations will start allocating each subband to transmission and reception to each station. Usually the process is started at the location where more transmissions are originated. In our case the process begins with the station of Brasilia. The initial decision is to use the high subband for transmission and the lower band for reception. This choice identifies the station in Brasilia as a "B station". This choice conditions the subbands for the rest of stations as described in Figure A.11.

The initial choice for transmission in Brasilia will be the first frequency of the subband, f_1, with vertical polarization. The second radio channel (backup channel of

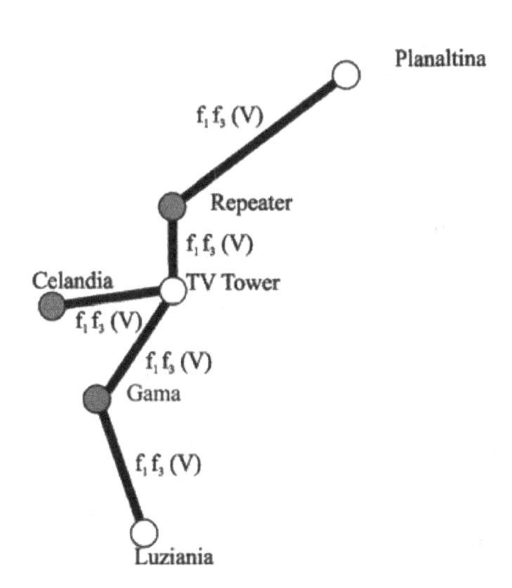

FIGURE A.11 Initial frequency plan. (B stations in white, A stations in grey).

TABLE A.29 Initial Frequency Plan

		Type A (Gama, Celandia, Sobradinho)		Type B (Luziania, TV Tower Planaltina)	
	Channel Number	Frequency (Mhz)	Pol.	Frequency (Mhz)	Pol.
TX1	1	6460	V	6800	V
TX1 Backup	3	6540	V	6880	V
TX1	1	6800	V	6460	V
RX1 Backup	3	6880	V	6540	V

the 1+1 configuration) will use then f_3, also with vertical polarization. Table A.29 summarizes the initial channel arrangement.

This first proposal will create interference problems in the nodal station of Brasilia. In fact, an experimental radio planner would directly skip this initial design and find directly alternative arrangements. In our case, this is the starting point to propose further refinements to the plan, evaluating the impact of countermeasures that reduce interferences.

Next section contains a detailed interference calculation example. If the C/I values are below the thresholds, the frequency plan should be modified and even different antennas might be required. Where the frequency use is not intensive in the area or the antenna choice is fixed, the usual procedure consists of selecting a different radio channel. If the frequencies are scarce, the techniques will focus on using high performance (HP) antennas and work with different polarizations on the interfering and interfered link, as well as adjusting carefully the power levels of the stations implied in the interference problem.

3.5 Interference Analysis

The objective of this study is the calculation of the C/I ratios at all the stations of the link and verify that these values are above the thresholds. The study would have to be carried out considering any potential interfering and interfered station 120 km around each station.

This task requires a radio environment map that might be available through the local, regional or national spectrum regulator. In any case, it is a good practice to check in the field the absence of potential interfering signals using appropriate radio frequency (RF) equipment. In our example we will assume that there will not be external interferences on our link and also that our link will not produce interferences into other systems. The interference analysis will be then an intrasystem interference analysis. Table A.30 summarizes the potential interference problems that might arise on our example.

If any of the link hops can contribute to the interference, the number of the interfering hops N can be expressed as a function of the total number of hops of the link, m:

$$N = 2m(m - 1) = 2 \cdot 5 \cdot 4 = 40 \tag{A.19}$$

TABLE A.30 Potential Interference Situations Associated with the Initial Channel Arrangement Plan

Interfered Station	Interfering Station					
	Luziania	Gama	Celandia	TV Tower	Sobradinho	Planaltina
Luziania	–	Desired signal	Co-channel overshoot interference NLOS path, significant terrain obstruction	Adjacent channel overshoot interference NLOS path, significant terrain obstruction	Co-channel overshoot interference NLOS path, significant terrain obstruction	Adjacent channel overshoot interference NLOS path, significant terrain obstruction
Gama	Co-channel nodal interference	–	Adjacent channel overshoot interference NLOS path, significant angular discrimination possible	Co-channel nodal interference Desired signals coming from the Luziania station	Adjacent channel. overshoot interference NLOS path, significant terrain obstruction	Co-channel overshoot interference NLOS path, significant terrain obstruction
Celandia	Co-channel overshoot interference NLOS path, significant terrain obstruction	Adjacent channel overshoot interference. NLOS path, significant angular discrimination	–	Desired signal	Adjacent channel overshoot interference. NLOS path, significant angular discrimination	Adjacent channel overshoot interference NLOS path, long distance between interference source and interfered receiver
TV Tower	Adjacent channel overshoot interference NLOS path with significant terrain obstruction	Co-channel nodal interference	Co-channel nodal interference	–	Co-channel nodal interference	Adjacent channel overshoot interference NLOS path with significant terrain obstruction

(Continued)

TABLE A.30 (*Continued*)

Interfered Station	Interfering Station					
	Luziania	Gama	Celandia	TV Tower	Sobradinho	Planaltina
TV Tower	Adjacent channel. overshoot interference NLOS path with significant terrain obstruction	Co-channel nodal interference	Co-channel nodal interference	–	Co-channel nodal interference	Adjacent channel overshoot interference NLOS path with significant terrain obstruction expected
Sobradinho	Co-channel overshoot interference NLOS path, significant terrain obstruction Angular discrimination possible	Adjacent channel overshoot interference NLOS path, significant terrain obstruction Angular discrimination possible	Adjacent channel overshoot interference NLOS path, significant angular discrimination possible	Co-channel nodal interference (desired signal from Planaltina)	–	Desired signal. Co-channel nodal interference with desired signal from TV tower
Planaltina	Adjacent channel overshoot interference Long distance between interference source and interfered receiver	Co-channel overshoot interference Long distance between interference source and interfered receiver	Co-channel overshoot interference NLOS path, significant terrain obstruction	Adjacent channel overshoot interference NLOS path, significant terrain obstruction	Desired signal	–

NLOS = Non Line Of Sight

In principle, cochannel interferences will have the highest impact on the system performance will occur at the nodal interference situations. The nodal stations where this problem occurs are TV Tower (Brasilia) with three hops sharing the site, and the repeaters in Gama and Sobradinho where two hops are connected at each of them.

In this phase of the design, the evaluation is based on the "worst case" criteria. We will assume that we are using standard performance parabolic antennas and the power levels are not modified. In addition, we will suppose that the interfering signal will not suffer any degree of variable fading.

Under that assumptions, the following expression from Chapter 8 (equation 8.55), provides the power received at a generic station B coming from a station C. The interfering path only considers free space attenuation.

$$L_{\mathrm{bC}} = 20\log_{10}\left(\frac{4\pi d}{\lambda}\right) \tag{A.20}$$

$$\begin{aligned} P_{\mathrm{IB}} = {} & P_{\mathrm{TC}} - (L_{\mathrm{TTC}} + L_{\mathrm{TTB}}) - L_{\mathrm{FC}} + G_{\mathrm{TC}} - A_{\mathrm{C}}(\beta) - L_{\mathrm{bC}} \\ & + G_{\mathrm{RB}} - A_{\mathrm{B}}(\alpha) - L_{\mathrm{TRB}} - L_{\mathrm{FB}} \end{aligned} \tag{A.21}$$

The radiation pattern of the antennas are described in Figure A.12 This figure contains the envelope pattern of different versions of the same antenna series, i.e.,

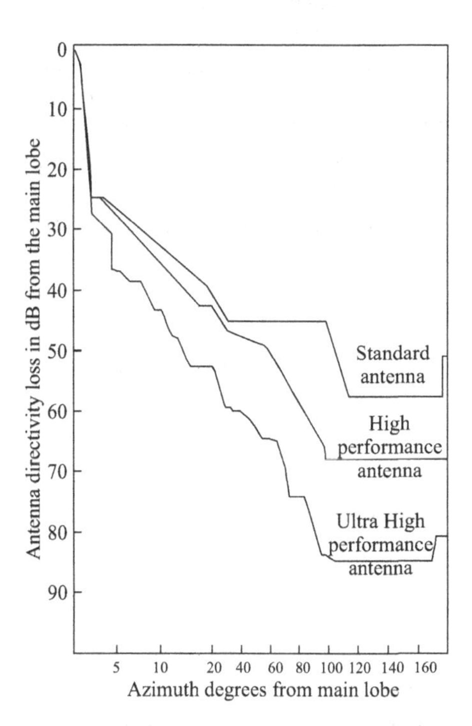

FIGURE A.12 Envelope radiation pattern of the antenna series used in the example.

TABLE A.31 Single Polarized Standard Antenna Characteristics

Diameter (ft/m)	Gain (dBi)			Beamwidth (Degrees)	XPD (dB)	VSWR Max. Return Losses (dB)
	Low	Mid-band	Top			
6/1.8	38.7	38.8	39.0	1.8	30	1.06/30.07
8/2.4	41.0	40.8	41.0	1.3	30	1.06/30.07
10/3.0	43.4	43.6	43.8	1.0	30	1.06/30.07
12/3.7	44.9	45.3	45.5	0.8	30	1.06/30.07

VSWR, voltage standing wave ratio

standard performance, high performance and others. Table A.31 contains more details about the single polarized Standard Antenna (the initial choice).

The initial choice is a 1.8 m antenna. The gain of this model is 38.7 dBi at the lowest frequency of the band. The exact antenna gain values associated with the interference calculations will not be the nominal maximum antenna gain, but the antenna gain at different directions, each one associated with the azimuth of the interfering–interfered path. In consequence, it will be also necessary to calculate the geometry of the interference problems: angles formed between each hop pair. Table A.32 and Figure A.13 display these data. All angles are given in north azimuth (degrees).

Once the antenna gain values associated with interference path angles, the interference power level can be calculated using equation (A.21). Table A.33 shows the results associated with the node TV Tower node.

Using the values in Table A.33, C/I results for unfaded desired signal reception are shown in Table A.34. The results have been obtained using receiver thresholds associated with both 10^{-3} and 10^{-6} BER values.

As an illustrating example the interference created by the route Celandia–TV Tower over the Sobradinho–TV Tower route is detailed in the following paragraphs. The interfering power is given by equation (A.22). Using the values from previous tables the interfering power results (dBm):

$$P_{\mathrm{I(Celandia-TV\ Tower)}} = 32 - 4 - 0 + 38.7 - 0 - 20\log_{10}\left(\frac{4\pi \cdot 22916}{3 \cdot 10^{11}/6460 \cdot 10^{9}}\right) + 38.7 - 45 - 0 \quad \text{(A.22)}$$

TABLE A.32 Azimuth Angles and Hop Distances

Route Origin–Destination		Azimuth (Degrees)	Hop Distance (km)
Luziania	Gama	342.41	31.9
Gama	TV Tower	32.98	25.0
Celandia	TV Tower	83.28	22.9
TV Tower	Sobradinho	358.39	16.3
Sobradinho	Planaltina	52.89	40.6

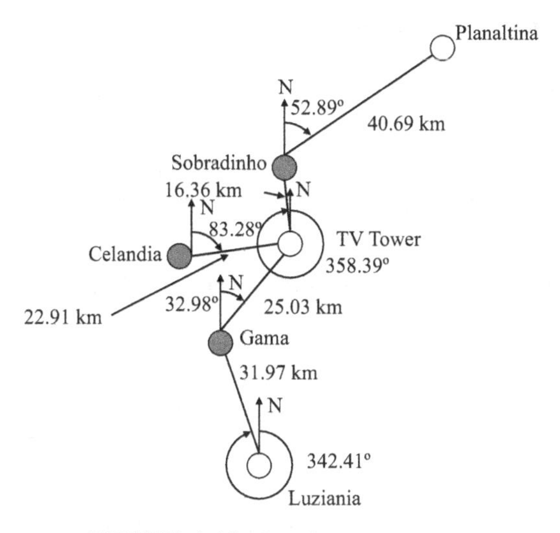

FIGURE A.13 Interference geometry.

The additional losses from IF filtering at both stations (L_{FC} and L_{FB}) are zero, as the interference is cochannel. The antennas of the hop Celandia–TV Tower are pointed toward the interfered system and the discrimination losses $A_C(\beta)$ toward the TV Tower, are practically also zero. The angle of the route Celandia–TV Tower from the TV Tower – Sobradinho, α, is (see Figure A.14).

$$\alpha = 180° - 83.28° - (360° - 358.39°) = 180° - 83.28° - 1.61° = 95.11° \quad (A.23)$$

The direction of 95.11° is associated with an antenna discrimination value of 44 dB. The interference level from the Celandia–TV Tower hop is −74.45 dBm.

Now, the C/I will be obtained using this value and the desired signal level (C) in dBm. This value can be obtained for faded and unfaded conditions, and for BER $= 10^{-3}$ and BER $= 10^{-6}$ thresholds. The nominal conditions will be associated with unfaded propagation. In this case the nominal received power on the

TABLE A.33 Interference Levels for all Routes Sharing the TV Tower Node (Values in dBm)

	Interfering Route			
Interfered Route	Sobradinho–TV Tower	Celandia–TV Tower	Gama–TV Tower	Overall Interference Level
Sobradinho–TV Tower	–	−74.45	−88.22	−74.27
Celandia–TV Tower	−71.52	–	−75.22	−69.98
Gama–TV Tower	−84.52	−74.45	–	−74.04

TABLE A.34 C/I Ratios for all Routes Sharing the TV Tower Node

Interfered Route	C/I (unfaded)	C/I @ BER = 10^{-3}	C/I @ BER = 10^{-6}
Sobradinho–TV Tower	46.74	1.27	3.27
Celandia–TV Tower	39.53	−3.02	−1.02
Gama–TV Tower	48.82	1.04	3.04

Sobradinho–TV Tower (at the TV Tower receiver) is in dBm:

$$P_{D(\text{Sobr.}-\text{TV Tower})} = 32 - 4 + 38.7 - 20\log_{10}\left(\frac{4\pi \cdot 16368}{3 \cdot 10^{11}/6460 \cdot 10^{9}}\right) + 38.7 = -27.53 \qquad \text{(A.24)}$$

The resulting unfaded C/I is then 46.92 dB. If now we consider the fading occurrences associated with 10^{-3} and 10^{-6} BER, the ratios obtained are 1.45 and 3.45 respectively. These C/I values are far below the minimum specified C/I limit (18 dB for BER = 10^{-3} and 23 dB for BER = 10^{-6}). In consequence, countermeasures will be required.

If similar calculations are performed for the interfering signals coming from the hop Gama–TV Tower onto the receiver of the hop Sobradinho–TV Tower, the interference level is −88.22 dBm. Now if we combine the interferences from the Celandia–TV Tower and Gama–TV Tower on the receiver of the Sobradinho–TV Tower link the composite value will be (dBm):

$$P^{Total}_{\text{Sobradinho}-\text{TV Tower receiver}} = 10\log\left(10^{P_{I(\text{Celandia}-\text{TV Tower})}/10} + 10^{P_{I(\text{Gama}-\text{TV Tower})}/10}\right) = -74.27 \qquad \text{(A.25)}$$

The composite unfaded C/I signal is 60.70 dB. The faded C/I associated with the BER 10^{-3} and BER 10^{-6} thresholds are15.22 dB and 17.22 dB respectively. The remaining interference calculation results at the Gama and Sobradinho repeaters are summarized in Table A.35.

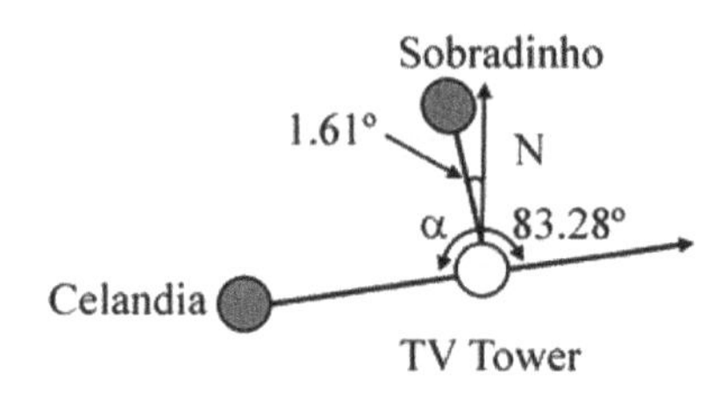

FIGURE A.14 Angle of route Celandia–TV Tower from the TV Tower–Sobradinho.

TABLE A.35 C/I Values at Gama and Sobradinho Repeater Stations

Interfering Route	Interfered Route	C/I (unfaded)	C/I @ BER = 10^{-3}	C/I @ BER = 10^{-6}
Luziania–Gama	TV Tower–Gama	59.13	17.79	19.79
TV Tower–Gama	Luziania–Gama	54.87	15.66	17.66
Planaltina–Sobradinho	TV Tower–Sobradinho	64.91	19.88	21.88
TV Tower–Sobradinho	Planaltina–Sobradinho	49.09	11.97	13.97

The results in faded conditions are again lower than the minimum requirements (18 and 23 dB for 10^{-3} and 10^{-6}) and thus the design needs to be refined.

3.6 Interference Reduction Design Techniques

There are several techniques that can be used in order to increase the C/I at those stations where the value is below the threshold. Each one of those techniques is evaluated in the following sections. In all cases, the worst case condition of unfaded interfering path has been applied.

3.6.1 Adjustment of Transmitted Output Power Levels The first technique related to power level adjustment is the use of ATPC techniques. The ATPC module provides the optimum radiated power within the power adjustment range in order to ensure the required reception level as a function of the instantaneous fading associated with different propagation factors.

The output power at short links can be also decreased permanently. This might not be always possible as the Gross Margin will be also reduced permanently and thus the availability and error performance values might be degraded below the thresholds. In summary, any decrease of the output power requires a validation of the system performance according to availability and error performance criteria.

A usual practice consists of adjusting the output levels so the received power by all the receivers at the nodal station is equal to the values associated with the longest hop, under unfaded conditions.

In our example, if the received powers at the TV Tower are equal to the path from Gama. The output transmitted powers at Sobradinho, Celandia, and Gama would result 28.31 dBm, 31.23 dBm and 32 dBm respectively. Considering these values, the new C/I ratios are shown in Table A.36.

TABLE A.36 C/I Ratios for Hops Sharing the TV Tower Node (After Transmitted Power Reduction)

Interfered Route	C/I (unfaded)	C/I @ BER = 10^{-3}	C/I @ BER = 10^{-6}
Sobradinho–TV Tower	43.79	2.00	4.00
Celandia–TV Tower	40.99	−0.79	1.20
Gama–TV Tower	43.79	2.00	4.00

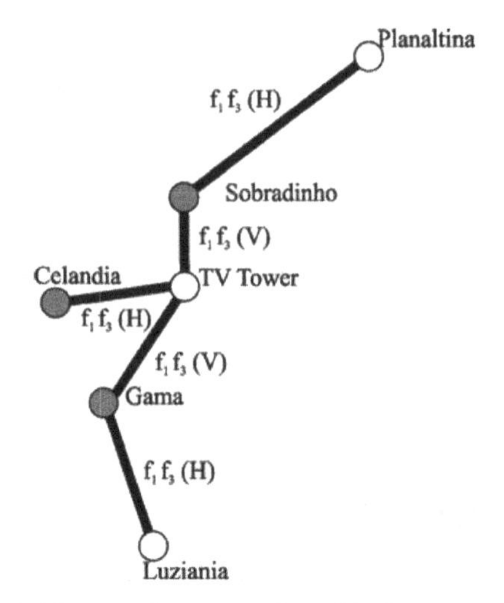

FIGURE A.15 New frequency plan with alternate polarizations.

The fact of balancing power levels to the ones associated with the longest hop involves the same C/I ratios at two of the hops (rows 1 and 3 in Table A.36). If these values are compared to the initial situation (Table A.34), the results are slightly better in the case of faded C/I but still far from the objectives. This measure will then be used to optimize the interference results once the design has achieved the specifications related to interference, availability, and error performance.

3.6.2 Cross-Polarization Operation Another technique usually employed to increase C/I ratios consists of alternating polarizations in paths that share a common node. In our case, we will change the polarization to the path suffering the highest interference: Celandia–TV Tower. On the other two nodes the change will be done to the paths Luziania–Gama and Sobradinho–Planaltina. The resulting polarization configuration is described in Figure A.15. The link budgets for interfering paths will be decreased by 30 dB (cross-polar discrimination value of the Standard Antenna). As an example, the interfering route Celandia–TV Tower on the interfered Sobradinho–TV Tower (power values in dBm). The resulting C/I values are summarized in Table A.37.

TABLE A.37 C/I Ratios for Hops Sharing the TV Tower Node (After Alternate Polarization)

Interfered Route	C/I (unfaded)	C/I @ BER = 10^{-3}	C/I @ BER = 10^{-6}
Sobradinho–TV Tower (V)	60.59	15.12	17.12
Celandia–TV Tower (H)	69.53	26.98	28.98
Gama–TV Tower (V)	53.26	11.48	13.48

TABLE A.38 C/I Ratios on the Repeater Stations Gama and Sobradinho (After Alternate Polarization)

Interfering Route	Interfered Route	C/I (unfaded)	C/I @ BER = 10^{-3}	C/I @ BER = 10^{-6}
Luziania–Gama	TV Tower–Gama	89.13	47.79	49.79
TV Tower–Gama	Luziania–Gama	84.87	65.66	67.66
Planaltina–Sobradinho	TV Tower–Sobradinho	94.91	49.88	51.88
TV Tower–Sobradinho	Planaltina–Sobradinho	79.09	41.97	43.97

$$P_{\mathrm{I(Celandia-TV)}} = 32 - 4 - 0 + 38.7 - 0 - 20\log_{10}\left(\frac{4\pi \cdot 22916}{3 \cdot 10^{11}/6460 \cdot 10^{9}}\right) + 38.7 - 45 - 30 \quad (A.26)$$

$$P_{\mathrm{I(Celandia-TV)}} = -104.45 \quad (A.27)$$

The paths Sobradinho–TV Tower and Celandia–TV Tower still have C/I ratios under the thresholds for faded conditions. If the horizontal and vertical polarization assignments are changed among the three paths the results are very similar. In consequence this countermeasure does solve the interference problems at the TV Tower. On the contrary, polarization solves the problems on the repeater stations of Gama and Sobradinho. The results for those nodes are shown in Table A.38.

3.6.3 Use of High Performance Antennas The use of HP antennas is another candidate technique for solving interference problems. The gain and discrimination values will be higher and thus the interfering levels will be reduced and the desired signal levels might be enhanced. Obviously, it will imply an increase of the cost of the link. The calculations in this section include HP antennas for all the links sharing the TV Tower site. The same frequencies and single polarization (vertical is assumed for all hops). The HP antenna specifications are summarized in Table A.39, and the envelope pattern was already presented in Figure A.12. The new C/I ratios

TABLE A.39 High Performance Antenna Data (Single Polarization)

Diameter (ft/m)	Gain (dBi) Low	Gain (dBi) Mid-Band	Gain (dBi) Top	Beamwidth (Degrees)	XPD (dB)	VSWR Max. Return Losses (dB)
6/1.8	39.0	39.5	39.8	1.7	30	1.06/30.07
8/2.4	41.9	42.3	42.8	1.3	30	1.06/30.07
10/3.0	43.1	43.4	43.8	1.0	27	1.06/30.07
12/3.7	45.2	45.6	46.1	0.8	30	1.06/30.07

VSWR, voltage standing wave ratio

TABLE A.40 C/I Ratios for Hops Sharing the TV Tower Node (HP antenna)

Interfered Route	C/I (unfaded)	C/I @ BER = 10^{-3}	C/I @ BER = 10^{-6}
Sobradinho–TV Tower	65.98	20.21	22.21
Celandia–TV Tower	48.28	−3.02	−1.02
Gama–TV Tower	48.82	1.04	3.04

are shown in Table A.40 and Table A.41. Once again the values do not meet the specifications.

3.6.4 New Frequency Assignments None of the previous techniques has provided enough C/I ratios. The final option will be always to find other radio channels and use more frequencies to make new assignments to the stations of each hop of the link. This section discusses the use of new frequencies and the calculations are based on using Standard Antennas. A possible set of new assignments is summarized in Figure A.16 and Table A.42.

The interference calculations will now include the filter discrimination. This value is the additional adjacent channel attenuation provided by the IF filters. The manufacturer specifies a value not lower than 45 dB. As shown in Figure A.16, the new frequency assignments are combined with alternate polarization at the node TV Tower.

The calculations for the interference level created by the link Celandia–TV Tower (interfering) to Sobradinho–TV Tower (interfered) are detailed now as an example. The calculation will be based on equation (A.28) that describes the adjacent channel interference level on a nodal interference situation.

$$
\begin{aligned}
P_{\mathrm{I(Celandia-TV)}} &= 32 - 2 - 0 + 38.7 - 0 - 135.84 + 38.7 - 5.3 - 2 - 30 - 45 \\
&= -149.44 \qquad \text{(A.28)}
\end{aligned}
$$

The value obtained in equation (A.28) expresses the power in dBm created by the Celandia Transmitter on the TV Tower receiver associated with the Sobradinho–TV Tower link. The desired received level from Sobradinho is −27.57 dBm, thus, the C/I ratio on unfaded condition is 128.86 dBm. Similar calculations would be performed to obtain the interference created on the same receiver by the Gama–TV Tower path.

TABLE A.41 C/I Ratios on the Repeater Stations Gama and Sobradinho (HP antenna)

Interfering Route	Interfered Route	C/I (unfaded)	C/I @ BER = 10^{-3}	C/I @ BER = 10^{-6}
Luziania–Gama	TV Tower–Gama	65.74	20.20	22.21
TV Tower–Gama	Luziania–Gama	48.28	−3.02	−1.02
Planaltina–Sobradinho	TV Tower–Sobradinho	48.82	1.04	3.04
TV Tower–Sobradinho	Planaltina–Sobradinho	46.74	1.27	3.27

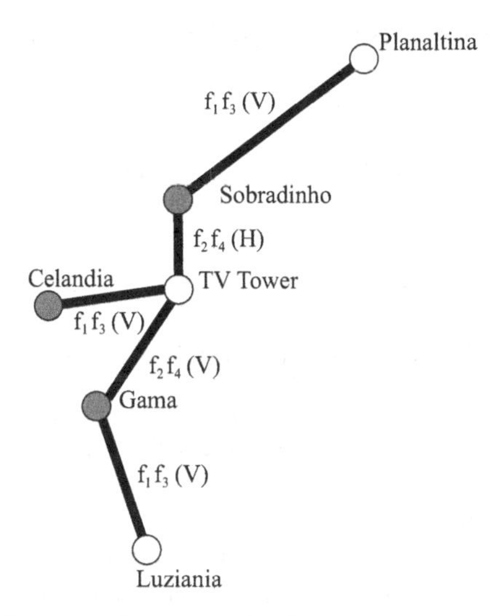

FIGURE A.16 Network diagram showing the new frequency assignments.

In this case the interfering signal is −118.27 dBm and the C/I is 90.69 dB. If both interference sources are aggregated, the overall interfering level is −118.26 dBm and the C/I is 90.68 dBm. Note that aggregated values will be almost equal to the highest interferer if the amplitudes of the interferers are higher than 10 dB. This analysis would be required for the other two interfered routes Sobradinho–TV Tower and Gama–TV Tower. The results are summarized in Table A.43.

The values satisfy the C/I specifications under fading conditions at the TV Tower node. The calculations of the repeater nodes are skipped, as the ratios there already were satisfactory using alternate polarizations.

TABLE A.42 New Channel Arrangement Plan

Route		TX1			TX1 Backup		
TX	RX	Channel Number	F (Mhz)	Pol.	Channel Number	F (Mhz)	Pol.
TV Tower	Sobradinho	f_2	6840	H	f_4	6920	H
Sobradinho	TV Tower	f_2	6500	H	f_4	6580	H
TV Tower	Celandia	f_1	6800	V	f_3	6880	V
Celandia	TV Tower	f_1	6460	V	f_3	6540	V
TV Tower	Gama	f_2	6840	V	f_4	6920	V
Gama	TV Tower	f_2	6500	V	f_4	6580	V
Gama	Luziania	f_1	6460	V	f_3	6540	V
Luziania	Gama	f_1	6800	V	f_3	6880	V
Sobradinho	Planaltina	f_1	6460	V	f_3	6540	V
Planaltina	Sobradinho	f_1	6800	V	f_3	6880	V

TABLE A.43 C/I Ratios for Hops Sharing the TV Tower Node (New Frequency Assignments)

Interfered Route	C/I (unfaded)	C/I @ BER = 10–3	C/I @ BER = 10–6
Sobradinho–TV Tower	90.68	45.26	47.26
Celandia–TV Tower	89.81	47.26	49.26
Gama–TV Tower	82.08	40.35	42.35

3.7 Availability and Error Performance Objectives

The availability and EPOs that apply to this example are obtained from ITU-R F.1668 y ITU-R F.1703 respectively, in line with the discussions provided in Chapter 5.

The link is part of the access section of the national portion of a hypothetical reference digital path . The maximum number of hops of a E1 frame is two. This is the case of the links from the nodal Switching station in Brasilia and local switching centers in Planaltina and Luziania. The Celandia–Brasilia route is composed by a single hop. The maximum hop is Planaltina–Brasilia (57.066 km = 40.698 + 16.368). The bit rate is 150.336 Mbps (STM-1).

3.7.1 Availability Objectives (ITU-R F.1703)

The availability objective is calculated using the data provided in Chapter 5, on Table 5.19. (SDH link, STM-1, access section, national portion):

$$AR = 1 - \left(B_6 \frac{L_{\text{link}}}{L_{\text{R}}} + C_6\right) = 1 - \left(0 \frac{57.066}{2500} + 4 10^{-4}\right) = 0.9996\ (99.96\%) \tag{A.29}$$

In the case of links with two different hops the objective is allocated homogeneously to each one of the hops: $[100 - (4\ 10^{-2}/2)] = 99.98\%$.

3.7.2 Error Performance Objectives (ITU-R F.1668)

The EPOs will be calculated using the reference data provided in Table 5.16 (Chapter 5). Table A.44 summarizes the procedure.

Again, considering two hops, the EPO values will be allocated 50% to each hop. In our example, the final EPO would be and ESR equal to 0.15%, SESR equal to 0.0075 % and $3.75\ 10^{-4}$ for background block error ratio (BBER).

TABLE A.44 EPO Calculations

	EPO
ESR	$0.04 \times C = 0.04 \times 0.075 = 0.003$ (0.3 %)
SESR	$0.002 \times C = 0.002 \times 0.075 = 0.00015$ (0.015 %)
BBER	$1 \times 10^{-4} \times C = 1 \times 10^{-4} \times 0.075 = 7.5 \times 10^{-6}$ (7.5×10^{-4} %)

TABLE A.45 Link Budgets

	Equipment					
	Luiziania Gama	Gama TV Tower	Celandia TV Tower	TV Tower Sobradinho	Sobradinho Planaltina	Units
(1) Reference frequency	6540	6580	6540	6580	6540	MHz
(2) Max. TX output power	32	32	32	32	32	dBm
(3) Branching losses	4	4	4	4	4	dB
(4) TX antenna gain (1.8 m)	38.7	38.7	38.7	38.7	38.7	dB
(5) RX antenna gain (1.8 m)	38.7	38.7	38.7	38.7	38.7	dB
(6) Threshold 10^{-3} T_{h3}	−73	−73	−73	−73	−73	dBm
(7) Threshold 10^{-6} T_{h6}	−71	−71	−71	−71	−71	dBm
(8) Threshold 10^{-12} T_{h12}	−68	−68	−68	−68	−68	dBm
(9) Link distance	31.978	25.037	22.916	16.368	40.698	km
Free Space Loss	138.9	136.8	136.0	133.1	140.9	dB
Gas absorption	Assumed negligible for frequencies below 10 GHz					dB
Field margin	2	2	2	2	2	dB
(10) Total fixed loss	136.9	134.8	134.0	131.1	138.9	dB
System Gains and Margins						
Nominal receiver input: (2)–(3)+(4)+(5)–(10)	−31.5	−29.4	−28.6	−25.7	−33.5	dBm
System gain (T_{h3})	105	105	105	105	105	dB
System gain (T_{h6})	103	103	103	103	103	dB
System gain (T_{h12})	100	100	100	100	100	dB
M_3	41.5	43.6	44.4	47.3	39.5	dB
M_6	39.5	41.6	42.4	45.3	37.5	dB
M_{12}	36.5	38.6	39.4	42.3	34.5	dB

3.8 Link Budget

Table A.45 summarizes the link budget calculations, including the gross margin values. The reference data have been obtained from Table A.25, the frequency plan using four channels (see Figure A.16) and the antenna characteristics already provided in Table A.31.

3.9 Availability Validation

The manufacturer specifies a mean time between failures (MTBF) equal to 50 000 h. Considering that the time to repair would be on the average of 5 h (typical data from the operator), the overall unavailability due to equipment failures and maintenance will be close to 10^{-8}. This value suggests that equipment will not be a relevant cause for unavailability.

Following ITU-R recommendations we will assume that anomalous refraction effects will not be a major cause of unavailability. This applies to short links not

TABLE A.46 Rain Attenuation Calculations

	Luiziania Gama	Gama TV Tower	Celandia TV Tower	TV Tower Sobradinho	Sobradinho Planaltina	Units
Rain rate $R_{0.01}$	145	145	145	145	145	mm/h
Polarization	Calculations made only for horizontal polarization					
Reference distance d_0	4.0	4.0	4.0	4.0	4.0	km
Distance factor r	0.1	0.1	0.1	0.2	0.1	–
Effective distance d_{eff}	3.5	3.4	3.4	3.2	3.6	km
Constant k	0.0024	0.0024	0.0024	0.0024	0.0024	–
Constant α	1.3214	1.3224	1.3214	1.3224	1.3214	–
γ_R	1.7	1.7	1.7	1.7	1.7	dB/km
$A_{0.01}$	6.0	6.0	5.8	5.6	6.2	dB
Availability objective (p %)	0.002	0.002	0.004	0.002	0.002	%
A_p	10.4	10.4	8.0	9.7	10.7	dB
Unallocated margin (M_3–A_p)	31.1	33.2	36.5	37.6	28.8	dB

being affected by intense duct propagation. According to the 1% value of the gradient refractivity in Brasilia (−83.74), we will suppose that the area is not affected by ducts. In consequence we will suppose that multipath will only influence error performance and rain will be the only cause for unavailability. The rain attenuation statistics are summarized on Table A.46.

TABLE A.47 Error Performance Validation Calculations (BER 10^{-6}, SESR)

	Luiziania Gama	Gama TV Tower	Celandia TV Tower	TV Tower Sobradinho	Sobradinho Planaltina	Units
Geoclimatic and Refraction Data						
dN1	−83.74	−83.74	−83.74	−83.74	−83.74	N units
Geoclimatic factor K	4.23E–05	4.23E–05	4.23E–05	4.23E–05	4.23E–05	–
Multipath occurrence factor p_0 (%)						%
Initial planning	1.019	0.352	0.265	0.091	1.648	%
Detailed planning	3.957	1.331	0.976	0.303	7.149	%
Multipath activity factor						–
Initial planning	0.006	0.003	0.002	0.001	0.009	–
Detailed planning	0.018	0.008	0.006	0.003	0.027	–
Average time delay	0.392	0.285	0.254	0.164	0.536	ns
Reference time delay	6.3	6.3	6.3	6.3	6.3	ns
A_t	25.0	24.5	24.3	23.7	25.3	dB
M_6	39.5	41.6	42.4	45.3	37.5	dB
Flat fading (Initial plan)	0.000113	0.000024	0.000015	0.000003	0.000296	%
Flat fading (Detailed plan)	0.000439	0.000092	0.000056	0.000009	0.001284	%
Selective fading (Initial plan)	0.000071	0.000017	0.000011	0.000002	0.000190	%
Selective fading (Detailed plan)	0.000195	0.000046	0.000029	0.000005	0.000567	%
Composite outage	0.000634	0.000138	0.000085	0.000014	0.001852	%

TABLE A.48 Error Performance Validation Calculations (BER 10^{-12} and BBER)

	Luiziania Gama	Gama TV Tower	Celandia TV Tower	TV Tower Sobradinho	Sobradinho Planaltina	Units
Geoclimatic and Refraction Data						
dN1	−83.74	−83.74	−83.74	−83.74	−83.74	N units
Geoclimatic factor K	4.23E–05	4.23E–05	4.23E–05	4.23E–05	4.23E–05	–
Multipath occurrence factor p_0 (%)						%
Initial planning	1.019	0.352	0.265	0.091	1.648	%
Detailed planning	3.957	1.331	0.976	0.303	7.149	%
Multipath activity factor						–
Initial planning	0.006	0.003	0.002	0.001	0.009	–
Detailed planning	0.018	0.008	0.006	0.003	0.027	–
Average time delay	0.392	0.285	0.254	0.164	0.536	ns
Reference time delay	6.3	6.3	6.3	6.3	6.3	ns
A_t	25.0	24.5	24.3	23.7	25.3	dB
M_{12}	36.5	38.6	39.4	42.3	34.5	dB
Flat fading (Initial plan)	0.000225	0.000048	0.000030	0.000005	0.000591	%
Flat fading (Detailed plan)	0.000876	0.000183	0.000111	0.000018	0.002563	%
Selective fading (Initial plan)	8.499E–05	2.033E–05	1.305E–05	2.431E–06	2.279E–04	%
Selective fading (Detailed plan)	2.338E–04	5.495E–05	3.462E–05	6.020E–06	6.788E–04	%
Composite outage	0.001110	0.000238	0.000146	0.000024	0.003242	%

3.10 Error Performance Validation

The EPOs are evaluated according to the 10^{-6} and 10^{-12} thresholds provided by the equipment manufacturer. As described in Chapter 8, the first BER target will be associated with the SESR and the 10^{-12} to BBER. Subsequently, the Tables A.47 and A.48 summarize the calculation procedure already described in detail for the cellular access network example.

INDEX

Microwave Line of Sight Link Engineering, First Edition. Pablo Angueira and Juan Antonio Romo.

Printed in the United States
By Bookmasters